MW00474556

Final Exam
mostly first 4 sections + La Grange
formula for remainder

X4   9.50
UNIVERSITY SHOP
A, No Refund if Removed
$42.75

{ find radius of conv
  v'ing endpoints
{ differ & integrating
  series expansions ⓑ 233
    ⓑ 229

4/30 8:00 –
      10:00

10 problems

La Grange formula for remainder

No – proof that func. is analytic

INTRODUCTORY ANALYSIS: THE THEORY OF CALCULUS

Introductory Analysis

The Theory of Calculus

J. A. Fridy

Kent State University

Harcourt Brace Jovanovich, Publishers

and its subsidiary, Academic Press

San Diego · New York · Chicago · Austin

London · Sydney · Tokyo · Toronto

Copyright © 1987 by Harcourt Brace Jovanovich, Inc.

All rights reserved. No part of this publication may be reproduced or transmitted in any form or by any means, electronic or mechanical, including photocopy, recording, or any information storage and retrieval system, without permission in writing from the publisher.

Requests for permission to make copies of any part of the work should be mailed to: Permissions, Harcourt Brace Jovanovich, Publishers, Orlando, Florida 32887.

ISBN: 0-15-501845-0

Library of Congress Catalog Card Number: 86-70759

Printed in the United States of America

To my family

PREFACE

Introductory Analysis: The Theory of Calculus contains a theoretical development of the calculus concepts that are presented intuitively in the typical freshman-sophomore calculus course. It is intended as a text for a theory course, usually taken in the junior or senior year. Such a theory course is typically called Advanced Calculus, Introduction to Analysis, or some variation. *Introductory Analysis: The Theory of Calculus* contains sufficient material for a two-semester course presented at a comfortable pace. A one-semester or two-quarter course could be taught by covering Chapters 1–8 plus a few other chapters at the discretion of the instructor. In an honors class, this book may be used to supplement a problem-solving calculus text for students with high mathematical aptitude.

The material is held at a reasonable level for undergraduates by presenting most of the theory in the familiar real number system as opposed to a more general or abstract setting. Thus, a student who is learning for the first time to discover and write mathematical proofs can more readily call upon his or her experience for examples that illustrate the situation at hand. After the student has some experience with the development of a mathematical proof in a concrete setting, the abstract theories will be more accessible.

One of the student's objectives at this level should be to develop skill in writing mathematics; therefore, the expository style of the book is intended to illustrate formal mathematical writing as opposed to the informality frequently found in freshman calculus texts. This is true not only of the formal statements and proofs, but also of the general discussions, which are neither casual nor chatty. In keeping with the level and style of exposition, there are only a few figures; this is typical and—I think—desirable in an advanced textbook on mathematical theory. At this stage, the student must learn that a pictorial argument does not constitute a proof. The figures in the book, as well as others that the student should be encouraged to draw, must be regarded as merely motivational and not an integral part of the logic. The exercise sets, which are located at the end of nearly every section, provide a means for the student to experience both the discovery and the writing of mathematics. Some routine problems appear at the beginning of most exercise sets, but the more beneficial exercises are those that ask for a proof or a detailed example. Although such exercises are not necessary for the proofs of later results, they add detail to the theory and provide experience for the student.

Because the student should be acquainted with the concepts and notation of elementary set theory, common set theory symbols are used without further explanation. In addition to union and intersection, these symbols include $\sim S$ for the complement of S and $T \sim S$ for the relative complement: $T \sim S = T \cap (\sim S)$.

The development of the theory begins in Chapter 1 with the definition of the ordering of the real numbers $\mathbb{R}$ and the introduction of the Least Upper Bound Axiom (all the arithmetic properties of the field $\mathbb{R}$ are *assumed*). The particular feature that marks this treatment of the subjects is the ordering of the various limit concepts. In Chapter 2 the first limit theory presented is that of convergent number sequences; it is the easiest to master in the rigorous "ε–δ manipulations," and it is easy to motivate by examples. Sequential limits are then *used* to develp the other convergence concepts, which include continuity, function limits, uniform continuity, derivatives, integrals, infinite series, function sequences, and power series. Along with the completeness of $\mathbb{R}$, which is the topic of Chapter 3, these limit concepts constitute the content of Chapters 1–12. Another unusual approach is found in Chapter 4, where the concept of a continuous function is studied first, and then continuity is used to define function limits. This ordering of topics is chosen because, to a student, continuous functions seem more natural than discontinuous functions that exhibit a limit. In addition to the Riemann integral, which is studied in Chapter 7, the Riemann-Stieltjes integral is presented in Chapter 10. In Chapter 11 the Weierstrass Approximation Theorem is studied as an example of uniform convergence of a function sequence.

The final four chapters are devoted to multidimensional theory, and they begin with a study of metric spaces. This is the only topic that is presented in an abstract setting, and even in this case the metric space theory is strongly slanted toward Euclidean spaces. In Chapters 14 and 15 continuity and differentiability are studied in n-dimensional Euclidean space, but in the final chapter the study of multidimensional integrals is limited to a two-dimensional theory. Here the approach uses Jordan content via inner and outer area of sets in the plane; this provides some useful background for those students who will go on to study Lebesgue measure theory.

A book of this type will naturally be influenced by the author's experiences through many years as a student and teacher of the subject. It would be impossible to cite the professors and references that have made a significant impact on this work without inadvertently slighting some of them. Therefore I must offer a general expression of gratitude to all my colleagues, students, and teachers who have helped in numerous ways over the years as this work progressed from handwritten course notes to book form. I am deeply indebted to all of you. I wish to thank Julia Froble for her superior work in typing the manuscript. The following reviewers provided very helpful comments on earlier versions of these materials: John Ingram, California State University–Sacramento; William Sledd, Michigan State University; Frank Cleaver, University of South Florida; W. R. Hintzman, San Diego State University; Michael Stein, Northwestern University; Joel Westman, University of California, Irvine; Walter K. Mallory, University of Delaware; Ralph Lakness, San Francisco State University; Eli Passow, Temple University; David Hallenback, University of Delaware; Terence Gaffney, Northeastern University; Michael B. Gregory, University of North Dakota; Esteban Poffald, Wabash College; William Armacost, California State University–Dominguez Hills; Martin Billik, San Jose State University; Barbara Shabell, California State Polytechnic University–Pomona; Douglas Hall, Michigan State University; and Timothy Sorenson, Kent State University, who did the final accuracy review. Finally, I want to thank the editors and staff of Academic Press and Harcourt Brace Jovanovich for their assistance and encouragement throughout the period of metamorphosis from course notes to book.

CONTENTS

INTRODUCTION: MATHEMATICAL STATEMENTS AND PROOFS

Types of Mathematical Statements

The subject matter of this book makes up what is known as a "mathematical theory," and the objective is to present this theory and develop it as a set of logically interrelated statements. This is in contrast to the more familiar undergraduate mathematics course, in which the primary objective is to teach a set of techniques for solving certain types of problems. Ideally, every mathematics course should present some theory, but in problem-solving courses, students are held responsible for very little. Plane geometry is possibly the only course prior to this one that is devoted entirely to the development of a mathematical theory. Here, as in geometry, we encounter various types of statements that are the elements of the theory. The label of each statement says something about its role in the theory.

First, there are *definitions*, statements that describe the objects, properties, or concepts to be studied. Next, there are *axioms*, which state something that is *assumed* to be true in this particular theory. The definitions and axioms do not have to be proved or justified in any way, although in some cases a motivating discussion is given in order to show how the ensuing theory parallels or extends something in our past experience. Another name for an axiom is *postulate*, a term commonly used in Euclidean geometry but not used here.

In addition to definitions and axioms, there is a hierarchy of formal statements that must be deduced using other statements in

1

which the truth is either assumed or previously proved. The most important of these proved assertions are called *theorems*. Next in significance are *corollaries*, a title that implies that the statement can be deduced as an additional consequence of another statement that has been proved previously. Thus it is customary to state a corollary immediately after the theorem from which it is deduced. Another class of deduced statements are the *lemmas*, preliminary results that are used to prove a theorem or theorems. Usually the statement of the lemma is not of general significance by itself, but it is given a title as a convenient way of subdividing a long proof of a theorem. Last in the hierarchy are statements known as *propositions*. This label is used when the assertion to be proved is not significant enough to be called a theorem, and calling it either a lemma or a corollary would be inappropriate. In the past, particularly in classical Euclidean geometry, the title "proposition" was used as we now use the title "theorem."

The Structure of Proofs

Most of the mathematical assertions that require proof are conditional; that is, something is asserted to be true *if* another thing is known or given to be true. The part that is given or assumed is called the *hypothesis*, and that which is deduced from the hypothesis is the *conclusion*. The simplest form of this conditional statement is "if H then C," where H denotes the hypothesis and C denotes the conclusion. Various other wordings of this that mean exactly the same thing are (i) "H implies C," (ii) "H only if C," (iii) "C is a necessary condition for H," and (iv) "H is sufficient for (implying) C." These are the most common forms, but there are others. In many conditional statements the hypothesis and conclusion consist of two or more parts; for example, "If H_1, H_2, and H_3, then C_1 and C_2."

If the roles of hypothesis and conclusion in a conditional statement are reversed, the resulting statement is called the *converse* of the original. The statements "if A then B" and "if B then A," for example, are converses of one another. Sometimes it is asserted that both a statement *and* its converse are valid, and then the two statements may be combined into one, such as "A if and only if B." In order to prove such a *biconditional* statement, one must prove *both* implications: "A implies B" *and* "B implies A." In this case we say that A and B are *equivalent*. There are other wordings of the biconditional statement. One of these variants that is found in much of the mathematical literature is "*A* is a necessary and sufficient condition for *B*." Another wording that is convenient for cases in which *A* and *B* are fairly complicated is

> "the following statements are equivalent:
>   (i) *A*;
>   (ii) *B*."

This form is certainly the simplest to use if there are more than two equivalent statements.

The arguments employed to prove an implication such as "if *H* then *C*" also have several forms. The simplest form (which is not necessarily the easiest to establish) is the *direct argument*. To prove that *H* implies *C*, one simply assumes that the truth of *H* is given and deduces from it the truth of *C*. An *indirect argument* can be made in either of two ways. The *contrapositive argument* is made by assuming that *C* is false and deducing that *H*, too, must be false. In the *contradiction,* or *reductio ad absurdum, argument,* one assumes that the statement to be proved is false and then reasons from that assumption until an absurdity is reached. The absurdity may be some blatantly false statement such as "0 = 1," or it may be the negation of an axiom or previously proved statement. Students who are writing proofs for the first time are often quite fond of this contradiction form, but they should be aware that there is a greater risk of overlooking some faulty reasoning when one argues from a hypothetically false premise. Thus the direct form or the contrapositive form of proof is preferred.

Although the implication "if *H* then *C*" appears brief and simple, both *H* and *C* can be very complicated compound statements. We illustrate this with an implication that has a two-part hypothesis and a single-part conclusion: "if H_1 and H_2 then *C*." The contrapositive of this is "if not *C*, then not (both) H_1 and H_2." Therefore a contrapositive proof could be achieved by assuming that *C* is false and H_1 is true, then deducing that H_2 is false. Alternatively, we could assume that *C* is false and H_2 is true, and then show that H_1 must be false. This discussion has barely scratched the surface of formal logic, but students should recognize these forms as they occur in the development of the mathematical theory.

This preliminary discussion closes with some remarks about word usages that are found in mathematical writing. The strict form of the implication "if *H* then *C*" would be very tedious if it were employed with no variation, so other ways of saying the same thing are used. For example, a well-known property of the real numbers may be stated, and restated, as follows:

(i) if x is a real number, then $x^2 \geq 0$

(ii) for every real number x, $x^2 \geq 0$

(iii) for each real number x, $x^2 \geq 0$

(iv) for any real number x, $x^2 \geq 0$

These statements mean the same thing, but the different wordings of the hypothesis allow one to emphasize subtle differences of connotation. Do not worry at this time about such matters of style and exposition; for now it is enough to know that such statements are equivalent. Similarly, the phrase "there exists" means exactly the same as "there is," but the former is used frequently for the sake of emphasis.

1

ORDERING
OF THE
REAL NUMBERS

1.1. The Order Axiom

The set of real numbers is denoted by $\mathbb{R}$. We assume all the familiar arithmetic and algebraic properties of the field of real numbers and choose—quite arbitrarily—to begin by defining the positive numbers $\mathbb{P}$ as a particular subset of $\mathbb{R}$.

ORDER AXIOM. The subset $\mathbb{P}$ of $\mathbb{R}$, called the *positive numbers,* satisfies the properties

(a) if x and y are in $\mathbb{P}$, then $x + y$ and xy are in $\mathbb{P}$; ← *2 positive elements*
(b) for each x in $\mathbb{R}$, exactly one of the following is true:
  (i) $x = 0$
  (ii) x is in $\mathbb{P}$
  (iii) $-x$ is in $\mathbb{P}$

In case $-x$ is in $\mathbb{P}$, we call x a *negative number.* Note that we do not use a symbol for the set of negative numbers. The natural choice is the symbol $\mathbb{N}$, but that is used almost universally to denote the natural numbers $\{1, 2, 3, \ldots\}$, and we, too, use $\mathbb{N}$ for the natural numbers.

The rules of arithmetic of positive and negative numbers are also assumed; for example, the product of two real numbers is negative if and only if one is positive and the other is negative.

DEFINITION 1.1. The inequality relation "less than" is defined as follows: $x < y$ ("x is less than y") means that x and y are in $\mathbb{R}$ and $y - x$ is in $\mathbb{P}$.

Several variations of the preceding relation are common and convenient. The statement "$y > x$" ("y is greater than x") means that $x < y$. The statement "$x \leq y$" ("x is less than or equal to y") means that $y \not< x$, that is, the negation of $y < x$. Similarly, "$y \geq x$" ("y is greater than or equal to x") means that $y \not< x$. Finally, the betweenness inequality "$x < y < z$" ("x is less than y, which is less than z") means that $x < y$ and $y < z$.

We now demonstrate some simple proofs using the Order Axiom and the inequality relation.

PROPOSITION 1.1. If $x < y$ and z is in $\mathbb{R}$, then $x + z < y + z$.

Proof. The hypothesis $x < y$ means that $y - x$ is in $\mathbb{P}$, and $y - x$ is the same as $(y + z) - (x + z)$. Thus $(y + z) - (x + z)$ is in $\mathbb{P}$, which means that $x + z < y + z$.

PROPOSITION 1.2. If $x < y$ and z is in $\mathbb{P}$, then $xz < yz$.

Proof. We assume that $x < y$, so $y - x$ is in $\mathbb{P}$. Also assume that z is in $\mathbb{P}$, so the closure of $\mathbb{P}$ under multiplication—property (a) of the Order Axiom—ensures that $(y - x)z$ is in $\mathbb{P}$. By the distributive law, this is the same as "$yz - xz$ is in $\mathbb{P}$"; hence, $xz < yz$.

PROPOSITION 1.3. If $x < y$ and $-z$ is in $\mathbb{P}$, then $xz > yz$.

Proof. See Exercise 1.1.1 (Exercise 1 at the end of Section 1.1).

PROPOSITION 1.4. If $x < y < z$, then $x < z$.

Proof. See Exercise 1.1.2.

PROPOSITION 1.5. If $0 < x < y$, then $x^2 < y^2$.

Proof. Since $x < y$, we can use Proposition 1.2 to multiply both sides by either x or y. (They are both positive by hypothesis and Proposition 1.4.) These multiplications yield $x^2 < xy$ and $xy < y^2$, respectively. Thus by Proposition 1.4, we conclude that $x^2 < y^2$.

In Propositions 1.6 and 1.7, we extend property (a) of the Order Axiom to show that $\mathbb{P}$ contains the sum and product of any finite set of its elements. This type of extension is encountered throughout mathematics. The proof uses the Principle of Mathematical Induction (see Appendix A).

$Prop$: $\text{If } N \in P, \text{ then } N > 0 \; (\text{iff})$

$N \in P \Rightarrow N - 0 \in P$

PROPOSITION 1.6. If $\{x_0, x_1, \ldots, x_n\} \subseteq \mathbb{P}$, then $x_0 + x_1 + \cdots + x_n$ is in $\mathbb{P}$.

Proof. First observe that if $n = 1$, the assertion is true, because it is part of property (a) of the Order Axiom. Next, we assume that the assertion is true when $n = k$ for some arbitrary k in $\mathbb{N}$. Thus

$$\begin{aligned} &\text{if } x_0, x_1, \ldots, x_k \text{ are } any \; k + 1 \text{ elements of } \mathbb{P}, \\ &\text{then } x_0 + x_1 + \cdots + x_k \in \mathbb{P}. \end{aligned} \tag{1}$$

Consider an arbitrary set of $k + 2$ elements of $\mathbb{P}$, say $\{x_0, x_1, \ldots, x_{k+1}\} \in \mathbb{P}$, and write their sum as

$$x_0 + x_1 + \cdots + x_{k+1} = (x_0 + \cdots + x_k) + x_{k+1}.$$

The right-hand side is the sum of two elements of $\mathbb{P}$: $(x_0 + \cdots + x_k)$ is in $\mathbb{P}$ by (1), and x_{k+1} is one of the set of $k + 2$ positive numbers. Therefore by property (a), the sum of these two positive numbers is in $\mathbb{P}$. Hence, by the Principle of Mathematical Induction, the assertion is true for any finite set of positive numbers.

PROPOSITION 1.7. If $\{x_0, x_1, \ldots, x_n\} \subseteq \mathbb{P}$, then $x_0 x_1 \ldots x_n$ is in $\mathbb{P}$.

Proof. This proof is similar to the proof of Proposition 1.6 and is left as an exercise (see Exercise 1.1.3).

PROPOSITION 1.8. If $0 < x < y$ and n is a positive integer, then $x^n < y^n$.

use $x^n < y^n$ $y^n - x^n > 0$ Either $\frac{1}{x} \in P$
$-\frac{1}{x} \in P$

Proof. See Exercise 1.1.4. $\frac{1}{x} = 0$

Prop. 1.9 $x \in P \Rightarrow \frac{1}{x} \in P$

Exercises 1.1

1. Prove Proposition 1.3.
2. Prove Proposition 1.4.
3. Prove Proposition 1.7. — MI   MI, factorization
4. Prove Proposition 1.8.
5. Prove: If $x < y$ and $a < b$, then $x + a < y + b$.   variation of 1.5
6. Prove: If x and y are in $\mathbb{P}$ and $x^2 < y^2$, then $x < y$.

Define $[x] = $ greatest integer $\leq x \Rightarrow [x] \leq x < [x] + 1$

7. Prove the Well-Ordering Principle: If S is a nonempty subset of $\mathbb{N}$, then S contains a smallest element. (*Hint:* Suppose S has no smallest element and let $T = \mathbb{N} \sim S$. Then use the Principle of Mathematical Induction to show that $T = \mathbb{N}$. Thus if S contains no smallest element, S is empty.)

1.2. Least Upper Bounds

We now consider the concept of bounded subsets of $\mathbb{R}$. A number set S is said to be *bounded* provided that it is contained in an interval $[a, b]$; that is, every element of S must satisfy $a \leq s \leq b$. In this case, the number a is called a *lower bound* of S, and b is called an *upper bound* of S. For example, the set $\{-1, 2, 5/2, 7\}$ is contained in the interval $[-1, 7]$ and therefore is bounded. The set of "proper" fractions, that is, $\{p/q: 0 < p < q$ and $p, q \in \mathbb{N}\}$, is contained in the interval $[0, 1]$, so it is bounded. Indeed, the interval $[0, 1]$ is itself a bounded set, as is any interval. On the other hand, the set $\mathbb{R}$ is not bounded, because no matter what interval $[a, b]$ may be chosen, it fails to contain $a - 1$ and $b + 1$, so it cannot contain $\mathbb{R}$.

If the number set S has an upper bound (but perhaps no lower bound), then S is said to be *bounded above*. Similarly, S is *bounded below* in case S has a lower bound (but perhaps no upper bound). A simple example is the set of positive numbers $\mathbb{P}$, which is bounded below by zero but is not bounded above.

If the set S is contained in the interval $[a, b]$, then a and b are not the only lower and upper bounds, respectively, for S. For if c is any number less than a, then for every s in S, $c < a \leq s$. So by Proposition 1.4, $c < s$, and c is therefore a lower bound of S. Similarly, the upper bound of S is not unique. The situation is sometimes quite different if we try to find an upper bound that is *smaller* than b or a lower bound that is *greater* than a. In the preceding paragraph we noted that the set $\{-1, 2, 5/2, 7\}$ is contained in $[-1, 7]$, but it is obvious that no upper bound can be less than 7 and no lower bound can be greater than -1. This suggests the concept of *least upper bound*.

Definition 1.2. The number β is said to be the *least upper bound* of the set S provided that

(i) β is an upper bound of S, and
(ii) if b is any upper bound of S, then $\beta \leq b$.

The least upper bound of S is abbreviated lub S.

The *greatest lower bound* of S is defined similarly and is abbreviated glb S.

It is important to realize that we are not requiring lub S to be an element of S. For example, the interval (a, b) consists of all numbers a that satisfy $a < s < b$. It is easy to show that no number less than b can be an upper bound of (a, b), and that no number greater than a can be a lower bound of (a, b). Thus $b = $ lub S and $a = $ glb S, although neither a nor b is in the set. These remarks lead us to the final, and perhaps most subtle, axiom of the real number system:

LEAST UPPER BOUND AXIOM. If S is a nonempty number set that is bounded above, then S has a least upper bound.

Following the LUB Axiom, we may expect a dual statement guaranteeing the existence of the greatest lower bound. It is not necessary, however, to postulate this as another axiom. We can deduce this property from those we have already assumed.[†]

COROLLARY 1.1: GREATEST LOWER BOUND PROPERTY. If S is a nonempty number set that is bounded below, then there is a greatest lower bound of S.

Proof. Let S^* denote the set $\{-s: s$ is in $S\}$. First we assert that if a is any lower bound of S, then $-a$ is an upper bound of S^*, because $a \leq s$ implies that $-s \leq -a$, by Proposition 1.3. Therefore S^* is bounded above, so by the LUB Axiom, there is a lub S^*, say $-\alpha$. This means that for every $-s$ in S^* and any upper bound $-a$ of S^*, $-s \leq -\alpha \leq -a$. By Proposition 1.3, this implies that $a \leq \alpha \leq s$; hence α is a lower bound of S that is as large as any lower bound of S; that is, $\alpha = $ glb S.

The importance of the LUB Axiom can hardly be overstated, for it is this property that makes possible the limiting concepts on which calculus is based and that guarantees that there is no "hole" in the real number line. If there were some point on the line to which no real number corresponded, then the set of all numbers corresponding to points lying to the left of the hole would have no least upper bound, although they would be bounded by any num-

---

[†]This illustrates the central spirit of mathematics: the goal is not the accumulation of a set of statements that are held to be true, but rather, it is the demonstration that the truth of some of these statements follows from that of others. Therefore the better theory is the one that assumes fewer axioms or proves more theorems.

the rational #'s do not have the lub property (& IRRATIONALS)

ber whose point lies to the right of the hole. Our present interest is more analytic than geometric, so we demonstrate the power of the LUB Axiom by proving two theorems that are at once intuitive and profound. The first deals with the set $\mathbb{N}$ of natural numbers. It is plain that no member of $\mathbb{N}$ can be an upper bound of $\mathbb{N}$, for if n is in $\mathbb{N}$, then $n + 1$ is also in $\mathbb{N}$. It is not obvious, however, that no noninteger can be an upper bound of $\mathbb{N}$. To prove this we use the LUB Axiom.

by contradiction

THEOREM 1.1. The set $\mathbb{N}$ of natural numbers is not bounded above.

Proof. Suppose that $\mathbb{N}$ did have an upper bound. Then by the LUB Axiom, $\mathbb{N}$ would have a least upper bound, say β. Thus for every n in $\mathbb{N}$, $n \le \beta$. But since $n + 1$ is also in $\mathbb{N}$, this implies that $n + 1 \le \beta$. Using Proposition 1.1 to rewrite the last inequality, we have $n \le \beta - 1$ for every n in $\mathbb{N}$. Hence $\beta - 1$ is another upper bound of $\mathbb{N}$, and it is *less than the least* upper bound β. From this contradiction we conclude that our original supposition was false; that is, $\mathbb{N}$ cannot have an upper bound.

1.3. The Density of the Rational Numbers

The final theorem of this chapter concerns a property of the rational numbers called *density*. For future use, the symbol $\mathbb{Q}$ denotes the rational numbers; that is, $\mathbb{Q} = \{n/m \colon n, m \text{ integers and } m \ne 0\}$. The number set S is said to be *dense* in $\mathbb{R}$ provided that for any two real numbers, say $x < y$, there is a member of S that satisfies $x < s < y$. If one pictures this on the number line, the density of S means that there is no interval that fails to contain an element of S.

THEOREM 1.2. The set of rational numbers is dense in $\mathbb{R}$.

Proof. Let x and y be any two real numbers, say $x < y$. Then $0 < y - x$, and by Theorem 1.1, $1/(y - x)$ is not an upper bound of $\mathbb{N}$. Choose a positive integer m such that $m > 1/(y - x)$, so $1/m < y - x$. Now let n be the positive integer such that $mx < n$; that is,

$$n - 1 \le mx < n,$$

or

$$\frac{n}{m} - \frac{1}{m} \le x < \frac{n}{m}. \tag{1}$$

Combining the left-hand inequality in (1) with $1/m < y - x$, we get

$$\frac{n}{m} \le x + \frac{1}{m} < x + (y - x) = y. \qquad (2)$$

Now (1) and (2) combine to give $x < n/m < y$.

Exercises 1.3

For each set, either find its least upper bound or show that it is not bounded above.   *alternately, could solve* $\{1 - \frac{1}{m}\}$ $1 - \frac{1}{n} < 1$

1. $A = \left\{ 1 - \frac{1}{n} : n \in \mathbb{N} \right\}$.   *Ans using greatest integer function as n gets larger value becomes closer to 1*

2. $B = \left(2 - \frac{1}{2}, 2 + \frac{1}{2} \right) \cup \left(3 - \frac{1}{3}, 3 + \frac{1}{3} \right) \cup \ldots$

$$\cup \left(n - \frac{1}{n}, n + \frac{1}{n} \right) \cup \ldots$$

3. $C = (0, 1) \cup \left(1, \frac{3}{2} \right) \cup \left(\frac{3}{2}, \frac{7}{4} \right) \cup \left(\frac{7}{4}, \frac{15}{8} \right) \cup \ldots$.

4. $D = \left\{ \frac{1}{2}, -\frac{1}{2}, \frac{3}{4}, -\frac{3}{4}, \frac{7}{8}, -\frac{7}{8}, \ldots \right\}$.

5. $E = \left\{ \frac{1}{2}, -\frac{1}{2}, 1, -1, \frac{3}{2}, -\frac{3}{2}, 2, -2, \ldots \right\}$.

6. $F = \left\{ n - \frac{1}{n} : n \in \mathbb{N} \right\}$.

7. Let A and B be nonempty subsets of $\mathbb{R}$ such that if $a \in A$ and $b \in B$, then $a < b$. Prove that lub $A \le$ glb B.

8. Assuming that it is known that $\sqrt{2}$ is irrational, prove that the set of irrational numbers is dense in $\mathbb{R}$.

Exercises 9–13 concern the concept of absolute value, which is defined as follows using the notation of this chapter:

$$|a| = \begin{cases} a, & \text{if } a \text{ is in } \mathbb{P}, \\ -a, & \text{if } a \text{ is in } \mathbb{R} \sim \mathbb{P}. \end{cases}$$

9. Prove that for every a in $\mathbb{R}$, $|a| \ge 0$.

10. Prove that for every a and b in $\mathbb{R}$, $|ab| = |a| \, |b|$.

11. Prove that for every a and b in $\mathbb{R}$, $|a + b| \le |a| + |b|$.

12. Prove that for every a and b in $\mathbb{R}$, $\big||a| - |b|\big| \leq |a - b|$.

13. Use Mathematical Induction to prove that for every a_1, $a_2, \ldots, a_n$ in $\mathbb{R}$,

$$|a_1 + a_2 + \cdots + a_n| \leq |a_1| + |a_2| + \cdots + |a_n|.$$

14. Prove the Dedekind Cut Theorem: Suppose that A and B are nonempty sets such that $A \cup B = \mathbb{R}$ and if $a \in A$ and $b \in B$, then $a < b$. Then there exists a "cut point" c such that if $x < c < y$, then $x \in A$ and $y \in B$.

1.11 $A \subset B$ and A not bounded f/ above (below)
$\Rightarrow B$ is not bounded b/ above (below)

1.12 $A \subset B$ and B is bounded f/ above (below)
$\Rightarrow A$ is bounded f/ above (below)

1.13 LUB of a set is unique
GLB of " " " "

Suppose α and β are lubs of a set A
α is an upper bound $\Rightarrow \beta \leq \alpha$
β " " " " $\Rightarrow \alpha \leq \beta$
$\Rightarrow \alpha = \beta$

1.14 $-c \leq A \leq c$ if $|A| \leq c$ $|A| = \begin{cases} A & A \geq 0 \\ -A & A < 0 \end{cases}$

$|A| = \sqrt{A^2}$

1.15 Every finite set is bounded above & below

1.16 If $A \subset B$ and both A and B are bounded above,
$\Rightarrow$ LUB $A \leq$ LUB B

2

SEQUENCE LIMITS

2.1. Convergent Sequences

A *function* can be defined as a collection (or set) of ordered pairs (x, y) such that no two pairs have the same first element. Then the *domain* of the function is the set of all first members of pairs in the collection, and its *range* is the set of all second members. In this chapter we study functions with a particular domain: A *sequence* is a function whose domain is an infinite subset of $\mathbb{N} = \{1, 2, 3, \ldots\}$, and a *number sequence* is a sequence whose range is a subset of $\mathbb{R}$. The only sequences that we consider here are number sequences, so we refer to them simply as sequences. It is sometimes convenient to think of a sequence as having $\{0, 1, 2, \ldots\}$ as its domain. This causes no substantive difference in the theory, and it is clear from the context whether the domain is $\mathbb{N}$ or $\{0\} \cup \mathbb{N}$.

If s is a sequence and n is in $\mathbb{N}$, it is customary to write s_n for the image of n under the function s, rather than the usual function notation $s(n)$. In this case, s_n is called the *nth term* of s.

The sequence s is said to be bounded provided that the range of s is a bounded set. This is equivalent to saying that there is a number B such that for every n in $\mathbb{N}$, $|s_n| \le B$. A *constant sequence* is a sequence whose range consists of a single number; that is, $s_n = c$ for every n in $\mathbb{N}$. A sequence is *eventually constant* if there is a number c and an integer N such that $n > N$ implies that $s_n = c$.

DEFINITION 2.1. The sequence s is said to *converge* to the number L provided that if $\varepsilon > 0$ then there is a number N such that

13

$$n > N \quad \text{implies} \quad |s_n - L| < \varepsilon.$$

In this case, s is called a *convergent sequence*, and we write $\lim_n s_n = L$.

It is sometimes convenient to describe a sequence by writing its first few terms to suggest the pattern of the remaining terms. Although this practice leads to occasional ambiguities, it is so widely used that one must get accustomed to it. It is good practice to try to write a formula for s_n whenever one encounters a sequence described by $\{s_1, s_2, s_3, \ldots\}$.

EXAMPLE 2.1. If $s = \{1, 1, 2, 2, 2, \ldots\}$, then

$$s_n = \begin{cases} 1, & \text{if } n \le 2, \\ 2, & \text{if } n > 2. \end{cases}$$

This sequence is eventually constant, and $\lim_n s_n = 2$.

EXAMPLE 2.2. If $s = \{a, \ldots, a, L, L, L, \ldots\}$, then

$$s_n = \begin{cases} a, & \text{if } n \le N, \\ L, & \text{if } n > N. \end{cases}$$

This is a more general eventually constant sequence, and it is convergent to L.

EXAMPLE 2.3. The *harmonic sequence* $\{1, 1/2, 1/3, \ldots\} = \{1/n\}_{n=1}^{\infty}$ is convergent to the limit zero, which we here prove in complete detail. Suppose $\varepsilon > 0$. By Theorem 1.1, $1/\varepsilon$ cannot be an upper bound for $\mathbb{N}$, so there exists an integer $N > 1/\varepsilon$. If $n > N$, then by Proposition 1.4, $n > 1/\varepsilon$; and by Proposition 1.2, $1/n < \varepsilon$; that is, $|(1/n) - 0| < \varepsilon$. Thus $\lim_n 1/n = 0$.

EXAMPLE 2.4. If $s = \{1, 0, 1/2, 0, 1/3, 0, \ldots\}$, then

$$s_n = \begin{cases} 1/k, & \text{if } n = 2k - 1, \\ 0, & \text{if } n = 2k, \end{cases}$$

for k in $\mathbb{N}$. Here, $\lim_n s_n = 0$, which can be shown by choosing $N > 2/\varepsilon$ and proceeding as in Example 2.3.

EXAMPLE 2.5. The *geometric sequence* is $\{r^n\}_{n=1}^{\infty}$, where $0 < r < 1$. (Here, r is restricted to the unit interval for the sake of convenience.) For such r, we prove that $\lim_n r^n = 0$. First note that $1/r > 1$ (by Propositions 1.4 and 1.2), so we can write $1/r = 1 + h$, for some positive h. Then

$$(1/r)^n = (1 + h)^n = 1 + nh + \frac{n(n-1)}{2}h^2 + \cdots + h^n > nh.$$

Therefore $r^n < 1/(nh)$. Now for a given $\varepsilon > 0$, choose $N > 1/(h\varepsilon)$ and proceed as in Example 2.3.

Examples 2.6–2.8 are nonconvergent sequences.

EXAMPLE 2.6. $\{1, 2, 3, \ldots\} = \{n\}_{n=1}^{\infty}$.

EXAMPLE 2.7. If $s = \{1, 0, 2, 0, 3, 0, \ldots\}$, then

$$s_n = \begin{cases} k, & \text{if } n = 2k - 1, \\ 0, & \text{if } n = 2k. \end{cases}$$

EXAMPLE 2.8. If $s = \{1, 0, 1, 0, 1, 0, \ldots\}$, then

$$s_n = \frac{1 + (-1)^{n+1}}{2}.$$

To show that none of these sequences satisfies the definition of convergence, we argue that for any value L to be the limit, all but a finite number of the terms must be between $L - 1/3$ and $L + 1/3$. (We are applying the definition using $\varepsilon = 1/3$.) But the only way this can occur for a sequence of integers is for the sequence to be eventually constant, and none of these examples is eventually constant.

It is inconvenient to have to rely solely on the definition to determine convergence of every example. Therefore it is to our advantage to prove some basic theorems about convergent sequences and their limits so that we can use them to investigate further examples. As an illustration of this, the next theorem could be used to deduce immediately that Examples 2.6 and 2.7 are nonconvergent.

THEOREM 2.1. If s is a convergent sequence, then s is bounded. $\star$ MEMORIZE!

$|a| - |b| \le |a + b|$

Proof. Suppose $\lim_n s_n = L$, and choose N so that $n > N$ implies $|s_n - L| < 1$, which is the same as $L - 1 < s_n < L + 1$. Thus $|s_n| < |L| + 1$ whenever $n > N$, which means that the number $|L| + 1$ is an upper bound for the set $\{|s_{N+1}|, |s_{N+2}|, \ldots\}$. Let B be the number given by

$|s_n - L| \ge |s_n| - |L|$

$$ B = \max\{|s_1|, |s_2|, \ldots, |s_N|, |L| + 1\}. $$

Then B is at least as great as each of the first N terms of $|s|$ as well as being an upper bound for the set $\{|s_{N+1}|, |s_{N+2}|, \ldots\}$. Hence for *every n,* $|s_n| \le B$.

The next theorem, although very important to our convergence theory, is one of those results that would be taken for granted by almost anyone other than a mathematician. It assures us that a convergent sequence cannot have more than one limit value.

THEOREM 2.2. If s is a convergent sequence, then its limit value is unique.

Proof. Assume that $\lim_n s_n = L$, and let M be a number not equal to L. We must show that s cannot converge to M. Suppose $\varepsilon = |L - M|/2$; that is, ε is half the distance between L and M. Since $\lim_n s_n = L$, there is an N such that $n > N$ implies that $L - \varepsilon < s_n < L + \varepsilon$. But if s_n is in the interval $(L - \varepsilon, L + \varepsilon)$ it cannot be in the interval $(M - \varepsilon, M + \varepsilon)$, because these intervals do not intersect one another. Therefore $|s_n - M| \not< \varepsilon$ whenever $n > N$, so s cannot converge to M.

Sometimes we wish to investigate the convergence of a sequence with a complicated formula that can be compared by inequalities to the formula of a known sequence. In this case, we wish to deduce convergence of the complicated sequence from the known one. For example, if $s_n = 2^{-n}[(n + 2)/(n + 3)]$, we know that the second factor is at most equal to 1, and therefore $s_n \le 2^{-n}$.

Since we know that $\lim_n 2^{-n} = 0$, this should tell us that $\lim_n s_n = 0$ also. This situation of "squeezing" the term s_n between the term of a convergent sequence and its limit is the essence of the next result.

"Squeeze Theorem"

PROPOSITION 2.1. If s and t are sequences such that $\lim_n t_n = L$ and for every n, $L \le s_n \le t_n$, then $\lim_n s_n = L$.

Proof. The inequalities $L \le s_n \le t_n$ imply that $|s_n - L| \le |t_n - L|$. Verify this and complete the proof in Exercise 2.1.12.

The next result is useful for approximating the limit value of a convergent sequence when there is not enough information to find its exact value.

PROPOSITION 2.2. If $\lim s_n = L$ and for every n, s_n is in the interval $[a, b]$, then L is also in $[a, b]$.

$\rightarrow$ by contradiction

Proof. First we show that $L \le b$. Suppose not, say $L > b$ and let $\varepsilon = L - b > 0$. Then the inequality $|s_n - L| < \varepsilon$ implies that

$$s_n > L - \varepsilon = L - (L - b) = b.$$

Thus s_n would not be in $[a, b]$. The remainder of the proof is required in Exercise 2.1.13.

Exercises 2.1

For each sequence in Exercises 1–11, determine convergence or nonconvergence and prove your conclusion.

1. $s_n = \dfrac{2n + 1}{n}$.

2. $s_n = (-1)^n$.

3. $s_n = \dfrac{1}{\sqrt{n}}$.

4. $s_n = \left[\dfrac{n + 1}{2}\right]$, where $[x]$ denotes the greatest integer not exceeding x.

5. $s_n = \left[\dfrac{1}{2} + \dfrac{1}{3} \sin n\right]$, where $[x]$ is as in Exercise 4.

6. $s_n = \dfrac{\cos n}{n}$.

7. $s_n = \begin{cases} 1/n, & \text{if } n \text{ is odd,} \\ 0, & \text{if } n \text{ is even.} \end{cases}$

8. $s_n = \begin{cases} 1/n, & \text{if } n \text{ is odd,} \\ 1, & \text{if } n \text{ is even.} \end{cases}$

9. $s_n = \begin{cases} 1/n, & \text{if } n \text{ is odd,} \\ 1/2^n, & \text{if } n \text{ is even.} \end{cases}$

10. $s_n = \left\{ \dfrac{1}{2}, \dfrac{1}{3}, \dfrac{1}{2^2}, \dfrac{1}{3^2}, \dfrac{1}{2^3}, \dfrac{1}{3^3}, \cdots \right\}.$

11. $s = \left\{ 0, 1, 0, \dfrac{1}{2}, 1, 0, \dfrac{1}{3}, \dfrac{2}{3}, 1, 0, \dfrac{1}{4}, \cdots \right\}.$

12. Prove Proposition 2.1.

13. Complete the proof of Proposition 2.2.

14. Show by an example that in Proposition 2.2 the stronger assumption that s_n is in (a, b) does *not* guarantee that L is also in (a, b).

15. Prove: For every real number L there exists a sequence q of rational numbers such that $\lim_n q_n = L$. (*Hint:* See Theorem 1.2.)

16. Prove: For every real number L there exists a sequence s of irrational numbers such that $\lim_n s_n = L$. (*Hint:* See Exercise 1.3.8.)

2.2. Algebraic Combinations of Sequences

Our next objective is to investigate the algebraic closure of the set of convergent sequences. For example, if s and t are sequences, we can form their sum $s + t$, which is the sequence whose nth term is $s_n + t_n$. Similarly, the difference $s - t$, the product st, and the quotient s/t are defined by the appropriate combination of nth terms of s and t. It would be useful to know that the convergence of s and t implies that these algebraic combinations of s and t are also convergent. That is the gist of the next two theorems. But before stating those results, we prove three lemmas that help prove the theorems that follow.

A sequence that converges to zero is called a *null sequence.* This subset of the convergent sequences can be used to characterize convergence to an arbitrary limit L, and that is the import of the first lemma.

Use to prove theorems

✳ **LEMMA 2.1.** If s is a sequence and L is a number, then

$$\lim_n s_n = L \quad \text{if and only if} \quad \lim_n (s_n - L) = 0.$$

Proof. The inequality $|s_n - L| < \varepsilon$ is obviously equivalent to the inequality $|(s_n - L) - 0| < \varepsilon$, so the assertion follows immediately from the definition of convergence.

→ *Product of a null sequence and bounded sequence = a null sequence*

✳ **LEMMA 2.2.** If s is a bounded sequence and t is a null sequence, then st is a null sequence.

Proof. Suppose $\varepsilon > 0$ and $|s_n| < B$ for every n in $\mathbb{N}$. Choose N so that $n > N$ implies $|t_n| < \varepsilon/B$. Then $n > N$ implies

$$|s_n t_n| = |s_n|\,|t_n| < B(\varepsilon/B) = \varepsilon.$$

Hence, $\lim_n s_n t_n = 0$.

✳ **LEMMA 2.3.** If t is a convergent sequence whose limit is not zero, and for every n in $\mathbb{N}$, $t_n \neq 0$, then $1/t$ is a bounded sequence.

Proof. Assume $\lim_n t_n = M \neq 0$, and choose N so that $n > N$ implies that t_n is between $M/2$ and $3M/2$. Then $n > N$ implies that $|t_n| > M/2$, or $|1/t_n| < 2/M$. Now choose $B = \max\{|1/t_1|, \ldots, |1/t_N|, 2/|M|\}$, and it is clear that for every n, $|1/t_n| \leq B$.

Armed with these lemmas, we are now prepared to prove that the set of convergent sequences is closed under addition, subtraction, multiplication, and division. (For division, it is necessary to make some further assumptions about t in order to avoid zeros in the denominator of s/t.)

✳ **THEOREM 2.3.** Suppose each of s and t is a convergent sequence and c is a number; then $s + t$, $s - t$, and cs are convergent. Also, if $\lim_n s_n = L$ and $\lim_n t_n = M$, then

$$\lim_n (s_n \pm t_n) = L \pm M$$

and

$$\lim_n cs_n = cL.$$

Proof. Suppose $\varepsilon > 0$. Choose N_s and N_t so that $n > N_s$ implies $|s_n - L| < \varepsilon/2$ and $n > N_t$ implies $|t_n - M| < \varepsilon/2$. Now define $N =$

$\max\{N_s, N_t\}$. Then $n > N$ implies both $n > N_s$ and $n > N_t$, which together imply that

$$|(s_n \pm t_n) - (L \pm M)| = |(s_n - L) \pm (t_n - M)|$$

$$\leq |s_n - L| + |t_n - M|$$

$$< \frac{\varepsilon}{2} + \frac{\varepsilon}{2}$$

$$= \varepsilon.$$

Hence $\lim_n(s_n \pm t_n) = L \pm M$.

To prove that $\lim_n cs_n = cL$, we first note that if $c = 0$, then the assertion is trivial, for cs is identically equal to zero. Assume $c \neq 0$ xnd $\varepsilon > 0$. Choose N so that $n > N$ implies $|s_n - L| < \varepsilon/|c|$. Now whenever $n > N$, we have

$$|cs_n - cL| = |c| \, |s_n - L| < |c| \, (\varepsilon/|c|) = \varepsilon.$$

Hence $\lim_n cs_n = cL$.

Theorem 2.4. Suppose each of s and t is a convergent sequence, say $\lim_n s_n = L$ and $\lim_n t_n = M$; then st is convergent and $\lim_n s_n t_n = LM$. If, in addition, t_n is never 0 and $M \neq 0$, then s/t is convergent and

$$\lim_n = \frac{s_n}{t_n} = \frac{L}{M}.$$

Proof. Consider the identity

$$s_n t_n - LM = (s_n t_n - Ms_n) + (Ms_n - LM)$$

$$= s_n(t_n - M) + M(s_n - L).$$

By Theorem 2.1, s is bounded, and by Lemma 2.1, $\lim_n(t_n - M) = 0$. Therefore by Lemma 2.2, $\lim_n s_n(t_n - M) = 0$. Also, by Lemma 2.1, $\lim_n(s_n - L) = 0$, so by Theorem 2.3, $\lim_n M(s_n - L) = 0$. Thus we have shown that $st - LM$ is the sum of two null sequences, and therefore by Theorem 2.4, $st - LM$ is also a null sequence. Hence, by Lemma 2.1, $\lim_n s_n t_n = LM$.

To prove that $\lim_n s_n/t_n = L/M$, it is sufficient to show that $\lim_n 1/t_n = 1/M$, for we can apply the conclusion of the preceding paragraph to the product $(s_n)(1/t_n)$ to get the conclusion for the quotient. By Lemma 2.1, the assertion $\lim_n 1/t_n = 1/M$ is equivalent to

$$\lim_n \left(\frac{1}{t_n} - \frac{1}{M} \right) = 0,$$

which is the same as

$$\lim_n \frac{t_n - M}{t_n M} = 0. \tag{1}$$

By Lemma 2.3, $1/t$ is bounded, because it is convergent to a non-zero limit; and by Lemma 2.1, $\lim_n (t_n - M) = 0$. Therefore by Lemma 2.2, (1) holds and the proof is complete.

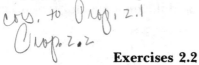

Exercises 2.2

1. Prove that $s_n = (2n + 3)/(n + 1)$ defines a convergent sequence by using the results of this section and writing s_n in the form

$$\frac{2 + (3/n)}{1 + (1/n)}.$$

2. Prove that $\{(n^2 - 2n)/(3n^2 + 1)\}_{n=1}^{\infty}$ is convergent by the method of Exercise 1.

3. In the part of Theorem 2.4 concerning the quotient, we assumed that (i) $t_n \neq 0$ for every n and (ii) $\lim_n t_n \neq 0$. Show by an example that both assumptions are needed, that is, that (i) does not imply (ii).

4. Show by an example that the sum or the difference of two nonconvergent sequences can be convergent.

5. Prove: If both the sum $s + t$ and the difference $s - t$ of two sequences are convergent, then s and t are themselves convergent.

6. Prove by Mathematical Induction that Theorem 2.3 can be extended to the sum of an arbitrary finite number of convergent sequences: if

$$\lim_n s_n^{(i)} = L_i \quad \text{for} \quad i = 1, 2, \ldots, k,$$

then

$$\lim_n \sum_{i=1}^{k} s_n^{(i)} = \sum_{i=1}^{k} L_i.$$

7. Prove by Mathematical Induction that Theorem 2.4 can be extended to the product of an arbitrary finite number of convergent sequences: if

$$\lim_n s_n^{(i)} = L_i \quad \text{for} \quad i = 1, 2, \ldots, k,$$

then

$$\lim_n s_n^{(1)} s_n^{(2)} \ldots s_n^{(k)} = L_1 L_2 \ldots L_k.$$

8. Let s be given by

$$s_n = \frac{a_k n^k + a_{k-1} n^{k-1} + \cdots + a_0}{b_k n^k + b_{k-1} n^{k-1} + \cdots + b_0},$$

where $a_k \neq 0$, $b_k \neq 0$, and the denominator is nonzero for every n. Prove that

$$\lim_n s_n = \frac{a_k}{b_k}.$$

2.3. Infinite Limits

There are cases in which a sequence is unbounded but its behavior is regular enough to be described with a pseudo-limit. This useful concept is the topic of this section.

DEFINITION 2.2. If s is a sequence, then the statement "s tends to infinity," denoted $\lim_n s_n = \infty$, means that if B is any number, then there is a number N such that

$$n > N \quad \text{implies} \quad s_n > B.$$

Similarly, the statement "s tends to negative infinity," denoted by $\lim_n s_n = -\infty$, means that if B' is any number, then there is a number N' such that

$$n > N' \quad \text{implies} \quad s_n < B'.$$

It is important to realize that such a sequence is *not* convergent, and therefore these infinite limits are not subject to the conclusions of Theorems 2.1–2.4. It is also important to observe that not every unbounded sequence tends to infinity. Consider, for example, the sequence $\{1, 0, 2, 0, 3, 0, \ldots\}$.

PROPOSITION 2.3. Suppose that s is a sequence such that for each n, $s_n > 0$; then $\lim_n s_n = \infty$ if and only if $\lim_n 1/s_n = 0$.

Keep this in mind when doing problems

Proof. Assume $\lim_n s_n = \infty$ and $\varepsilon > 0$. Apply Definition 2.2 with $B = 1/\varepsilon$ to get a number N such that

$$n > N \quad \text{implies} \quad s_n > \frac{1}{\varepsilon}. \tag{1}$$

By Proposition 1.2, this is equivalent to

$$n > N \quad \text{implies} \quad \frac{1}{s_n} < \varepsilon. \tag{2}$$

Therefore $\lim_n 1/s_n = 0$. Since (2) implies (1), the converse implication is proved similarly.

Exercises 2.3

1. Prove that $\lim_n (n^2 - 5n + 1) = \infty$.

2. Prove that $\lim_n (n - 7\sqrt{n}) = \infty$.

3. Prove that $\lim_n (-n + \sin n) = -\infty$.

4. Prove that $\lim_n s_n = \infty$ if and only if $\lim_n (-s_n) = -\infty$.

5. Prove: If each of s and t is a sequence that tends to infinity, then $s + t$ tends to infinity.

6. Prove: If each of s and t is a sequence that tends to infinity, then st tends to infinity.

7. Show by an example that two sequences can tend to infinity and their difference *not* tend to infinity.

8. Show by an example that two sequences can tend to infinity and their quotient *not* tend to infinity.

2.4. Subsequences and Limit Points

The concept of a subsequence is both natural and simple. Since a sequence is a function and every function is a collection of ordered pairs, we can display it as

$$\{(1, s_1), (2, s_2), \ldots, (n, s_n), \ldots\}.$$

If t is an infinite subset of this collection, then t is called a *subsequence* of s. Note that t is itself a sequence because it is a function whose domain is an infinite subset of $\mathbb{N}$. The domain of t consists of those n's whose pairs are in the subcollection making up t. These n's can be thought of as an increasing sequence of positive integers, and therefore they are usually denoted by n_1, n_2, ..., n_k, ... Thus the first term of t is denoted by s_{n_1}, the kth term by s_{n_k}, and in general

$$t = \{s_{n_1}, s_{n_2}, \ldots, s_{n_k}, \ldots\} = \{s_{n_k}\}_{k=1}^{\infty}.$$

Sometimes the subscript-of-a-subscript notation is too cumbersome, and we write $s_{n(k)}$ to denote the kth term of a subsequence of s. If a formula for n_k or $n(k)$ is known and simple, we may write that in the subscript position. For example,

$$\{s_{2k}\}_{k=1}^{\infty} = \{s_2, s_4, s_6, \ldots\}$$

is called the *subsequence of even-indexed terms*. Similarly,

$$\{s_{2k-1}\}_{k=1}^{\infty} = \{s_1, s_3, s_5, \ldots\}$$

is called the *subsequence of odd-indexed terms*.

It should be emphasized that the terms of the subsequence must retain the same order as in the original sequence. This is implied by the fact that the subscripts n_1, n_2, ... are chosen in *increasing* order. Thus if $s = \{1/n\}_{n=1}^{\infty}$, then

$$\left\{\frac{1}{2}, \frac{1}{4}, \frac{1}{6}, \frac{1}{8}, \ldots\right\} \text{ is a subsequence of } s,$$

but

$$\left\{\frac{1}{2}, \frac{1}{6}, \frac{1}{4}, \frac{1}{8}, \ldots\right\} \text{ is } not \text{ a subsequence of } s.$$

PROPOSITION 2.4. If s is a bounded sequence, then every sub-sequence of s is also bounded.   *iff*

Converse is also true

Proof. See Exercise 2.4.1.

PROPOSITION 2.5. If s is a sequence that converges to L, then every subsequence of s also converges to L. *Converse also true*

Proof. See Exercise 2.4.2.

Next we introduce the concept of *limit point*, a concept that is related to subsequences and is an extension of the idea of the limit of a convergent sequence.

P29 in notes

DEFINITION 2.3. The number λ is called a limit point of the sequence s provided that s has a subsequence that converges to λ.

Some authors use the terms *cluster point* or *accumulation point* instead of limit point. Indeed, one must be careful not to confuse a limit point of a sequence with the limit of a sequence. For example, the sequence

$$s = \left\{1, \frac{1}{2}, 1, \frac{1}{3}, 1, \frac{1}{4}, \dots\right\}$$

has two limit points, 0 and 1, because

$$\lim{}_k s_{2k} = \lim{}_k \frac{1}{k+1} = 0$$

and

$$\lim{}_k s_{2k-1} = \lim{}_k 1 = 1.$$

But s does not have a limit, because it is nonconvergent. This example suggests the following general result.

PROPOSITION 2.6. If the sequence s is convergent, then s has exactly one limit point. *Converse is false*

Proof. See Exercise 2.4.4.

Proposition 2.6 gives us a very convenient method of showing that a sequence is nonconvergent, namely, to show that there are two convergent subsequences with limits that are not equal.

Exercises 2.4

1. Prove Proposition 2.4.

2. Prove Proposition 2.5. (*Hint:* Since the subscripts n_k of the subsequence $\{s_{n_k}\}$ are increasing, it follows that $n_k \geq k$ for $k = 1, 2, \dots$.)

3. Give an example of an unbounded sequence that has a convergent subsequence.

4. Prove Proposition 2.6.

In Exercises 5–8, find all limit points of s.

5. $s_n = \dfrac{(-1)^n n + 1}{n}$.

6. $s = \left\{0, 1, 0, \dfrac{1}{2}, 1, 0, \dfrac{1}{4}, \dfrac{1}{2}, \dfrac{3}{4}, 1, \ldots\right\}$.

7. $s = \{0, 1, 0, 0, 1, 0, 0, 0, 1, \ldots\}$.

8. $s_n = \begin{cases} \dfrac{2n + 1}{n}, & \text{if } n = k^2 \text{ for some } k \text{ in } \mathbb{N}, \\ (-1)^n, & \text{if } n \text{ is not a square.} \end{cases}$

9. Prove: The sequence s has no limit point if and only if $\lim_n |s_n| = \infty$.

2.5. Monotonic Sequences

In the discussion of subsequences we noted that the subscripts n_k were "increasing." Familiarity with the concepts of increasing and decreasing functions is assumed here. These are now applied to sequences.

DEFINITION 2.4. The sequence s is *increasing* (respectively, *decreasing*) provided that $s_n < s_{n+1}$ (respectively, $s_n > s_{n+1}$) for every n. Similarly, s is *nondecreasing* (respectively, *nonincreasing*) provided that $s_n \le s_{n+1}$ (respectively, $s_n \ge s_{n+1}$) for every n. If s satisfies any of these four criteria, then s is called a *monotonic sequence*.

Because of their very regular behavior, monotonic sequences are particularly easy to work with in determining convergence. This fact is brought out in the next theorem, which is a fundamental result (see Exercise 2.5.12).

THEOREM 2.5: MONOTONIC SEQUENCE THEOREM. A monotonic sequence is convergent if and only if it is bounded.

Proof. First note that the implication in one direction is already proved: By Theorem 2.1, if s is unbounded, then it cannot be convergent. To prove the converse implication, consider first the case in which s is nondecreasing and bounded above. (Note that this case includes the increasing sequences. Also, we do not have to assume that a nondecreasing sequence is bounded below; its first term must be a lower bound.) By the LUB Axiom, there is a least upper bound of the range of s, say, $\beta = \text{lub}\{s_n : n \in \mathbb{N}\}$. For any positive ε, $\beta - \varepsilon$ is not an upper bound of s, so there is an N such that $s_N > \beta - \varepsilon$. Since s is nondecreasing and β is an upper bound of s, $n > N$ implies

$$\beta - \varepsilon < s_N \leq s_n \leq \beta < \beta + \varepsilon.$$

Thus $n > N$ implies $|s_n - \beta| < \varepsilon$, so $\lim_n s_n = \beta$. In a similar fashion, we can prove that if s is nonincreasing and bounded below, then $\lim_n s_n = \text{glb}\{s_n : n \in \mathbb{N}\}$, which completes the proof.

In the proof of Theorem 2.5, we actually proved more than the theorem states. We proved that if s is nondecreasing and bounded above, then s converges *to the least upper bound of its range.* It is sometimes useful to use the Monotonic Sequence Theorem in this form because it gives a precise value to the limit of s.

Exercises 2.5

1. Find a monotonic subsequence of $\left\{n + \dfrac{(-1)^n}{n}\right\}_{n=1}^{\infty}$.

2. Find an increasing subsequence of

$$s = \left\{0, 1, 0, \frac{1}{2}, 1, 0, \frac{1}{4}, \frac{1}{2}, \frac{3}{4}, 1, 0, \ldots\right\}.$$

3. Find a nonincreasing subsequence of $s = \{0, 1, 0, 2, 0, 3, 0, \ldots\}$.

4. For the proof of Theorem 2.5, give the details of the proof for the case of a nonincreasing sequence that is bounded below.

In Exercises 5–7, prove that s is monotonic.

5. $s_n = \dfrac{n+1}{n}$.

6. $s_n = r^n$, where $r \geq 0$.

7. $s_n = \dfrac{2^n}{n!}$.

8. Prove: If each of s and t is a nondecreasing sequence, then $s + t$ is nondecreasing.

9. Give an example of two monotonic sequences whose sum is not monotonic.

10. Prove: If s is an unbounded sequence, then s has a monotonic subsequence.

11. Prove: If s is a monotonic sequence of positive numbers, then $1/s$ is a monotonic sequence.

12. Prove that the Monotonic Sequence Theorem implies the Least Upper Bound Axiom.

13. Prove that the Dedekind Cut Theorem of Exercise 1.3.14 implies the Least Upper Bound Axiom.

3

COMPLETENESS
OF THE
REAL NUMBERS

3.1. The Bolzano-Weierstrass Theorem

In this chapter we investigate some concepts and results that are directly related to the LUB Axiom. Indeed, the major theorems of this chapter consist of four statements that are equivalent to the LUB Axiom on $\mathbb{R}$; that is, any one of these four statements could have been assumed as the axiom and the others deduced from it. We have already encountered two such results: namely, the Dedekind Cut Theorem (Exercise 1.3.14) and the Monotonic Sequence Theorem (Theorem 2.5). In Exercises 2.5.12 and 2.5.13, it was asserted that each of these theorems implies the LUB Axiom. In developing the foundations of the real number system, it is very important to establish the equivalence of these various properties. But our objective is to develop limit and convergence theories in $\mathbb{R}$, so we must be content in this chapter to prove only that the LUB Axiom implies the new properties.

The first of the four completeness theorems of this chapter concerns bounded sequences. It is plausible to assert the following observation: A bounded sequence has infinitely many terms in a finite interval, so the terms must cluster somewhere, which means that there is a limit point or a convergent subsequence. We can prove this easily once we have established the following lemma.

LEMMA 3.1. Every sequence has a monotonic subsequence.

Proof. We first introduce the collection of "tail end" subsequences of a sequence s: $S_1 = \{s_1, s_2, \ldots\}$, $S_2 = \{s_2, s_3, \ldots\}, \ldots,$

$S_N = \{s_N, s_{N+1}, \ldots\}, \ldots$. There are two cases to consider. First, suppose that some S_N has no largest term. Then it is clear that S_N contains an increasing subsequence, because for each term that is selected there must follow a term that is even bigger. Thus s has an increasing subsequence because any subsequence of S_N is also a subsequence of s. Now consider the case in which every S_N has a largest term. Let $s_{k(1)}$ be the largest term of S_1 (= s). (In case more than one term has the same largest value, we choose $s_{k(1)}$ to be the first such term.) Now consider $S_{1+k(1)} = \{s_{1+k(1)}, s_{2+k(1)}, \ldots\}$, and let $s_{k(2)}$ be its largest term. Since $S_{1+k(1)}$ is a subsequence of S_1, its largest term cannot exceed that of S_1; that is, $s_{k(1)} \geq s_{k(2)}$. Next let $s_{k(3)}$ be the largest term of $S_{1+k(2)}$, so $s_{k(2)} \geq s_{k(3)}$. Continuing in this manner we can always select a next term $s_{k(n+1)}$ that is the largest term of $S_{1+k(n)}$, which gives $s_{k(n)} \geq s_{k(n+1)}$. Hence, $\{s_{k(n)}\}_{n=1}^{\infty}$ is a non-increasing subsequence of s. Therefore, in either case, s has a monotonic subsequence.

Before applying this lemma to bounded sequences, we should note the total generality of the implication. We have proved that *every* sequence—bounded, unbounded, convergent, or what have you—has a monotonic subsequence. So although the proof of the lemma involves a lengthy construction argument, we receive a lot in return.

THEOREM 3.1: BOLZANO-WEIERSTRASS THEOREM. If s is a bounded sequence, then s has a convergent subsequence.

Proof. By Lemma 3.1, s has a monotonic subsequence t, and t must be bounded because s is bounded. Therefore, by the Monotonic Sequence Theorem, t is convergent.

3.2. Cauchy Sequences

Our next goal is to develop a criterion for determining convergence of a sequence without referring to the limit value of the sequence. To do this we introduce the concept of a Cauchy sequence.

DEFINITION 3.1. The sequence s is called a *Cauchy sequence* provided that if $\varepsilon > 0$, there is a number N such that

$$m > N \quad \text{and} \quad n > N \quad \text{imply} \quad |s_m - s_n| < \varepsilon. \tag{1}$$

The last line (1) of this definition is usually written more compactly as

$$m, n > N \quad \text{implies} \quad |s_m - s_n| < \varepsilon.$$

One may think of this as saying that the terms of *s* get arbitrarily *close to one another*. Contrast this with the definition of convergence, which says that the terms of *s* get arbitrarily *close to some number L*. We can prove that these two concepts are equivalent for real number sequences, but first we must develop some facts about Cauchy sequences.

If *s* is a Cauchy sequence, it is easy to see that $\lim_n (s_{n+1} - s_n) = 0$; for in (1) the number *m* could be $n + 1$, which yields $|s_{n+1} - s_n| < \varepsilon$ whenever $n > N$. Similarly, *m* could be replaced by $n + 2$, $n + 3$, or $n + k$ for any positive integer *k*. This shows that if *s* is a Cauchy sequence and *k* is in $\mathbb{N}$, then $\lim_n (s_{n+k} - s_n) = 0$. It is important to be aware that this last property is not strong enough to imply that *s* must satisfy the Cauchy criterion of Definition 3.1.

EXAMPLE 3.1. If $s = \{0, 1/2, 1, 2/3, 1/3, 0, 1/4, 1/2, 3/4, 1, 4/5, \ldots\}$, then for any *k*, $\lim_n (s_{n+k} - s_n) = 0$. But *s* is not a Cauchy sequence. (The details of this example are left to the reader in Exercise 3.2.2.)

Next we prove a property of Cauchy sequences that is used in the proof of the theorem that follows it.

LEMMA 3.2. If *s* is a Cauchy sequence, then *s* is bounded.

Proof. Let *s* be a Cauchy sequence and apply Definition 3.1 for the case $\varepsilon = 1$. Thus there is an *N* such that $m, n > N$ imply $|s_m - s_n| < 1$. Since this holds for *all* $m > N$, we may replace *m* by $N + 1$ and say that

$$|s_{N+1} - s_n| < 1 \quad \text{whenever} \quad n > N.$$

This is equivalent to

$$s_{N+1} - 1 < s_n < s_{N+1} + 1 \quad \text{whenever} \quad n > N.$$

Therefore

$$n > N \quad \text{implies} \quad |s_n| < |s_{N+1}| + 1.$$

Now that we have the tail end of *s* bounded, we can define the number *B* by

$$B = \max\{|s_1|, \ldots, |s_N|, |s_{N+1}| + 1\},$$

and it is clear that $|s_n| \leq B$ for every n.

THEOREM 3.2: CAUCHY CRITERION FOR CONVERGENCE. The sequence s is convergent if and only if it is a Cauchy sequence.

Proof. First assume that s is convergent, say $\lim_n s_n = L$, and suppose $\varepsilon > 0$. Then we can choose N so that $n > N$ implies $|s_n - L| < \varepsilon/2$. This is the same as saying that $m > N$ implies $|s_m - L| < \varepsilon/2$. Thus for $m, n < N$ we have

$$|s_m - s_n| = |(s_m - L) - (s_n - L)|$$
$$\leq |s_m - L| + |s_n - L|$$
$$< \varepsilon.$$

Therefore s is a Cauchy sequence.

To prove the converse, assume that s is a Cauchy sequence; therefore by Lemma 3.2, s is bounded. By Theorem 3.1, s has a convergent subsequence, say $\lim_n s_{k(n)} = L$. If $\varepsilon > 0$, choose N so that $m, n > N$ implies $|s_m - s_n| < \varepsilon/2$. Since $\lim_n s_{k(n)} = L$, we can choose some term of this subsequence for which $k(n) > N$ and $|s_{k(n)} - L| < \varepsilon/2$. Now $n > N$ implies

$$|s_n - L| = |(s_n - s_{k(n)}) + (s_{k(n)} - L)|$$
$$\leq |s_n - s_{k(n)}| + |s_{k(n)} - L|$$
$$< \varepsilon/2 + \varepsilon/2$$
$$= \varepsilon.$$

Hence s converges to L.

Upon closer examination we see that once again we have proved more than the theorem states explicitly. In the first part of the proof, we used only the definitions of convergence and Cauchy sequence; therefore that implication does *not* depend on the LUB Axiom. Thus even in a number system like $\mathbb{Q}$, which, as is shown in Example 3.2, does not satisfy the LUB Axiom or Theorems 2.5 or 3.1, a convergent sequence is also a Cauchy sequence. It is the converse implication that is deeply connected to the LUB Axiom. This implication guarantees that for any sequence that satisfies the Cauchy Criterion there must be a limit value to which it converges. A system in which this implication holds is called *complete*. It is best to illustrate this with an example of a number system that is *not* complete.

EXAMPLE 3.2. Consider $\mathbb{Q}$, the rational numbers. By Exercise 1.3.2, there is a sequence r in $\mathbb{Q}$ that converges to $\sqrt{2}$, which is not in $\mathbb{Q}$. By the first part of Theorem 3.2, r is a Cauchy sequence. But there is no limit *in* $\mathbb{Q}$ to which r converges. Therefore $\mathbb{Q}$ is incomplete.

3.3. The Nested Intervals Theorem

The next theorem on the completeness of $\mathbb{R}$ is concerned with sequences of intervals. Therefore we should briefly review some notation and terminology. If a and b are real numbers such that $a \leq b$, then each of the following sets is called an interval:

$$(a, b) = \{r \in \mathbb{R}: a < r < b\},$$

$$[a, b) = \{r \in \mathbb{R}: a \leq r < b\},$$

$$(a, b] = \{r \in \mathbb{R}: a < r \leq b\},$$

$$[a, b] = \{r \in \mathbb{R}: a \leq r \leq b\}.$$

The first type, (a, b), is called an *open interval,* and the fourth type, $[a, b]$, is called a *closed interval.* The other two are called either *half-open* or *half-closed intervals.*

A sequence of intervals $\{I_n\}_{n=1}^{\infty}$ can be described by the sequences of their endpoints, say $I_n = [a_n, b_n]$. If, for each n, $I_{n+1} \subseteq I_n$, then $\{I_n\}_{n=1}^{\infty}$ is called a *nested sequence of intervals.* In this case it is obvious that $\cap_{n=1}^{N} I_n = I_N$. We are interested in the intersection $\cap_{n=1}^{\infty} I_n$, which consists of those numbers r that lie in every one of the I_n's. It is possible that this intersection is empty even though the intervals are nested (see Exercise 3.3.6). But this cannot happen if the sequence consists of closed intervals. As we see in the proof of the next theorem, this assertion depends on the LUB Axiom via the Monotonic Sequence Theorem.

THEOREM 3.3: NESTED INTERVALS THEOREM. If $\{I_n\}_{n=1}^{\infty}$ is a nested sequence of closed intervals, then $\cap_{n=1}^{\infty} I_n \neq \emptyset$.

Proof. Let I_n be the interval $[a_n, b_n]$. It is clear that the endpoint sequences are monotonic: $\{a_n\}_{n=1}^{\infty}$ is nondecreasing and

$\{b_n\}_{n=1}^{\infty}$ is nonincreasing. Moreover, for any k and n, $a_k \le b_n$; for, if $k < n$, then $a_k \le a_n \le b_n$, and if $k > n$, then $a_k \le b_k \le b_n$. Therefore $\{a_k\}_{k=1}^{\infty}$ is bounded above by every b_n, so by Theorem 2.5, $\{a_k\}_{k=1}^{\infty}$ is convergent. Also, by Proposition 2.2, $\alpha \equiv \lim_k a_k \le b_n$ for every n. Hence, for every n, $a_n \le \alpha \le b_n$; that is, α is in $\cap_{n=1}^{\infty} I_n$, which shows that the intersection is nonempty. (Of course, we could have shown that $\lim_n b_n$ is in $\cap_{n=1}^{\infty} I_n$ by a similar argument, but one point is sufficient.)

Exercise 3.4.6 requests you to show by an example that the implication of Theorem 3.3 is not valid unless the intervals are closed.

3.4. The Heine-Borel Covering Theorem

In the next theorem we give a criterion that expresses the completeness of $\mathbb{R}$ by using open intervals. If S is a subset of $\mathbb{R}$ and $\mathcal{G}$ is a collection of open intervals, then $\mathcal{G}$ is said to be an *open cover* of S provided that every element of S is in at least one member of $\mathcal{G}$. In set theoretic symbols, $S \subseteq \cup_{I \in \mathcal{G}} I$. This is also described by the phrase "$\mathcal{G}$ covers S." In the following discussion we place no restriction on the number of intervals in the collection $\mathcal{G}$. Thus $\mathcal{G}$ may have infinitely many members, even an uncountable number; that is, $\mathcal{G}$ may have so many members that they cannot be put in one-to-one correspondence with $\mathbb{N}$ (see Appendix B).

In some cases $\mathcal{G}$ may contain more intervals than are needed to cover a given set. For example, if S has only a finite number of elements, then it takes at most that finite number of intervals to cover S. For a less trivial example, consider the following situation:

EXAMPLE 3.3. Let S be the interval $[0, 1)$ and let $\mathcal{G}$ consist of the intervals $\{(0, 1/n)\}_{n=1}^{\infty}$ as well as $(-1/2, 1/2)$ (see Figure

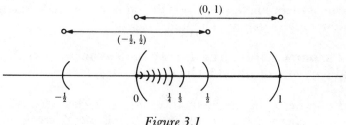

Figure 3.1

3.1). It is obvious that $\mathcal{G}$ is an open cover of S, but $\mathcal{G}$ covers S very inefficiently. Only a finite number of the members of $\mathcal{G}$ are needed to cover S:

$$S \subset (-1/2, 1/2) \cup (0, 1).$$

The two-interval covering of S in Example 3.3 is called a (finite) *subcover* of $\mathcal{G}$. This property of being able to cover S with a finite subcover is the theme of the next result.

THEOREM 3.4: HEINE-BOREL COVERING THEOREM. If J is a closed interval and $\mathcal{G}$ is an open cover of J, then there is a finite subcollection of $\mathcal{G}$ that covers J.

Proof. Suppose the assertion were false. Let J be the closed interval $[a, b]$, and let $\mathcal{G}$ be an open cover of J that cannot be reduced to a finite subcover. Consider the two subintervals $[a, (a + b)/2]$ and $[(a + b)/2, b]$. At least one of these two subintervals cannot be covered by a finite number of members of $\mathcal{G}$, for if both could be so covered then the two finite subcovers could be combined as a finite subcover of $[a, b]$. Let J_1 be one of these two subintervals such that J_1 cannot be covered by any finite subcollection of $\mathcal{G}$, and bisect J_1 into two subintervals of equal length $(b - a)/4$. As before, at least one of these two subintervals cannot be covered by any finite subcollection of $\mathcal{G}$; call that subinterval J_2. Continue bisecting the subintervals in this manner: J_n is bisected into two subintervals of length $(b - a)/2^{n+1}$, and at least one of these cannot be covered by any finite subcollection of $\mathcal{G}$. Call that subinterval J_{n+1}.

The above construction does not terminate, so it yields a nested sequence of closed intervals $\{J_n\}_{n+1}^{\infty}$. By Theorem 3.3, there is a number μ that is in every J_n. But μ is also in J, so μ must be covered by some I in $\mathcal{G}$, say $\mu \in I = (c, d)$. Since $\lim_n (b - a)/2^n = 0$, there is an N that is sufficiently large to yield $(b - a)/2^N < \min\{\mu - c, d - \mu\}$. Thus the length of J_N is less than the distance between μ and either endpoint of I. Since μ is in J_N, we assert that J_N must lie entirely in I. To prove this inclusion, let p be an arbitrary element of J_N. Then

$$\mu - p \le | \mu - p| \le (b - a)2^N < \mu - c,$$

which yields $c < p$; also

$$p - \mu \le |p - \mu| \le (b - a)/2^N < d - \mu,$$

which yields $p < d$. Thus $c < p < d$, which means that p is in I. Hence $J_N \subseteq I$. In the construction process, J_N was chosen so that it could not be covered by any finite subcollection of $\mathcal{G}$. But we have now shown that J_N is covered by a single interval in $\mathcal{G}$. Both of these contradictory statements are valid deductions from the original supposition, so we conclude that our original supposition is false. Hence the assertion of the theorem must be true, and the proof is complete.

The foregoing proof is admittedly difficult, both in subtlety of logic and detail of construction. At this stage, the student is asked to accept on faith the claim that this result is worth the effort. As we see in several subsequent proofs, the Heine-Borel Theorem is a tool of great power and wide applicability.

As in the case of the Nested Intervals Theorem, the interval in question in Theorem 3.4 must be closed and bounded; otherwise it will not have the Heine-Borel covering property. This is demonstrated in Exercises 3.4.9 and 3.4.10.

Exercises 3.4

1. Prove: The bounded sequence s is convergent if and only if s has exactly one limit point. (*Hint:* See Proposition 2.6.)

2. Let s be a sequence satisfying the following property: if $\varepsilon > 0$ and $k \in \mathbb{N}$, there is an N such that $n > N$ implies $|s_{n+k} - s_n| < \varepsilon$. Show that this does not imply that s is a Cauchy sequence by considering the following counterexample:

$$s = \left\{ 0, \frac{1}{2}, 1, \frac{2}{3}, \frac{1}{3}, 0, \frac{1}{4}, \frac{1}{2}, \frac{3}{4}, 1, \frac{4}{5}, \ldots \right\}.$$

3. Write a formula for s_n, the nth term of the sequence in Exercise 2.

4. Let $U = (0, 1]$. Find a Cauchy sequence in U that does not converge to a limit in U.

5. If s is given by

$$s_n = 1 + \frac{1}{2} + \frac{1}{2} + \frac{1}{3} + \frac{1}{3} + \frac{1}{3} + \frac{1}{4} + \frac{1}{4} + \frac{1}{4} + \frac{1}{4}$$

$$+ \cdots + \frac{1}{n} + \cdots + \frac{1}{n},$$

is s a Cauchy sequence?

6. Show by an example that a nested sequence of intervals can have empty intersection if the intervals are not closed. (*Hint:* $I_n = (0, 1/n]$.)

7. Is it possible for a nested sequence of *open* intervals to have a nonempty intersection?

8. Show by an example that closed sets of the form $[a_n, \infty)$ cannot be substituted for closed intervals in the Nested Intervals Theorem; that is, give a sequence sets $[a_1, \infty) \supseteq [a_2, \infty) \supseteq \ldots$ that has empty intersection.

9. Let J be the interval $(0, 1)$ and find an open cover $\mathcal{G}$ such that no finite subcollection of $\mathcal{G}$ covers J.

10. Let $\mathbb{P} = (0, \infty)$ and find an open cover $\mathcal{G}$ such that no finite subcollection of $\mathcal{G}$ covers $\mathbb{P}$.

11. Let S be an arbitrary unbounded set in $\mathbb{R}$ and find an open cover $\mathcal{G}$ such that no finite subcollection of $\mathcal{G}$ covers S.

12. Suppose that s is a convergent sequence such that for each n, $s_n \neq L = \lim_n s_n$, and let S denote the range of s. Find an open cover $\mathcal{G}$ of S such that no finite subcollection of $\mathcal{G}$ covers S.

4

CONTINUOUS
FUNCTIONS

4.1. Continuity

In the previous chapters we have, for the most part, restricted our attention to functions whose domains are $\mathbb{N}$, that is, sequences. In a first calculus course, one deals with functions whose domains are made up of intervals, half-lines, or all of $\mathbb{R}$. We are now ready to develop a limit theory for such functions, and the background that we have gained in sequential convergence will be a great help in reducing the amount of "epsilon-delta" work that is usually encountered.

Throughout the subsequent discussion, f is a function whose domain and range are subsets of $\mathbb{R}$. The statement "the number a is in the interior of the domain of f" means that there is an open interval $(a - \delta, a + \delta)$ that is contained entirely in the domain of f. This ensures that $f(x)$ is defined whenever x is sufficiently close to a; this also implies that a itself is in the domain of f.

DEFINITION 4.1. Let f be a number function; we say that f is *continuous at a* provided that the number a is in the interior of the domain of f and for every positive number ε there exists a positive number δ such that

$$|f(x) - f(a)| < \varepsilon \quad \text{whenever} \quad |x - a| < \delta.$$

If, for every a in some set D, f is continuous at a, then we say, "f is continuous on D." In case f is continuous at each number a in its domain, we say simply, "f is continuous." When f fails to be continuous—either at a or on D—we say that f is *discontinuous*.

The above definition describes the phenomenon, familiar to the calculus student, of a function "approaching its given function value as x approaches a point in its domain." Most of the functions that are encountered in elementary calculus are continuous, and these are the functions whose graphs are most easily sketched.

Although the property of continuity is a natural one that is encountered in any standard first course in calculus, it is worthwhile to reacquaint ourselves with it by giving rigorous ε–δ verifications of continuity in some examples.

EXAMPLE 4.1. If $f(x) = mx + b$, then f is continuous (on $\mathbb{R}$). Let a be any real number and suppose $\varepsilon > 0$. Consider the factorization

$$|f(x) - f(a)| = |(mx + b) - (ma + b)| = |m|\,|x - a|.$$

If $m = 0$, there is nothing more to do, for then $|f(x) - f(a)| = 0 < \varepsilon$ for *every* x. If $m \neq 0$, we define $\delta = \varepsilon/|m|$; then $|x - a| < \delta$ implies

$$|f(x) - f(a)| = |m|\,|x - a| < |m| \cdot \frac{\varepsilon}{|m|} = \varepsilon.$$

(See Figure 4.1.)

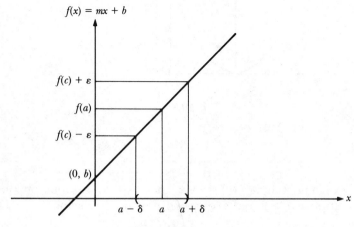

Figure 4.1

EXAMPLE 4.2. If $f(x) = x^2$, then f is continuous. Suppose $\varepsilon > 0$ and let a be any real number. Consider the factorization

$$|f(x) - f(a)| = |x^2 - a^2| = |x + a|\,|x - a|.$$

Define $\delta = \min\{1,\ \varepsilon/(1 + 2|a|)\}$. Thus when $|x - a| < \delta$, we have $|x - a| < 1$, so $a - 1 < x < a + 1$. This implies that

$$|x + a| \leq |x| + |a| < |a| + 1 + |a| = 1 + 2|a|.$$

Also, $|x - a| < \delta$ implies that $|x - a| < \varepsilon/(1 + 2|a|)$, so whenever $|x - a| < \delta$, we have

$$|f(x) - f(a)| = |x + a|\,|x - a| < (1 + 2|a|) \cdot \frac{\varepsilon}{1 + 2|a|} = \varepsilon.$$

(See Figure 4.2.)

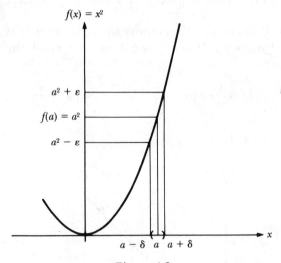

Figure 4.2

EXAMPLE 4.3. If $f(x) = \sqrt{x}$, then f is continuous on $(0, \infty)$. Suppose $a > 0$ and $\varepsilon > 0$; consider the following:

$$|f(x) - f(a)| = |\sqrt{x} - \sqrt{a}|$$

$$= \frac{|\sqrt{x} - \sqrt{a}|\,|\sqrt{x} + \sqrt{a}|}{|\sqrt{x} + \sqrt{a}|}$$

$$= \frac{1}{\sqrt{x} + \sqrt{a}}\,|x - a|$$

$$\leq \frac{1}{\sqrt{a}}\,|x - a|.$$

Define $\delta = \varepsilon\sqrt{a}$. Thus $|x - a| < \delta$ implies that

$$|f(x) - f(a)| = |\sqrt{x} - \sqrt{a}| < \frac{1}{\sqrt{a}}\,(\varepsilon\sqrt{a}) = \varepsilon.$$

(See Figure 4.3.)

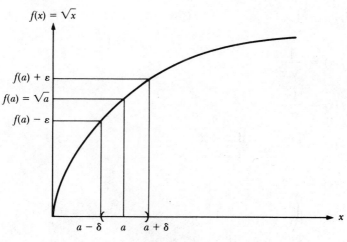

Figure 4.3

There is obviously a pattern to these examples. One first writes $|f(x) - f(a)|$ in a factored form, say $|f(x) - f(a)| = |g(x)|$ $|x - a|$. Then δ is chosen small enough so that $g(x)$ is bounded, say, $|g(x)| \leq B$ whenever $|x - a| < \delta$, and simultaneously δ is chosen less than ε/B. In this way $|x - a| < \delta$ implies that $|g(x)|\,|x - a| < B(\varepsilon/B)$ $= \varepsilon$. Although this is a routine procedure, it is a good way to develop an understanding of the "given ε, choose δ" aspect of the definition. Therefore one should review it by doing some exercises on similar functions.

Exercises 4.1

1. Prove that $f(x) = 1/x$ is continuous at 2.

2. Prove that $f(x) = 1/(3x - 2)$ is continuous at 1.

3. Prove that $f(x) = 1/\sqrt{x}$ is continuous at 4.

4. Prove that $f(x) = 1/x^2$ is continuous at 2.

Prove that each of the following functions is continuous.

5. $f(x) = x^2 + x - 1$.

6. $f(x) = x^3$.

7. $f(x) = x^n$, where n is a positive integer.

8. $f(x) = \dfrac{1}{x}$.

9. $f(x) = \dfrac{1}{\sqrt{x}}$.

10. $f(x) = \dfrac{1}{x^2}$.

11. $f(x) = \begin{cases} x^2, & \text{if } x < 2, \\ x + 2, & \text{if } x \geq 2. \end{cases}$

12. $f(x) = \dfrac{x}{x + 1}$.

13. $f(x) = |x|$.

14. $f(x) = \begin{cases} x^3, & \text{if } x < 1, \\ 2x - 1, & \text{if } x \geq 1. \end{cases}$

15. $f(x) = \sqrt{x^2 + 1}$.

4.2. The Sequential Criterion for Continuity

After introducing any limit concept, one seeks to prove basic properties such as those expressed in Theorems 2.1 through 2.4 for sequential limits. We could prove such results by manipulating ε's and δ's similar to the proofs in Chapter 2, but it is possible to avoid much of that detail by proving one theorem that allows us to make use of the sequential limit theory we have already developed.

THEOREM 4.1: SEQUENTIAL CRITERION FOR CONTINUITY. Let f be a function and a be a number in the interior of the domain of f. Then the following are equivalent:

(i) f is continuous at a;
(ii) if s is a sequence in the domain of f such that $\lim_n s_n = a$, then $\lim_n f(s_n) = f(a)$.

Proof. Assume (i) holds, suppose $\varepsilon > 0$, and let s be any sequence in the domain of f satisfying $\lim_n s_n = a$. Using (i) we choose $\delta > 0$ so that

$$|f(x) - f(a)| < \varepsilon \quad \text{whenever} \quad |x - a| < \delta. \tag{1}$$

Since $\lim_n s_n = a$, we can choose N so that $n > N$ implies $|s_n - a| < \delta$, and because of (1) this implies that $|f(s_n) - f(a)| < \varepsilon$. Hence $\lim_n f(s_n) = f(a)$, and we have proved that (i) implies (ii).

The converse implication can be proved by an indirect argument. We assume that (i) is false and show that (ii) must also be false by constructing a sequence s such that $\lim_n s_n = a$ but $\lim f(s_n) \neq f(a)$. Since we are assuming that f is not continuous at a, the defining implication fails no matter how small δ is chosen. Thus there is some $\varepsilon^* > 0$ such that for any positive δ, say $\delta = 1/n$,

$$|x - a| < 1/n \quad \text{does } not \text{ imply} \quad |f(x) - f(a)| < \varepsilon^*.$$

Therefore there is some value of x, say $x = s_n$, such that

$$|s_n - a| < 1/n \quad \text{but} \quad |f(s_n) - f(a)| \geq \varepsilon^*. \tag{2}$$

By choosing such a value s_n for each n in $\mathbb{N}$, we have defined a sequence s such that $\lim_n s_n = a$ (because $|s_n - a| < 1/n$), while $\lim_n f(s_n) \neq f(a)$ because $|f(s_n) - f(a)| \geq \varepsilon^*$ for each n. Hence we have shown that if (i) is false then (ii) is also false, and the proof is complete.

Although the Sequential Criterion for Continuity (hereafter SCC) is a bit awkward to use for proving that a particular function is continuous, it is a very handy tool for showing that a given function is discontinuous. This is illustrated in the next example.

EXAMPLE 4.4. If

$$f(x) = \begin{cases} 1/x, & \text{if } x \neq 0, \\ M, & \text{if } x = 0, \end{cases}$$

then f is discontinuous at 0 (regardless of the value of M). We take $s_n = 1/n$, so that $\lim_n s_n = 0$, but $\lim_n f(s_n) = \lim_n n = \infty$. Thus f fails to satisfy (ii) of the SCC.

This example suggests a general observation, which we prove next.

COROLLARY 4.1a. If the function f is continuous at a, then there is an open interval I containing a such that f is bounded on I.

Proof. Suppose the stated conclusion is false, and assume that a is in the domain of f. Then f is unbounded on every open interval containing a, so for each n, f is unbounded on the open interval $I_n = (a - 1/n, a + 1/n)$. Since the number n cannot be an upper bound for $|f(x)|$ on I_n, we can choose some s_n in I_n such that $|f(s_n)| > n$. This determines a sequence s such that $\lim_n s_n = a$ (because $|s_n - a| < 1/n$), but $\{f(s_n)\}$ cannot converge, because it is unbounded. Hence, f fails to satisfy (ii) of the SCC at a and is therefore discontinuous at a.

In the next example we use the SCC to establish the discontinuity of a bounded function.

EXAMPLE 4.5. If

$$f(x) = \begin{cases} \sin(1/x), & \text{if } x \neq 0, \\ M, & \text{if } x = 0, \end{cases}$$

then f is discontinuous at 0 (regardless of the value of M). We choose $s_n = 1/(n\pi)$ and $t_n = 1/\{2n\pi + (\pi/2)\}$. Then $\lim_n s_n = 0$ and $\lim_n t_n = 0$, but for each n,

$$f(s_n) = \sin n\pi = 0 \quad \text{and} \quad f(t_n) = \sin(2n\pi + (\pi/2)) = 1.$$

Thus $\lim_n f(s_n) = 0$, while $\lim_n f(t_n) = 1$. Since $f(0)$ cannot equal both of these limit values, at least one of the sequences s or t shows that f fails to satisfy (ii) of the SCC.

As before, this example suggests a general conclusion, which we state as the next result. The proof is self-evident and is left as an exercise (Exercise 4.2.8).

COROLLARY 4.1b. Let f be a function with the number a in the interior of its domain. If there exist two sequences s and t both converging to a such that $\lim_n f(s_n) \neq \lim_n f(t_n)$, then f is discontinuous at a.

This corollary makes it particularly easy to establish a discontinuity of the following type, which is known as a *jump discontinuity*.

EXAMPLE 4.6. Let f be the "bracket function" or "greatest integer function": $f(x) = [x] =$ the greatest integer n such that $n \leq x$. Then f is discontinuous at each integer. We take $s_k = n - 1/k$ and $t_k = n + 1/k$; then $\lim_k s_k = n = \lim_k t_k$, but for every $k > 1$, $f(s_k) = n - 1$ and $f(t_k) = n$. Therefore

$$\lim_k [s_k] = n - 1 \quad \text{and} \quad \lim_k [t_k] = n,$$

so by Corollary 4.1b, f is discontinuous at n.

4.3. Combinations of Continuous Functions

It is now time to combine the SCC with the sequential limit theory of Chapter 2. The result concerns the limits of algebraic combinations of continuous functions.

THEOREM 4.2. If the functions f and g are both continuous at a, then so are $f + g$, $f - g$, and fg. Furthermore, if $g(a) \neq 0$, then f/g is continuous at a.

Proof. Let s be a sequence in the domains of both f and g such that $\lim_n s_n = a$. By the SCC, we have $\lim_n f(s_n) = f(a)$ and $\lim_n g(s_n) = g(a)$. Therefore by Theorem 2.3,

$$\lim{}_n[f(s_n) \pm g(s_n)] = f(a) \pm g(a),$$

and by Theorem 2.4,

$$\lim{}_n f(s_n)\ g(s_n) = f(a)g(a).$$

Also, if $g(a) \neq 0$, then by Theorem 2.4,

$$\lim{}_n \frac{f(s_n)}{g(s_n)} = \frac{f(a)}{g(a)}.$$

Since s is an arbitrary sequence converging to a, we have shown that property (ii) of the SCC holds for each of the functions $f + g$, fg, and f/g. Hence each is continuous at a.

In our treatment of the quotient f/g we glossed over a small but necessary point. In order to show that property (ii) of the SCC holds for f/g, it is necessary to consider only sequences *in the domain of f/g* that converge to a. This means that for each n, $g(s_n) \neq 0$, and this is what is required in the hypothesis of Theorem 2.4, which we used. The question of $g(x)$ being nonzero is of no concern for the other three combination functions, so it was not raised until the end of the proof. The next result guarantees the existence of such a sequence s with nonvanishing function values.

LEMMA 4.1. If the function g is continuous at a and $g(a) > 0$, then there is an open interval I containing a such that $g(x) > 0$ for every x in I.

Proof. Suppose the conclusion is false. Then every open interval containing a must contain some number x for which $g(x) \leq 0$. In particular, for each n, the interval $(a - 1/n, a + 1/n)$ contains some s_n such that $g(s_n) \leq 0$. If g is continuous at a, then the SCC implies that $\lim{}_n g(s_n) = g(a)$, but Exercise 2.2.3 implies that $g(a) \leq 0$. Thus if the conclusion is false, the hypotheses cannot both hold, and this proves the assertion.

THEOREM 4.3. If the function f is continuous at a and $f(a) > c$ (respectively, $f(a) < c$), then there is an open interval I containing a such that $f(x) > c$ (respectively, $f(x) < c$) for every x in I.

Proof. If f satisfies the hypotheses of the theorem, we let $g(x) = f(x) - c$. Then g satisfies the hypotheses of Lemma 4.1, so there is an open interval I throughout which $0 < g(x) = f(x) - c$, that is,

$f(x) > c$ for every x in I. To prove the case in which $f(a) < c$, we apply Lemma 4.1 to the function $h(x) = c - f(x)$.

In the proof of Theorem 4.3 it was assumed as known that the constant function $\varphi(x) = c$ is continuous (this was proved in Example 4.1 with $m = 0$). Then Theorem 4.2 ensures that the combination $f(x) - c$ is also continuous, and it was the continuity of $f(x) - c$ that was used in the proof of Theorem 4.3.

Next we present a brief discussion of composite functions. This topic is thoroughly treated in elementary calculus, and it was mentioned briefly in connection with the concept of subsequence in Chapter 2. Nevertheless, it is worth a little time to review some notation and terminology. If each of f and g is a function, then the *composition* of g with f, denoted $g \circ f$, is the function consisting of the following collection of ordered pairs:

$$g \circ f = \{(x, y) : (x, f(x)) \in f \quad \text{and} \quad (f(x), y) \in g\}.$$

Thus the image $f(x)$ must be in both the range of f and the domain of g. If (x, y) is in $g \circ f$, then we write

$$(g \circ f)(x) = y = g(f(x)).$$

The above composition can be viewed as a binary operation in which two functions are combined to yield another function. In Theorem 4.2 we encountered binary operations on number functions that were given by algebraic combinations of the two functions. According to that theorem, the resulting combinations exhibited the same continuity property that the original two functions possessed. This is the nature of the next theorem, in which it is shown that continuity is preserved under the composition of functions. The proof of this result makes direct use of the definition of continuity without recourse to the SCC.

THEOREM 4.4. If the function f is continuous at a and the function g is continuous at $f(a)$, then $g \circ f$ is continuous at a.

Proof. Suppose $\varepsilon > 0$. Since g is continuous at $f(a)$, there is a positive number δ' such that

$$|g(z) - g(f(a))| < \varepsilon \quad \text{whenever} \quad |z - f(a)| < \delta'. \tag{1}$$

Since f is continuous at a, there is a positive number δ such that

$$|f(x) - f(a)| < \delta' \quad \text{whenever} \quad |x - a| < \delta. \tag{2}$$

Therefore whenever $|x - a| < \delta$, implication (2) ensures that $f(x)$ satisfies the condition required of z in (1), so that $|g(f(x)) - g(f(a))| < \varepsilon$. Hence $|x - a| < \delta$ implies that $|(g \circ f)(x) - (gf)(a)| < \varepsilon$, so $g \circ f$ is continuous at a.

4.4. One-Sided Continuity

A frequently useful variation of continuity is the concept of *one-sided continuity*. The definition is very similar to that of regular continuity, but it focuses on points in the domain of the function that lie on only one side of the point at which continuity is asserted.

DEFINITION 4.2. The function f is *continuous from the left* (respectively, *right*) at a provided that an interval $(a - c, a]$ (respectively, $[a, a + c)$) is in the domain of f and for every positive number ε there exists a positive number δ such that $|f(x) - f(a)| < \varepsilon$ whenever $0 \leq a - x < \delta$ (respectively, whenever $0 \leq x - a < \delta$).

Note that in the last phrase of the definition, "$0 \leq a - x < \delta$" says that x is close to a and *less than a*, while "$0 \leq x - a < \delta$" says that x is close to a and *greater than a*.

One-sided continuity is useful in describing the behavior of certain functions that fail to be continuous but still behave well from one side.

EXAMPLE 4.6a. The bracket function $f(x) = [x]$ is continuous from the right at each integer, although as we have seen in Example 4.6, $[x]$ is discontinuous there.

EXAMPLE 4.7. The function $f(x) = \sqrt{2 - x}$ is continuous from the left at 2, but it fails to be continuous there because 2 is not in the interior of its domain.

Exercises 4.4

1. Prove that $(|x + 2|)/(x + 2)$ is discontinuous at -2.

2. Use the Principle of Mathematical Induction to extend The-

orem 4.2 to an arbitrary finite number of terms or factors; that is, prove that if each of $f_1, \ldots, f_n$ is continuous at a, then so are $\sum_{k=1}^{n} f_k$ and $\prod_{k=1}^{n} f_k$.

3. Prove that every polynomial is continuous (on $\mathbb{R}$).

4. Prove that $\psi(x) = \begin{cases} 1, & \text{if } x \in \mathbb{Q}, \\ 0, & \text{if } x \in \mathbb{R} \sim \mathbb{Q}, \end{cases}$ is discontinuous everywhere.

5. Prove that $f(x) = \begin{cases} x, & \text{if } x \in \mathbb{Q} \cap [0, 1], \\ 1 - x, & \text{if } x \in [0, 1] \sim \mathbb{Q}, \end{cases}$ is continuous at $1/2$ but nowhere else.

6. Let f be defined on $[0, 1]$ as follows: If x is irrational or zero, then $f(x) = 0$; if $x = p/q$, where p and q are positive integers with no common factor (that is, the fraction p/q is reduced to lowest terms), then $f(p/q) = 1/q$. Prove that f is discontinuous at each rational number and continuous at each irrational number in $[0, 1]$.

7. Use Theorem 4.4 to prove that $f(x) = \sqrt{x^2 + 1}$ is continuous.

8. Prove Corollary 4.1b.

9. Prove: If f is a nonnegative continuous function, then $h(x) = \sqrt{f(x)}$ is continuous.

10. Prove: If f is continuous at a and $\varepsilon > 0$, then there is an open interval I containing a such that for any two points x_1 and x_2 in I, $|f(x_1) - f(x_2)| < \varepsilon$.

11. Prove: If f is continuous on $[a, b]$, then f is bounded there. (*Hint:* Use Corollary 4.1a and the Heine-Borel Theorem.)

4.5. Function Limits

In many cases it is necessary to study the behavior of a function at points that are close to some point that is not in the domain of the function. For example, the derivative of a function, a theory that is developed in Chapter 5, is based on just such a relationship. Our present task is to lay the groundwork for that theory. It is convenient to introduce at this time some notation for the type of set that is frequently encountered in the ensuing discussion. If a is a number and $\delta > 0$, let Δ_a be the set given by

$$\Delta_a = (a - \delta, a) \cup (a, a + \delta).$$

DEFINITION 4.3. Let f be a function and L and a be numbers; then f has limit L at a provided that the domain of f contains Δ_a for some positive δ and there exists a function $\bar{f}$ such that

(i) $\bar{f}(x) = f(x)$ for every x in Δ_a,
(ii) $\bar{f}$ is continuous at a, and
(iii) $\bar{f}(a) = L$.

In this case we write $\lim_{x \to a} f(x) = L$.

When it is clear which letter denotes the variable in the domain of f, the above limit is written more briefly as $\lim_a f(x) = L$.

In order to become acquainted with this definition, let us consider some possibilities. If f is continuous at a, then we can take $\bar{f}$ to be f itself, and $\lim_a f(x) = L$ then becomes $\lim_a f(x) = f(a)$. Indeed, this is the way that continuity at a is usually *defined*. But the concept of function limit is broader than that of continuity, so we ask the question: In what situations can $\lim_a f(x) = L$ occur even though f is discontinuous at a? The simplest such example is one in which f meets all the criteria for continuity at a except that $f(a)$ is not defined; that is, a is not in the domain of f. This is illustrated by the following function:

EXAMPLE 4.8. If $f(x) = (x^2 - 1)/(x - 1)$, then $\lim_1 f(x) = 2$. It is obvious that 1 is not in the domain of f, but for any other value of x we see that $f(x)$ reduces to $x + 1$. So we take $\bar{f}(x) = x + 1$, which we know to be continuous (see Exercise 4.4.3). Thus $\lim_1 f(x) = \bar{f}(1) = 2$.

Another illustration of the existence of a function limit for a discontinuous function can be given by starting with a given continuous function $\bar{f}$ and changing its value at a from $\bar{f}(a)$ to some other number, say M. If we call the new function f, then f is discontinuous at a because $M \neq \bar{f}(a)$, but $\lim_a f(x) = \bar{f}(x)$. This is illustrated by the following functions:

EXAMPLE 4.9. Let

$$\bar{f}(x) = x^2 \quad \text{and} \quad f(x) = \begin{cases} x^2, & \text{if } x \neq 0, \\ 1, & \text{if } x = 0. \end{cases}$$

It is clear that $\lim_0 f(x) = 0 \neq f(0)$.

It may seem that the last two examples are contrived; in both cases the function f is so nearly continuous that we had to do something that seems unnatural in order to prevent its being continuous. That is precisely the impression that one should have, for function limits occur only when the function is "nearly continuous." The ambiguous phrase "nearly continuous" needs to be clarified. A quick study of the definition of $\lim_a f(x) = L$ shows us that the given function f differs from the continuous function $\bar{f}$ only at the point a itself—and not even there if f is continuous. Therefore we are quite correct in saying that $\lim_a f(x)$ exists if and only if it is possible to define $f(a)$—or *redefine* it, if necessary—so that the resulting function is continuous at a. Thus one can think of this as "removing the discontinuity" by properly defining $f(a)$. Indeed, such a phenomenon is called a *removable discontinuity*.

4.6. The Sequential Criterion for Function Limits

As with continuity, we wish to establish a connection between function limits and sequence limits that can aid us in our investigation of function limits. That is our next objective. In the following discussion, the phrase "f has a limit at a" means that there exists a number L such that f has limit L at a.

THEOREM 4.5: SEQUENTIAL CRITERION FOR FUNCTION LIMITS. Suppose f is a function such that the domain of f contains Δ_a for some number a and some $\delta > 0$; then the following statements are equivalent:

(a) f has a limit at a;
(b) if s is any sequence in Δ_a such that $\lim_n s_n = a$, then $\{f(s_n)\}$ is convergent.

Proof. First assume f has a limit at a, say $\lim_a f(x) = L$. Let $\bar{f}$ be the function continuous at a such that $\bar{f}(a) = L$ and $\bar{f}(x) = f(x)$ on Δ_a. Then $\bar{f}$ satisfies the SCC, so consider any sequence s in Δ_a such that $\lim_n s_n = a$. By property (ii) of the SCC, $\lim_n f(s_n) = \lim_n \bar{f}(s_n) = L$. Hence (a) implies (b).

Conversely, suppose (b) holds. We assert that there is a *unique* number L such that for every s in Δ_a that converges to a, the convergent sequence $\{f(s_n)\}$ has that number L as its limit. For, if s and t are two such sequences in Δ_a, let $\lim_n f(s_n) = L_s$ and $\lim_n f(t_n) = L_t$. Now consider the sequence $u = \{s_1, t_1, s_2, t_2, \ldots\}$. It is clear that each u_n is in Δ_a and $\lim_n u_n = a$, so by (b) we know that $\{f(u_n)\}$ is convergent. Consider the following two subsequences of $\{f(u_n)\}$:

$$\lim_k f(u_{2k-1}) = \lim_k f(s_k) = L_s$$

and

$$\lim_k f(u_{2k}) = \lim_k f(t_k) = L_t.$$

Since *all* subsequences of the convergent sequence $\{f(u_n)\}$ must converge to the same limit, it follows that $L_s = L_t$. Now define $\bar{f}(a)$ to be that common limit value: $\bar{f}(a) = L$; and define $\bar{f}(x) = f(x)$ on Δ_a. Then (b) says that $\bar{f}$ satisfies property (ii) of the SCC, so $\bar{f}$ is continuous at a. Hence f has limit L at a, and we have shown that (b) implies (a).

As with the SCC, it is convenient to abbreviate the Sequential Criterion for Function Limits; we hereafter refer to Theorem 4.5 as the SCL.

After introducing another limit concept, we raise the same questions that were addressed in Theorems 2.3, 2.4, and 4.2 concerning algebraic combinations of limits. By now we know that these previous results can be put to good use in the present setting, so the proofs are left as exercises.

THEOREM 4.6. If the functions f and g both have a limit at a, then so do $f + g$, $f - g$, and fg; and in this case

$$\lim_a (f \pm g)(x) = \lim_a f(x) \pm \lim_a g(x),$$

and

$$\lim_a (fg)(x) = [\lim_a f(x)][\lim_a g(x)].$$

Furthermore, if $\lim_a g(x) \neq 0$, then f/g has a limit at a and

$$\lim_a (f/g)(x) = \frac{\lim_a f(x)}{\lim_a g(x)}.$$

Proof. See Exercises 4.6.4, 4.6.5, and 4.6.6.

In our discussion of function limits we have deliberately avoided the customary ε–δ definition. It is important, however, to establish the equivalence of the usual definition of function limit with the one that we have used here. That is the content of the next result.

THEOREM 4.7. If f is a function and L and a are numbers, then the following statements are equivalent:

(a) $\lim_a f(x) = L$;
(b) if $\varepsilon > 0$, there is a positive number δ such that if $0 < |x - a| < \delta$, then x is in the domain of f and $|f(x) - L| < \varepsilon$.

Proof. First assume (a) and let $\bar{f}$ be continuous at a with $\bar{f}(a) = L$ and $\bar{f}(x) = f(x)$ throughout some set $\Delta_a^c = (a - c, a) \cup (a, a + c)$. If $\varepsilon > 0$, then there is a positive number $\delta < c$ such that

$$|\bar{f}(x) - \bar{f}(a)| = |\bar{f}(x) - L| < \varepsilon \quad \text{whenever} \quad |x - a| < \delta. \qquad (1)$$

By excluding $x = a$ in (1), we can replace $\bar{f}$ with f, which yields

$$|f(x) - L| < \varepsilon \quad \text{whenever} \quad 0 < |x - a| < \delta.$$

This proves that (a) implies (b).

Conversely, assume that (b) holds. Define $\bar{f}(x) = f(x)$ for $x \neq a$ and $\bar{f}(a) = L$. Then it is clear that $\bar{f}$ satisfies Definition 4.1, so $\lim_a f(x) = L$.

4.7. Variations of Function Limits

There are several common variations of the limit concept for functions. For example, one can define the "limit of $f(x)$ as x tends to infinity," denoted by $\lim_{x \to \infty} f(x) = L$, by the following:

if $\varepsilon > 0$, then there is a number N such that
$x > N$ implies $|f(x) - L| < \varepsilon$.

This is an exact analog of the sequential limit, since we have merely replaced n and s_n with x and $f(x)$, respectively. The net result of this notational change is to replace the sequential domain $\mathbb{N}$ with a function domain of the form (a, ∞). When $\lim_{x \to \infty} f(x) = L$, the line $y = L$ is a horizontal asymptote to the graph of f. One defines $\lim_{x \to -\infty} f(x)$ in a similar fashion.

Another familiar variant of the function limit is the one-sided limit. This can be defined by saying that the function f has a left-hand limit at a equal to L provided that the domain of f contains an open interval $(a - \delta, a)$ and there is a function $\bar{f}$ continuous from the left at a such that $\bar{f}(x) = f(x)$ on $(a - \delta, a)$ and $\bar{f}(a) = L$. This is denoted by either

$$\lim_{x \to a-} f(x) = L,$$
$$\lim_{a-} f(x) = L,$$

or simply

$$f(a-) = L.$$

We similarly define the right-hand limit at a, which is denoted by either

$$\lim_{x \to a+} f(x) = L, \quad \lim_{a+} f(x) = L, \quad \text{or} \quad f(a+) = L.$$

The results of Theorem 4.6 on algebraic combinations also hold for one-sided limits and limits as x tends to infinity. The proof of the one-sided limit variation of Theorem 4.6 is almost exactly the same as that of Theorem 4.6. For the case in which x tends to infinity, the proof is the same as those of Theorems 2.3 and 2.4. The details are requested in Exercises 4.7.7 and 4.7.8.

Exercises 4.7

1. Prove that if $f(x) = (x^3 - 8)/(x - 2)$, then f has a limit at 2.

2. Prove that if $f(x) = (x^2 + x - 2)/(x^2 - 4)$, then f has a limit at -2.

3. Prove that if $f(x) = (x^n - a^n)/(x - a)$, then f has a limit at a. (Here n, as usual, denotes a positive integer.)

4. Prove that if f and g both have limits at a, then
$$\lim_a (f \pm g)(x) = \lim_a f(x) \pm \lim_a g(x).$$

5. Prove that if f and g both have limits at a, then
$$\lim_a (fg)(x) = [\lim_a f(x)][\lim_a g(x)].$$

6. Prove that if f and g both have limits at a and $\lim_a g(x) \neq 0$, then
$$\lim_a (f/g)(x) = \frac{\lim_a f(x)}{\lim_a g(x)}.$$

7. State and prove an analog of Theorem 4.6 for one-sided limits.

8. State and prove an analog of Theorem 4.6 for limits as x tends to infinity.

9. Prove that $\lim_{x \to \infty} 1/x = 0$.

10. Prove that if $a_n \neq 0$ and $b_n \neq 0$, then

$$\lim_{x \to \infty} \frac{a_n x^n + \cdots + a_1 x + a_0}{b_n x^n + \cdots + b_1 x + b_0} = \frac{a_n}{b_n}.$$

11. Prove: If f is nondecreasing on $\mathbb{R}$, then for every a in $\mathbb{R}$, $f(a-)$ exists.

12. Prove: If f is monotonic on $\mathbb{R}$, then for every a in $\mathbb{R}$ both $f(a-)$ and $f(a+)$ exist.

13. Prove the Cauchy Criterion for limits as $x \to \infty$: $\lim_{x \to \infty} f(x) = L$ if and only if for every $\varepsilon > 0$ there is a number B such that $|f(x) - f(y)| < \varepsilon$ whenever $x > B$ and $y > B$.

Define $\lim_{x \to a+} f(x) = \infty$ as follows: If B is any number, then there is a positive number δ such that $(a, a + \delta)$ is in the domain of f and $f(x) > B$ whenever $0 < x - a < \delta$.

14. Prove that $\lim_{x \to 3+} \dfrac{1}{x - 3} = \infty$.

15. Prove that $\lim_{x \to 1+} \dfrac{2x}{x^2 - 1} = \infty$.

16. Prove: If $f(x) > 0$ for all x, then $\lim_{x \to a+} f(x) = \infty$ if and only if $\lim_{x \to a+} 1/f(x) = 0$.

5

CONSEQUENCES OF CONTINUITY

5.1. The Range of a Continuous Function

In this chapter we deduce properties of functions that are continuous on an interval, which is usually a closed interval. These theorems rely heavily on the completeness of $\mathbb{R}$ via the theorems in Chapter 3. In the first theorem, we assert the boundedness of a function, which means that its range is a bounded set. In particular, the function f is bounded on the set D provided that there is a number B such that for every x in D, $|f(x)| \le B$.

THEOREM 5.1. If the function f is continuous on the interval $[a, b]$, then f is bounded there.

Proof. Suppose that f is not bounded on $[a, b]$. We show that f is not continuous on $[a, b]$. Since no positive integer can be an upper bound for the range of f, for each n we can select a number s_n in $[a, b]$ such that $|f(s_n)| > n$. Then s is a bounded sequence because $a \le s_n \le b$ for every n. By the Bolzano-Weierstrass Theorem (Theorem 3.1), s has a convergent subsequence, say, $\lim_n s_{k(n)} = c$. Also, by Proposition 2.2, it follows that c is in $[a, b]$. But for each n, $|f(s_{k(n)})| > k(n)$, so $\{f(s_{k(n)})\}_{n=1}^{\infty}$ is unbounded and therefore nonconvergent. Hence, by the SCC (Theorem 4.1), f is not continuous at c, so f is not continuous on $[a, b]$.

In Theorem 5.1 it is necessary to assume that the interval in question is a *closed* interval; otherwise the implication may not be valid. Consider the following example.

EXAMPLE 5.1. Define $f(x) = 1/x$ on $(0, 1]$. By Theorem 4.2, f is continuous everywhere except zero, and therefore f is continuous on $(0, 1]$. But $f(1/n) = n$, so it is clear that f is not bounded on $(0, 1]$.

In the next theorem we show that if f is a continuous function on a closed interval, then not only is its range a bounded set, but the range actually contains its least upper bound and greatest lower bound. Again, it is necessary to assume that the interval is closed. For example, the identity function $f(x) = x$ is continuous on $(0, 1)$, but its range is $(0, 1)$ and obviously does not contain its least upper bound or its greatest lower bound.

THEOREM 5.2. If the function f is continuous on the closed interval $[a, b]$, then there are numbers c and d in $[a, b]$ such that for every x in $[a, b]$, $f(c) \le f(x) \le f(d)$; that is,

$$f(c) = \min\{f(x): x \in [a, b]\}$$

and

$$f(d) = \max\{f(x): x \in [a, b]\}.$$

Proof. Since f is continuous on $[a, b]$, Theorem 5.1 ensures that f is bounded there. By the LUB Axiom, we may define $M = \text{lub}\{f(x): x \in [a, b]\}$. We must show that there is a number d in $[a, b]$ such that $f(d) = M$. For each positive integer n, $M - 1/n$ cannot be an upper bound of the range of f, so we can select an s_n in $[a, b]$ satisfying

$$M - 1/n < f(s_n) \le M.$$

Since $a \le s_n \le b$, the Bolzano-Weierstrass Theorem guarantees that there is a convergent subsequence, say $\lim_n s_{k(n)} = d$; and Proposition 2.2 ensures that d is in $[a, b]$. Now for each positive integer n, we have

$$\frac{M - 1}{k(n)} < f(s_{k(n)}) \le M;$$

thus $\lim_n f(s_{k(n)}) = M$. But also $\lim_n s_{k(n)} = d$, and f is continuous at d; so by the SCC, $\lim_n f(s_{k(n)}) = f(d)$. Hence, by the uniqueness of sequence limits (Theorem 2.2), $f(d) = M$. The proof that glb $\{f(x): x \in [a, b]\}$ is in the range of f can be achieved by a similar argument (see Exercise 5.1.1).

Exercises 5.1

1. Give the details of the remainder of the proof of Theorem 5.2: If f is continuous on $[a, b]$, prove that there exists a number c in $[a, b]$ such that for every x in $[a, b]$, $f(x) \geq f(c)$.

2. Give an example to show that the maximum and minimum values of $f(x)$ that are guaranteed by Theorem 5.2 can be achieved at more than one point in $[a, b]$.

3. Give an example (or show that one cannot exist) of a function that is bounded and one-to-one on $[0, 1]$ but is not continuous there.

4. Prove that the function given by $f(x) = (x^3 - 5x + 3)/(x^2 - 4)$ assumes a maximum value and a minimum value on $[-1, 1]$.

5. Prove that the function given by $f(x) = x^4 - 2x^3 + x + 5$ assumes a minimum value on $\mathbb{R}$. (*Hint:* $\lim_{x \to \pm\infty} f(x) = \infty$.)

6. Prove: If P is a polynomial of even degree with a positive coefficient of the highest degree term, then P assumes a minimum value on $\mathbb{R}$. (*Hint:* See Exercise 5.)

5.2. The Intermediate Value Property

In the next theorem we prove a property of continuous functions that illustrates—perhaps better than any other property—the reason we choose the word *continuous* to describe these functions. When speaking about continuity to a beginning calculus student, one frequently resorts to the graphic description: The graph of a continuous function can be drawn by making a continuous curve without lifting the chalk or pencil. Thus, such a graph exhibits no "holes," "jumps," or "skipped portions." We express this property in precise language as follows: If μ is any number that is between two numbers in the range of f, then μ itself must be in the range of f. Such a number μ is called an *intermediate value*, and a function whose range contains all intermediate values is said to possess the *intermediate value property*. The next theorem asserts that functions that are continuous on an interval have this property.

THEOREM 5.3: INTERMEDIATE VALUE THEOREM. If the function f is continuous on the interval $[a, b]$, where $f(a) \neq f(b)$, and μ is a number between $f(a)$ and $f(b)$, then there is a number c in (a, b) such that $f(c) = \mu$.

Proof. Without loss of generality we may assume that $f(a) < \mu < f(b)$. Let S be the set defined by $S = \{x \in [a, b]: f(x) < \mu\}$. Then a is in S, so S is nonempty and bounded above by b. Define $c = \text{lub } S$. We assert that $f(c) = \mu$, and we prove this by showing that both $f(c) < \mu$ and $f(c) > \mu$ lead to contradictions. First suppose that $f(c) < \mu$. By Proposition 4.2, there is an interval $(c - \delta, c + \delta)$ throughout which $f(x) < \mu$. Then $f(c + \delta/2) < \mu$, so $c + \delta/2$ is in S; but $c + \delta/2 > c$, which contradicts the choice of c as an upper bound of S. Now suppose that $f(c) > \mu$. Again, Proposition 4.2 ensures that there is an interval $(c - d, c + d)$ throughout which $f(x) > \mu$. Then $x > c - d$ implies that $f(x) \geq \mu$, which says that any x greater than $c - d$ is not in S. Therefore $c - d$ is an upper bound of S, which contradicts the choice of c as the *least* upper bound of S. Since f is defined at c, and neither $f(c) < \mu$ nor $f(c) > \mu$ can hold, we conclude that $f(c) = \mu$.

In order to appreciate the full generality of the Intermediate Value Theorem, one must realize that if f is continuous on any interval (open, closed, or half-open), then we can choose any two points of the interval to serve as a and b in Theorem 5.3. For $[a, b]$ is then contained in the original interval, so f is continuous on the closed interval $[a, b]$. Thus we can conclude that for any two points a and b in an interval on which f is continuous, the range of f contains every intermediate value between $f(a)$ and $f(b)$.

As is nearly always the case when we have proved an implication such as Theorem 5.3, we must ask whether the converse implication also holds; that is, if a function has the intermediate value property, is it necessarily continuous? The answer is no, as is seen by the following counterexample.

EXAMPLE 5.2. Define $f(x) = \sin(1/x)$, if $x \neq 0$, and $f(0) = 0$. Then f is not continuous at zero, because $\lim_0 f(x)$ does not exist (Example 4.5). But with the oscillations of f near zero, we see that every value in the range of f (which is $[-1, 1]$) is taken on in every interval $[0, \varepsilon]$, no matter how small ε may be.

Theorems 5.1, 5.2, and 5.3 can be combined into one statement, which provides another perspective of the properties they assert.

COROLLARY 5.3. If f is continuous on a closed interval domain, then its range is also a closed interval.

In some elementary algebra courses, it is proved that $\sqrt{2}$ is not a rational number. This is accomplished by showing that the supposition $\sqrt{2} = m/n$, where m and n are in $\mathbb{N}$, leads to a contradiction. Although it is usually not stressed in an elementary course, this proof does not show that there exists a real number whose square is 2; it merely proves that $\mathbb{Q}$ contains no such number. With the aid of the Intermediate Value Theorem, we can prove that $\mathbb{R}$ contains such a number. We here prove that $\mathbb{R}$ contains unique positive roots of every order.

THEOREM 5.4. If $a > 0$ and n is in $\mathbb{N}$, then there exists a unique positive number c such that $c^n = a$; that is, $c = \sqrt[n]{a}$.

Proof. Consider the function f defined by $f(x) = x^n$. Then f is continuous on $[0, 1 + a]$; also, $f(0) = 0$ and

$$f(1 + a) = (1 + a)^n = 1 + na + \cdots + a^n > a.$$

By the Intermediate Value Theorem, there is a number c in $(0, 1 + a)$ such that $f(c) = a$; that is, $c^n = a$. To show that c is the only such positive number, suppose that $b > 0$ and $b^n = a$. Then $b^n = c^n$, so

$$
\begin{aligned}
0 &= b^n - c^n \\
&= (b - c)(b^{n-1} + b^{n-2}c + \cdots + bc^{n-2} + c^{n-1}).
\end{aligned}
$$

The second factor cannot be zero, since $b > 0$ and $c > 0$; therefore we conclude that $b = c$.

Exercises 5.2

1. Prove that $2x^4 - x^3 + x^2 - 1 = 0$ has a solution in $(0, 1)$.

2. Prove that F given by $F(x) = x^3 + 2x + 7$ has a real zero. (A "zero" of a function f is a number z such that $f(z) = 0$.)

3. Prove that if P is a polynomial of odd degree, then P has a real zero. (*Hint:* Compare Exercise 5.1.5.)

4. Let f be a continuous function on $[a, b]$ and let $[c, d] \subseteq [a, b]$. Prove that there exists a number μ in $[a, b]$ such that

$$f(\mu) = \frac{f(c) + f(d)}{2}.$$

very imp. problem

5. Prove the "Fixed-Point Theorem": if f is continuous on [0, 1] and its range is also [0, 1], then there is a "fixed point" c in [0, 1] such that $f(c) = c$. (*Hint:* Consider $g(x) = f(x) - x$.)

6. Prove: If each of f and g is continuous on $[a, b]$ and for every rational number r in $[a, b]$, $f(r) = g(r)$, then $f(x) = g(x)$ for every x in $[a, b]$.

5.3. Uniform Continuity

The concept of continuity that we have studied thus far is sometimes called *pointwise continuity*. The word *pointwise* is used to emphasize the fact that the continuity of the function is inherently dependent upon the particular point in the domain: $\lim_a f(x) = f(a)$. This dependence can be illustrated by a closer examination of the delta–epsilon verification of continuity in the following examples.

EXAMPLE 5.3. If $f(x) = x^2$, then we can show continuity at a by factoring

$$|f(x) - f(a)| = |x^2 - a| = |x + a|\,|x - a|;$$

then for any given $\varepsilon > 0$, we define $\delta = \min\{1, \varepsilon/(1 + |a|)\}$. With this choice of δ, it is easy to see that $|x - a| < \delta$ implies $|f(x) - f(a)| < \varepsilon$.

But the important detail at this time is that the choice of δ depends on a as well as ε. Now suppose that the domain of f is restricted to some interval, say, $[-5, 5]$. Then for any a in the domain of f, $|a| \le 5$. Now the choice of δ can be changed to $\delta = \min\{1, \varepsilon/6\}$. In this case, δ is dependent upon ε alone, regardless of the point of the domain at which continuity is being verified.

EXAMPLE 5.4. Suppose $f(x) = 1/x$. As usual, we start by factoring:

$$|f(x) - f(a)| = \left|\frac{1}{x} - \frac{1}{a}\right| = \frac{|x - a|}{|xa|}.$$

Given $\varepsilon > 0$, we define $\delta = \min\{|a|/2, \varepsilon a^2/2\}$. Then $|x - a| < \delta$ implies that $|x| > |a|/2$, and therefore

$$\frac{|x - a|}{|xa|} < \frac{\varepsilon a^2/2}{|a/2|\,|a|} = \varepsilon.$$

Again we see that the definition of δ depends on a as well as ε. We may describe the situation by observing that in order to guarantee that $|x - a|/|xa|$ is small, it is necessary first to bound $|xa|$ away from zero. This is achieved by requiring $\delta \leq |a|/2$; then $|x - a| < \delta$ implies that r is between $a/2$ and $3a/2$. Now suppose that the domain of f is restricted to $[1, \infty)$, so that all points in the domain of f are at least one unit away from zero. Then $|x - a|/|xa| \leq |x - a|$, because $|xa| \geq 1$. Therefore we can choose $\delta = \varepsilon$, and it is plain that $|1/x - 1/a| < \varepsilon$ whenever $|x - a| < \delta$.

The ordinary definition of continuity allows us to choose δ in a point-by-point manner, using different choices of δ for different points in the domain of f. But in each of the modified versions of the two preceding examples, we were able to choose one δ that worked uniformly well at any point a in the domain of f. This provides both the motivation and the name of the next concept.

DEFINITION 5.1. The function f is said to be *uniformly continuous* on the set D provided that if $\varepsilon > 0$, then there is a positive number δ such that for any x_1 and x_2 in D,

$$|x_1 - x_2| < \delta \quad \text{implies} \quad |f(x_1) - f(x_2)| < \varepsilon.$$

In Examples 5.3 and 5.4 we proved that $f(x) = x^2$ is uniformly continuous on $[-5, 5]$ and $f(x) = 1/x$ is uniformly continuous on $[1, \infty)$; in those arguments the roles of x_1 and x_2 in the definition of uniform continuity were played by x and a, respectively. Here is another example that is very simple, but it illustrates nicely the required uniformity of the choice of δ.

EXAMPLE 5.5. If $f(x) = mx + b$ and $\varepsilon > 0$, we define $\delta = \varepsilon/|m|$; then $|x_1 - x_2| < \delta$ implies

$$|f(x_1) - f(x_2)| = |(mx_1 + b) - (mx_2 + b)|$$

$$= |m(x_1 - x_2)|$$

$$< |m| \, (\varepsilon/|m|)$$

$$= \varepsilon.$$

Hence f is uniformly continuous on $\mathbb{R}$.

Note that uniform continuity on a set D implies uniform continuity on any subset of D. This is an immediate consequence of the definition. As we see next, more care must be taken if one tries to enlarge the domain on which a function is uniformly continuous.

EXAMPLE 5.6. If $f(x) = |x|/x$, then f is uniformly continuous on both of the sets $(-\infty, 0)$ and $(0, \infty)$, but f is not uniformly continuous on their union, because for any δ, the numbers $x_1 = \delta/4$ and $x_2 = -\delta/4$ differ by $\delta/2$, yet $|f(x_1) - f(x_2)| = 2$.

We have presented the concept of uniform continuity as a special—and stronger—version of pointwise continuity. This approach suggests that for a function to be uniformly continuous on a set, that function must be (pointwise) continuous at every point of that set. We state this formally in the following proposition.

PROPOSITION 5.1. If the function f is uniformly continuous on the set D, and a is in D, then f is continuous at a.

Proof. There is almost nothing to prove. Simply write the definition of uniform continuity replacing x_1 by x and x_2 by a: If $\varepsilon > 0$, there is a positive number δ such that for any x in D,

$$|x - a| < \delta \quad \text{implies} \quad |f(x) - f(a)| < \varepsilon. \tag{1}$$

Hence, by Definition 4.1, f is continuous at a.

In the preceding proof we glossed over a minor point, namely, in the definition of $\lim_a f(x)$, it was assumed that the point a is interior to the domain of f. Now we are saying only that a must be in the domain of f. This is not a serious problem, however, because we can agree that implication (1), which is the main idea for continuity at a, need be valid only when x is in the domain of f, that is, only for x's where the symbol $f(x)$ makes sense. This variation of the limit definition was used before in our brief discussion of one-sided limits and one-sided continuity.

Although Proposition 5.1 establishes a relationship between uniform and pointwise continuity, it raises a much deeper question: Is uniform continuity a stronger property than pointwise continuity, or does each one imply the other? In order to settle this question, it is necessary to take a closer look at what is meant by the failure of a function to be uniformly continuous on a set D. It must be realized that in discussing the examples $f(x) = x^2$ and $f(x) = 1/x$, we did not show that they were not uniformly continuous on $\mathbb{R}$ and $\mathbb{R} \sim \{0\}$, respectively. We merely demonstrated that *our* choice of δ was dependent upon the point a. This is not the same as showing that δ *could not* be defined independently of a. Thus far we have merely failed to prove that these functions are uniformly continuous on their domains, and our failing to prove something does not mean that it is false. In order to settle the issue, we first give a precise statement of the *negation of uniform continuity:*

DEFINITION 5.2. The function f *fails to be uniformly continuous on the set D* provided that there is a positive number ε^* such that for any positive number δ there are numbers x_1 and x_2 in D satisfying

$$|x_1 - x_2| < \delta \quad \text{and} \quad |f(x_1) - f(x_2)| \geq \varepsilon^*.$$

EXAMPLE 5.3a. To show that $f(x) = x^2$ is not uniformly continuous on $\mathbb{R}$, we use $\varepsilon^* = 1$ as follows: For any positive δ, consider the numbers $x_1 = 1/\delta$ and $x_2 = 1/\delta + \delta/2$. Then $|x_1 - x_2| = \delta/2 < \delta$, but

$$|f(x_1) - f(x_2)| = |x_1 + x_2|\,|x_1 - x_2|$$
$$= \left|\frac{2}{\delta} + \frac{\delta}{2}\right|\left|\frac{\delta}{2}\right|$$
$$> 1.$$

Hence f is not uniformly continuous on $\mathbb{R}$.

EXAMPLE 5.4a. To show that $f(x) = 1/x$ is not uniformly continuous on $(0, 1]$, we use $\varepsilon^* = 1/2$. If $0 < \delta \leq 1$ is given, consider $x_1 = \sqrt{\delta}$ and $x_2 = \sqrt{\delta} - \delta/2$. Note that since $\delta \leq \sqrt{\delta}$, we have $0 < x_2 < x_1 \leq 1$. Then $|x_1 - x_2| = \delta/2 < \delta$, but

$$|f(x_1) - f(x_2)| = \frac{|x_1 - x_2|}{|x_1 x_2|}$$

$$> \frac{\delta/2}{\sqrt{\delta}\,\sqrt{\delta}}$$

$$= \frac{1}{2}.$$

Hence f is not uniformly continuous on $(0, 1]$.

There are situations in which continuity at every point of a set D *does* imply uniform continuity on D. The next theorem gives such a condition, and it is the major result of this section. The depth of the assertion is seen by the fact that our proof uses the LUB Axiom via the Heine-Borel Theorem.

THEOREM 5.5. If f is continuous on the closed interval $[a, b]$, then f is uniformly continuous on $[a, b]$.

Proof. Let f be continuous at every c in $[a, b]$, and assume that $\varepsilon > 0$. For each c in $[a, b]$, choose $\delta_c > 0$ such that

$$|x - c| < \delta_c \quad \text{implies} \quad |f(x) - f(c)| < \frac{\varepsilon}{2}.$$

Now consider the collection of open intervals

$$\mathcal{G} = \left\{ \left(c - \frac{\delta_c}{2},\, c + \frac{\delta_c}{2} \right) : c \in [a, b] \right\}.$$

For every c in $[a, b]$, there is an interval in $\mathcal{G}$ centered at c, so $\mathcal{G}$ is obviously an open cover of $[a, b]$. By the Heine-Borel Theorem there is a finite subcollection of $\mathcal{G}$ that covers $[a, b]$, say,

$$[a, b] \subset \left(c_1 - \frac{\delta_1}{2},\, c_1 + \frac{\delta_1}{2} \right) \cup \ldots \cup \left(c_n - \frac{\delta_n}{2},\, c_n + \frac{\delta_n}{2} \right), \qquad (2)$$

where δ_i is used in place of δ_{c_i}. Define $\delta = \min\{\delta_1/2, \ldots, \delta_n/2\}$. Now if $|x_1 - x_2| < \delta$, then x_1 is in one of the n intervals in (2), so for some c_i, $|x_1 - c_i| < \delta_i/2$. Therefore

$$|x_2 - c_i| = |(x_2 - x_1) + (x_1 - c_i)|$$
$$\leq |x_2 - x_1| + |x_1 - c_i|$$
$$< \delta + \frac{\delta_i}{2}$$
$$\leq \delta_i.$$

Thus, by the choice of δ_i,

$$|f(x_1) - f(c_i)| < \frac{\varepsilon}{2} \quad \text{and} \quad |f(x_2) - f(c_i)| < \frac{\varepsilon}{2},$$

which yields

$$|f(x_1) - f(x_2)| \leq |f(x_1) - f(c_i)| + |f(x_2) - f(c_i)| < \varepsilon.$$

Hence f is uniformly continuous on $[a, b]$.

Up to this point, all our examples of uniformly continuous functions have had the property that their graphs had bounded slope on D. It is natural to conjecture that there is an inherent relationship between uniform continuity and the slope of the graph; the definition asks that the vertical difference $f(x_1) - f(x_2)$ be small whenever the horizontal difference $x_1 - x_2$ is sufficiently small. Indeed, there is such a relationship, which we prove when we discuss the derivative. But the relationship holds in one direction only; as we show next, it is not necessary for the graph of a function to have bounded slope in order to have uniform continuity.

EXAMPLE 5.7. Consider the function $f(x) = \sqrt{x}$ on $[0, 1]$. Although the graph of f has a vertical tangent line at $x = 0$, we still have uniform continuity on $[0, 1]$ because f is continuous at each point of the closed interval $[0, 1]$. It is a challenging exercise to prove this directly from the definition by finding a formula for δ in terms of ε (see Exercise 5.4.7).

5.4. The Sequential Criterion for Uniform Continuity

The Sequential Criterion for Continuity says that continuous functions are characterized by the property of mapping convergent sequences in the domain into convergent sequences in the range. There is an analogous characterization of uniform continuity that involves Cauchy sequences instead of convergent sequences.

THEOREM 5.6: SEQUENTIAL CRITERION FOR UNIFORM CONTINUITY. If f is a function and D is a bounded subset of $\mathbb{R}$, then the following statements are equivalent:

  (i) f is uniformly continuous on D;
  (ii) if s is a Cauchy sequence in D, then $\{f(s_n)\}_{n=1}^{\infty}$ is a Cauchy sequence.

Proof. Assume (i) is true, let s be a Cauchy sequence in D, and suppose $\varepsilon > 0$. Then there exists a positive number δ such that if x_1 and x_2 are in D,

$$|x_1 - x_2| < \delta \quad \text{implies} \quad |f(x_1) - f(x_2)| < \varepsilon.$$

Since s is a Cauchy sequence, we can choose N so that

$$m, n > N \quad \text{implies} \quad |s_m - s_n| < \delta.$$

Then

$$|f(s_m) - f(s_n)| < \varepsilon \cdot \quad \text{whenever} \quad m, n > N,$$

so $\{f(s_n)\}_{n=1}^{\infty}$ is a Cauchy sequence.

Conversely, suppose f is not uniformly continuous on the bounded set D. Then there is a positive number ε^* such that for each n in $\mathbb{N}$ there are numbers s_n and t_n in D satisfying

$$|s_n - t_n| < 1/n \quad \text{and} \quad |f(s_n) - f(t_n)| \geq \varepsilon^*.$$

Since D is bounded, the Bolzano-Weierstrass Theorem guarantees that s has a convergent subsequence, say $\lim_n s_{k(n)} = L$. Since $s - t$ is a null sequence and

$$t_{k(n)} = s_{k(n)} + (t_{k(n)} - s_{k(n)}),$$

it follows from Theorem 2.3 that $\lim_n t_{k(n)} = L$. (Note that L may not be in D, but this does not matter.) Now define the sequence u by

$$u = \{s_{k(1)}, t_{k(1)}, s_{k(2)}, t_{k(2)}, \ldots \}.$$

Then u has limit L, so by Theorem 3.2, u is a Cauchy sequence in D. But for each n,

$$|f(u_{2n-1}) - f(u_{2n})| = |f(s_{k(n)}) - f(t_{k(n)})| \geq \varepsilon^*,$$

and therefore $\{f(u_n)\}_{n=1}^{\infty}$ is not a Cauchy sequence. Hence (ii) does not hold, and the proof is complete.

In the preceding proof, the boundedness of D was not used in proving that (i) implies (ii). Therefore we may conclude that a uniformly continuous function *on any domain* will preserve Cauchy sequences. In proving that (ii) implies (i), however, the boundedness of D was needed to be able to apply the Bolzano-Weierstrass Theorem. More important, this implication is not valid on unbounded domains. For a simple counterexample, we can recall that $f(x) = x^2$ is not uniformly continuous on $\mathbb{R}$ (Example 5.3a). But f does map Cauchy sequences into Cauchy sequences, for if s is a Cauchy sequence, then s is convergent. By Theorem 2.4, s^2 is also convergent. Hence, $\{f(s_n)\}_{n=1}^{\infty} = \{s_n^2\}_{n=1}^{\infty}$ is a Cauchy sequence.

As we have just seen, the Sequential Criterion for Uniform Continuity is a convenient tool for demonstrating that a given function is not uniformly continuous.

Exercises 5.4

1. Prove that $f(x) = x^3$ is not uniformly continuous on $\mathbb{R}$.

2. Prove that $f(x) = \cos(1/x)$ is not uniformly continuous on $(0, 1]$. (*Hint:* Find a Cauchy sequence s in $(0, 1]$ such that $|f(s_n) - f(s_{n+1})| \geq 1$.)

3. Prove that $f(x) = 1/(2x - 1)$ is not uniformly continuous on $[0, 1/2)$.

4. Prove that $f(x) = \tan x$ is not uniformly continuous on $[0, \pi/2)$.

5. Prove that $1/x^2$ is uniformly continuous on $[1, \infty)$.

6. Prove $f(x) = \sqrt{x}$ is uniformly continuous on $[c, \infty)$ for any $c > 0$.

7. Prove that $f(x) = \sqrt{x}$ is uniformly continuous on $[0, \infty)$ by using only Definition 5.1 and choosing $\delta = \varepsilon^2/4$. (Compare Example 5.7.)

8. Prove: If f is uniformly continuous on (a, b), then f is bounded there.

9. Prove: If f is uniformly continuous on $[a, b]$ and uniformly continuous on $[b, c]$, then f is uniformly continuous on $[a, c]$.

10. Prove: If f and g are uniformly continuous on D, then $f + g$ is uniformly continuous on D.

11. Prove: If f is continuous on (a, b) and $\lim_{a+} f(x)$ and $\lim_{b-} f(x)$ both exist, then f can be defined (or redefined) at a and b so that f is uniformly continuous on $[a, b]$.

12. Prove: If f is uniformly continuous on (a, b), then $\lim_{a+} f(x)$ and $\lim_{b-} f(x)$ exist.

6

THE DERIVATIVE

6.1. Difference Quotients

Consider a number function f whose domain contains an open interval about the number a; that is, a is an interior point of the interval. Define a "slope function" Q_a by

$$Q_a(h) = \frac{f(a + h) - f(a)}{h},$$

where h is in an interval $(-\delta, \delta)$ in the domain of f. This slope function is called a *difference quotient* for f near a. The geometric motivations for this quantity are already familiar to the reader. Note that Q_a is not defined at zero, but, of course, this does not prevent the existence of its limit as h approaches zero.

DEFINITION 6.1. The function f is said to be *differentiable* at a provided that Q_a has a limit at zero. In this case we write

$$f'(a) = \lim_{h \to 0} Q_a(h).$$

The geometric manifestation of differentiability is seen in the smoothness of the graph of f (at a). For example, polynomial functions, whose graphs have no breaks or sharp corners, are differentiable everywhere (see Exercise 6.1.2). But functions such as $|x|$, $1/x$, and $\sqrt{|x|}$ fail to be differentiable at zero, where their graphs are not smooth. Before examining these functions in more detail,

70

we note that Definition 6.1 can be used to define a new "derived" function f' whose value at a is the number $f'(a)$. Thus the domain of f' is a subset of the domain of f. The function f' is called the *derivative* of f.

EXAMPLE 6.1. Let $f(x) = |x|$, and take $a = 0$. Then we have

$$Q_0(h) = \frac{|0 + h| - |0|}{h} = \frac{|h|}{h} = \begin{cases} 1, & \text{if } h > 0, \\ -1, & \text{if } h < 0. \end{cases}$$

Thus $\lim_{h \to 0} Q_0(h)$ fails to exist, because its right-hand and left-hand limits are unequal.

EXAMPLE 6.2. Let $f(x) = \sqrt{|x|}$. Again taking $a = 0$, we have

$$Q_0(h) = \frac{\sqrt{|0 + h|} - \sqrt{|0|}}{h} = \frac{\sqrt{|h|}}{h}.$$

When $h > 0$, this gives $Q_0(h) = \sqrt{h}/h = 1/\sqrt{h}$, and $\lim_{h \to 0} 1/\sqrt{h} = \infty$. Hence the right-hand limit of Q_0 fails to exist, so Q_0 cannot have a limit as h tends to zero. (The left-hand limit also fails to exist, but it is not necessary to show that, too.)

EXAMPLE 6.3. Let $f(x) = 1/x$. We see that $f(0)$ is not defined, and therefore $Q_0(h)$ is not defined for any h. Therefore $f'(0)$ cannot exist.

In the third example we see that it is necessary that the number a be in the domain of f in order to have differentiability. It is not enough, though, to have f defined arbitrarily at a. The following example illustrates this:

EXAMPLE 6.4. If

$$f(x) = \begin{cases} 2x + 1, & \text{if } x \neq 3, \\ 4, & \text{if } x = 3, \end{cases}$$

then

$$Q_3(h) = \frac{2(3+h) + 1 - 4}{h} = \frac{3}{h} + 2;$$

therefore

$$\lim_{h \to 0} Q_3(h) = \infty,$$

and f is not differentiable at 3.

This example tells us that the value $f(a)$ must be the "right one" in some sense. A brief re-examination of the definition of $f'(a)$ sheds some light on this. We wish to take the limit of a quotient as its denominator h tends to zero. Our past experience tells us that in such a situation the quotient will increase without bound (that is, tend to ∞) so long as the numerator stays away from zero. So to prevent this "blow up," we must have the numerator also approach zero:

$$\lim_{h \to 0} \{f(a + h) - f(a)\} = 0,$$

or equivalently,

$$\lim_{h \to 0} f(a + h) = f(a).$$

This final limit is equivalent to continuity at a, so our heuristic reasoning has led us to the first theorem of this chapter, which gives a relationship between differentiability and continuity.

THEOREM 6.1. If f is differentiable at a, then f is continuous at a.

Proof. Suppose $\varepsilon > 0$. Since $\lim_{h \to 0} Q_a(h) = f'(a)$, we can choose a positive number δ such that if $|h| < \delta$, then

$$|Q_a(h)| \le |f'(a)| + 1.$$

We can also choose δ even smaller, if necessary, so that

$$\delta < \frac{\varepsilon}{|f'(a)| + 1}.$$

Now $|h| < \delta$ implies that

$$|f(a + h) - f(a)| = \left|\frac{f(a + h) - f(a)}{h}\right| |h|$$

$$= |Q_a(h)| \, |h|$$

$$< \{|f'(a)| + 1\} \frac{\varepsilon}{|f'(a)| + 1}$$

$$= \varepsilon.$$

Hence, $\lim_{h \to 0} f(a + h) = f(a)$, or $\lim_{x \to a} f(x) = f(a)$.

It is easy to show that each power function (that is, $f(x) = x^n$) is differentiable (see Exercise 6.1.1a). Then, using Theorem 4.6 on products and sums of function limits, we can deduce that every polynomial is differentiable at every point of $\mathbb{R}$ (see Exercise 6.1.2).

For the purpose of producing examples, we assume that all the familiar differentiation formulas from elementary calculus have been proved.

EXAMPLE 6.5. Let

$$f(x) = \begin{cases} x \sin(1/x), & \text{if } x \neq 0, \\ 0, & \text{if } x = 0. \end{cases}$$

We assert that f is continuous everywhere. The only unusual point is the origin, and since $|\sin(1/x)| \leq 1$, it follows easily that $|f(x)| \leq |x|$; thus

$$\lim_{x \to 0} f(x) = 0 = f(0).$$

As for differentiability, if $x \neq 0$, then

$$f'(a) = \sin\frac{1}{x} + x\left(\frac{-1}{x^2}\right)\cos\frac{1}{x}$$

$$= \sin\frac{1}{x} - \frac{1}{x}\cos\frac{1}{x}.$$

If $a = 0$, then we get

$$Q_0(h) = \frac{h\sin\dfrac{1}{h} - 0}{h} = \sin\frac{1}{h}.$$

In every interval $(-\delta, \delta)$, $\sin(1/h)$ takes on every value between -1 and 1 infinitely many times, so $\lim_{h \to 0} Q_0(h)$ does not exist. Hence f is not differentiable at zero.

EXAMPLE 6.6. Let

$$f(x) = \begin{cases} x^2 \sin (1/x), & \text{if } x \neq 0, \\ 0, & \text{if } x = 0. \end{cases}$$

In this case, the amplitude factor x^2 is sufficient to make f differentiable at zero, because

$$f'(0) = \lim_{h \to 0} Q_0(h) = \lim_{h \to 0} h \sin \frac{1}{h} = 0.$$

When $x \neq 0$, we can rely on our elementary derivative formulas to compute

$$f'(x) = 2x \sin \frac{1}{x} - \cos \frac{1}{x}.$$

Now we can use these formulas to calculate the derivative of f', that is, the second derivative of f, denoted f''. There is no problem at points other than zero, but at $a = 0$, the difference quotient for f' is

$$Q_0'(h) = \frac{f'(0 + h) - f'(0)}{h} = \frac{f'(h)}{h}$$

$$= 2 \sin \frac{1}{h} - \frac{1}{h} \cos \frac{1}{h}.$$

It is clear that $Q_0'(h)$ is underbounded as h tends to zero, so the limit of $Q_0'(h)$ cannot exist. Hence f is differentiable at zero, but it does not have a second derivative there.

Exercises 6.1

1. For each of the following functions, derive the formula for $f'(x)$ by simplifying $Q_x(h)$ and evaluating its limit as h tends to zero:

(a) $f(x) = x^n$, where n is a positive integer.

(b) $f(x) = \dfrac{1}{x}$.

(c) $f(x) = \sqrt{x}$.

  (*Hint:* $(x + h) - x = (\sqrt{x + h} - \sqrt{x})(\sqrt{x + h} + \sqrt{x})$.)

(d) $f(x) = \dfrac{1}{\sqrt{x}}$.

(e) $f(x) = \dfrac{1}{x^2}$.

(f) $f(x) = \sqrt{2x + 1}$.

(g) $f(x) = \dfrac{x}{2x - 3}$. ~~find derivative using def.~~

2. Prove that for any polynomial $P(x) = a_n x^n + \cdots + a_1 x + a_0$, P is differentiable, and $P'(x) = na_n x^{n-1} + \cdots + 2a_2 x + a_1$.

3. Discuss the continuity and differentiability of f at zero, where ~~no 2nd derivative~~

$$f(x) = \begin{cases} x^2, & \text{if } x > 0, \\ 0, & \text{if } x \le 0. \end{cases}$$

4. Discuss the existence of $f'(0), f''(0), \ldots, f^{(n)}(0)$, where

$$f(x) = \begin{cases} x^n, & \text{if } x > 0, \\ 0, & \text{if } x \le 0, \end{cases}$$

  and n is a positive integer.

5. Prove that the derivative of a nondecreasing differentiable function is nonnegative. Similarly, prove that the derivative of a nonincreasing differentiable function is nonpositive.

6. Prove the following "rules" for differentiation of algebraic combinations of differentiable functions f and g:

(a) $(f + g)' = f' + g'$;

(b) $(fg)' = f'g + fg'$;

(c) $\left(\dfrac{f}{g}\right)' = \dfrac{f'g - fg'}{g^2}$.

7. Use Mathematical Induction to prove the Leibnitz Rule for the nth-order derivatives of products:

$$(fg)^{(n)} = f^{(n)}g + \binom{n}{1}f^{(n-1)}g' + \binom{n}{2}f^{(n-2)}g'' + \cdots + fg^{(n)},$$

where $f^{(k)}$ denotes the kth-order derivative of f, and $\binom{n}{k}$ denotes the binomial coefficient and $f^{(0)} = f$.

6.2. The Chain Rule

The next result is the so-called Chain Rule, which is used throughout elementary calculus. It allows us to compute the derivative of a composite function as the product of derivatives of the component functions. Recall from Section 4.3 that we write the composition of the functions f and g as $f \circ g$; that is, if x is in the domain of g and $g(x)$ is in the domain of f, then

$$(f \circ g)(x) = f[g(x)].$$

In the following discussion we assume that the domain of f contains an open interval about the point $g(a)$.

THEOREM 6.2: CHAIN RULE. If the function g is differentiable at a and the function f is differentiable at $g(a)$, then their composition $f \circ g$ is differentiable at a, and

$$(f \circ g)'(a) = f'[g(a)]g'(a).$$

Proof. Let $Q_a(h)$ denote the difference quotient for $f \circ g$ near a. In order to distinguish the difference quotients for f and g, we write

$$G_a(h) = \frac{g(a + h) - g(a)}{h}$$

and

$$F_{g(a)}(t) = \frac{f[g(a) + t] - f[g(a)]}{t}.$$

Let us examine $Q_a(h)$ as follows:

$$Q_a(h) = \frac{(f \circ g)(a + h) - (f \circ g)(a)}{h}$$

$$= \begin{cases} \dfrac{f[g(a + h)] - f[g(a)]}{g(a + h) - g(a)} \cdot \dfrac{g(a + h) - g(a)}{h}, & \text{if } g(a + h) \neq g(a), \\ 0, & \text{if } g(a + h) = g(a). \end{cases}$$

Now let t denote $g(a + h) - g(a)$, so that $g(a + h) = g(a) + t$. With this substitution, we have

$$Q_a(h) = \begin{cases} \dfrac{f[g(a) + h)] - f[g(a)]}{t} \cdot \dfrac{t}{h}, \\ 0, \end{cases} \tag{1}$$

$$= \begin{cases} F_{g(a)}(t)\, G_a(h), & \text{if } t \neq 0, \\ 0, & \text{if } t = 0. \end{cases}$$

Since we want to show that $Q_a(h)$ approaches $f'[g(a)]g'(a)$, it would be simpler if we needed only to use the upper line of Equation (1). For this reason, we separate the argument into two cases.

Case (i): Suppose $g(a + h) \neq g(a)$ for every h in some set $\Delta_0 = (-\delta, 0) \cup (0, \delta)$. Then the upper formula is the only one needed to get $\lim_{h \to 0} Q_a(h)$. Since g is differentiable at a, $\lim_{h \to 0} G_a(h) = g'(a)$. By Theorem 6.1, g is also continuous at a, so

$$0 = \lim_{h \to 0} \{g(a + h) - g(a)\} = \lim_{h \to 0} t.$$

Therefore

$$\lim_{h \to 0} F_{g(a)}(t) = \lim_{t \to 0} F_{g(a)}(t) = f'[g(a)].$$

As h approaches zero, the limit of the product $F_{g(a)}(t)\, G_a(h)$ equals the product of the limits; hence, $\lim_{h \to 0} Q_a(h) = f'[g(a)]\, g'(a)$.

Case (ii): Suppose that for every $\delta > 0$, the set $\Delta_0 = (-\delta, 0) \cup (0, \delta)$ contains some h^* such that $g(a + h^*) = g(a)$. For each such h^*, we have

$$G_a(h^*) = \frac{g(a + h^*) - g(a)}{h^*} = 0.$$

Therefore the difference quotient for g near a is equal to zero at all these h^*'s, which occur arbitrarily close to zero. Therefore zero is the only possible value for $\lim_{h \to 0} G_a(h)$, and the limit must exist because g is differentiable at a. Hence, $g'(a) = 0$. Examining (1), we conclude that, as h tends to zero, both the upper and lower formulas approach the limit value zero. This, of course, is equal to $f'[g(a)]\, g'(a)$, because $g'(a) = 0$, and we have proved the assertion for Case (ii).

Skip these ↓   **Exercises 6.2**

1. Prove: If g is a differentiable function and n is a positive integer, then $h(x) = [g(x)]^n$ is differentiable and $h'(x) = n[g(x)]^{n-1}g'(x)$. (*Hint:* Use Exercise 6.1.1a.)

2. Prove: If g is a differentiable function, then $h(x) = 1/g(x)$ is differentiable (wherever $g(x) \neq 0$). (*Hint:* Use Exercise 6.1.1b.)

3. Prove: If g is a positive-valued differentiable function, then $h(x) = \sqrt{g(x)}$ is differentiable. (*Hint:* Use Exercise 6.1.1c.)

4. Let f be a one-to-one function and ϕ be its inverse function; that is, $f[\phi(x)] = x$. Prove: If f and ϕ are differentiable, then

$$\phi'(a) = \frac{1}{f'[\phi(a)]}.$$

5. Use Mathematical induction to extend the Chain Rule to the derivative of the composition of n functions:

$$\phi(x) = (f_n \circ f_{n-1} \circ \ldots \circ f_1)(x) = f_n[f_{n-1}(\ldots f_2[f_1(x)]\ldots)].$$

6.3. The Law of the Mean

The result that gives this section its title asserts that the average (or mean) rate of change of a function over an interval must equal one of the values of the instantaneous rate of change in that interval. An interesting physical illustration of this can be given as follows: If you drive an automobile 100 miles in two hours, then your instantaneous speed—as shown on the speedometer—must be exactly 50 miles per hour at least once during the two hours.

The geometric interpretation of the Law of the Mean is familiar from elementary calculus. The average rate of change of the function f over the interval $[a, b]$ is the slope of the segment that connects the endpoints of the graph, and the instantaneous rate of change is the slope of the tangent line to the graph. The Law of the Mean asserts that there is some point on the graph where the tangent line is parallel to that segment (see Figure 6.1).

Of course we would not expect this property to be true for every function on $[a, b]$. For example, consider a step function whose tangent lines—wherever they exist—are necessarily horizontal. We see later that smoothness of the graph (that is, differentiability of the function) is needed to guarantee that this property holds. But first, we must do some preliminary work.

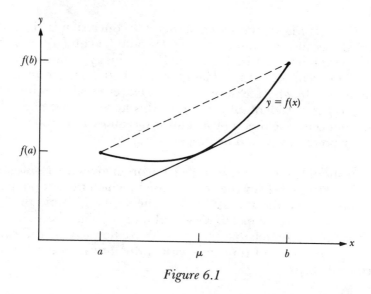

Figure 6.1

LEMMA 6.1. If the function f is differentiable on (a, b) and has a relative extremum at μ in (a, b), then $f'(\mu) = 0$.

Proof. For the sake of definiteness, suppose $f(\mu)$ is a maximum; then $f(\mu + h) \leq f(\mu)$, whenever $\mu + h$ is in a small interval about μ. Thus the numerator of $Q_\mu(h)$ is nonpositive when h is close to zero, which implies

$$Q_\mu(h) = \frac{f(\mu + h) - f(\mu)}{h} \begin{cases} \leq 0, & \text{if } h > 0, \\ \geq 0, & \text{if } h < 0. \end{cases}$$

Therefore as h tends to zero, the only possible limit value for $Q_\mu(h)$ is zero. Since we are assuming that f is differentiable (at μ), this limit exists:

$$f'(\mu) = \lim_{h \to 0} Q_\mu(h) = 0.$$

It is clear that a similar argument works in the case where $f(\mu)$ is a minimum.

LEMMA 6.2: ROLLE'S THEOREM. If the function f is differentiable on (a, b), continuous on $[a, b]$, and $f(a) = f(b) = 0$, then there is a point μ in (a, b) where $f'(\mu) = 0$.

Proof. If f is identically zero, then the conclusion is obviously satisfied, because *any* number in (a, b) could be chosen for μ. Assume that f is not identically zero on $[a, b]$. Since f is continuous on $[a, b]$, Theorem 5.2 ensures that f takes on both a maximum and a minimum in $[a, b]$. At least one of these extreme values is nonzero (because f is not identically zero), so this nonzero extreme value does not occur at either a or b. Thus f has an extreme value at an interior point, say μ, and by Lemma 6.1, $f'(\mu) = 0$.

It should be noted that Rolle's Theorem gives the result illustrated in Figure 6.1 for the special case in which the endpoints of the graph lie on the x-axis; for then the segment connecting the endpoints has zero slope, so we would want a point where the tangent line is horizontal, that is, $f'(\mu) = 0$. We now extend this result to the general case of functions with endpoint values that are not necessarily zero.

Theorem 6.3: Law of the Mean. If the function f is differentiable on (a, b) and continuous on $[a, b]$, then there is a point μ in (a, b) where

$$f'(\mu) = \frac{f(b) - f(a)}{b - a}. \tag{1}$$

Proof. Consider the function ϕ given by

$$\phi(x) = f(b) - f(x) - \left(\frac{f(b) - f(a)}{b - a} \right)(b - x). \tag{2}$$

In order to apply Rolle's Theorem to ϕ, we check to see that ϕ satisfies all the hypotheses of Rolle's Theorem. Since ϕ is a linear combination of constants, first-degree polynomials, and f, it follows that ϕ is continuous and differentiable wherever f is. Indeed, we can differentiate the formula in (2) to get

$$\phi'(x) = -f'(x) - \left(\frac{f(b) - f(a)}{b - a} \right)(-1). \tag{3}$$

Also, substituting first a then b for x in (2), we see that $\phi(a) = 0 = \phi(b)$. Hence, Rolle's Theorem assures us that for some μ in (a, b), $\phi'(\mu) = 0$. According to (3), this means

$$0 = -f'(\mu) + \frac{f(b) - f(a)}{b - a}, \tag{4}$$

which is obviously equivalent to (1).

Whenever a theorem of the form "if A and B, then C" is proved, it is desirable to give some examples to show that each part of the hypothesis is needed to guarantee the truth of the conclusion. Therefore we consider the following two functions.

EXAMPLE 6.7. Let $f(x) = |x|$ on the interval $[-1, 1]$. Although f is continuous on $[-1, 1]$, it is not differentiable on $(-1, 1)$; we see that the conclusion of the Law of the Mean does not hold, because there is no horizontal tangent (see Figure 6.2).

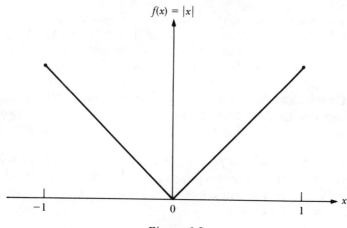

Figure 6.2

EXAMPLE 6.8. Let $g(x) = x - [x]$ on $[0, 1]$, where $[x]$ is the greatest integer function. Although g is differentiable on $(0, 1)$, it is not continuous on $[0, 1]$. Again, the lack of a horizontal tangent shows that the conclusion of the Law of the Mean does not hold (see Figure 6.3).

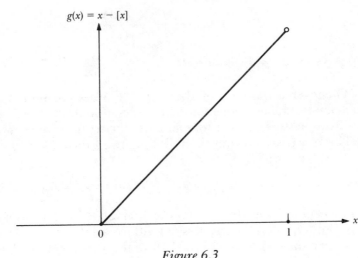

Figure 6.3

The importance of the Law of the Mean can hardly be over-stated. In the following corollaries and theorems, we give but a few of its many consequences. Corollaries 6.3a and 6.3b are the basis for the concept of the indefinite integral.

COROLLARY 6.3a. A differentiable function with an identically zero derivative throughout an interval must be constant there.

Proof. Suppose that f is differentiable but *nonconstant* on some interval. Then there are two points a and b in the interval such that $f(a) \neq f(b)$. Applying the Law of the Mean to f on $[a, b]$, we get a point μ in (a, b) where

$$f'(\mu) = \frac{f(b) - f(a)}{b - a} \neq 0.$$

Hence f' cannot be identically zero on the original interval.

COROLLARY 6.3b. Two differentiable functions with identical derivatives throughout an interval must differ by a constant there.

Proof. Let f and g be two functions such that $f' = g'$, and apply Corollary 6.3a to their difference, $f - g$.

The next corollary shows how the Law of the Mean plays a role in determining where a function is increasing and where it is decreasing.

COROLLARY 6.3c. If f is continuous on $[a, b]$, differentiable on (a, b), and f' does not change sign on (a, b), then f is monotonic on $[a, b]$.

Proof. Suppose that $f'(x) \geq 0$ for every x in (a, b), and let x_1 and x_2 be two points in $[a, b]$, say, $a \leq x_1 < x_2 \leq b$. By the Law of the Mean, there is a point μ in (x_1, x_2) such that

$$f(x_2) - f(x_1) = f'(\mu)(x_2 - x_1) \geq 0.$$

Hence f is nondecreasing on $[a, b]$. Similarly, if $f'(x) \leq 0$ on (a, b), then f is nonincreasing on $[a, b]$.

There are some notational variations of the Law of the Mean that are useful at times. For example, if f is differentiable in an open interval about a, we can apply the Law of the Mean to f on the interval $[a, x]$ (or $[x, a]$, in case $x < a$) and get a point μ between x and a where

$$f(x) = f(a) + f'(\mu)(x - a).$$

The import of this formula is that if the value of a function is known at a, and its rate of change is known for points near a, then this information can be used to compute the value of f at some point other than a.

There is a fairly standard type of problem in which the Law of the Mean is used to establish an inequality involving some differentiable function(s). This is illustrated in the following result.

PROPOSITION 6.1. If $x > 0$, then $\sin x < x$.

Proof. If $f(x) = x - \sin x$, then for some μ in $(0, x)$,

$$f(x) = f(0) + f'(\mu)(x - 0) = 0 + (1 - \cos \mu)(x).$$

Therefore

$$x - \sin x = x - x \cos \mu,$$

or

$$\sin x = x \cos \mu.$$

If $x \le 1$, then $0 < \mu < 1$, so $0 < \cos \mu < 1$. Thus $\sin x < x$. If $x > 1$, then it is obvious that $\sin x \le 1 < x$.

  **Exercises 6.3**

Use the Law of the Mean or Rolle's Theorem to prove the following assertions.

1. If $x > 0$, then $\log(1 + x) < x$.

2. If $x > -1$ and $x \neq 0$, then $\sqrt{1 + x} < 1 + (x/2)$.

3. If $x \neq 0$, then $\cos x > 1 - (x^2/2)$.

4. If $a < b$, then $e^a(b - a) < e^b - e^a < e^b(b - a)$.

5. If $x > 0$, then $\dfrac{x}{1 + x^2} < \arctan x < x$.

6. If $0 < x < 1$, then $x < \arcsin x < \dfrac{x}{\sqrt{1 - x^2}}$.

7. If $P(x) = a_n x^n + a_{n-1} x^{n-1} + \cdots + a_1 x + a_0$, and the coefficients satisfy

$$\frac{a_n}{n + 1} + \frac{a_{n-1}}{n} + \cdots + \frac{a_1}{2} + a_0 = 0,$$

then the equation $P(x) = 0$ has at least one root in $(0, 1)$.

8. If the polynomial equation $P(x) = 0$ has k distinct real roots, then the equation $P'(x) = 0$ has at least $k - 1$ distinct roots.

9. If $f(x) = \begin{cases} 1, & \text{if } x \ge 0, \\ 0, & \text{if } x < 0, \end{cases}$ then there is no function F such that $F'(x) = f(x)$ for every x in $\mathbb{R}$.

10. If the function f has a bounded derivative on the interval (a, b), then f is uniformly continuous there.

6.4. Cauchy's Law of the Mean

Theorem 6.4 has the same geometric interpretation as the Law of the Mean (Figure 6.1), but in this case the curve would be given by parametric equations, say, $x = g(t)$ and $y = f(t)$, where $a \le t \le b$.

THEOREM 6.4: CAUCHY'S LAW OF THE MEAN. Suppose that each of f and g is continuous on $[a, b]$ and differentiable on (a, b). If $g'(t) \neq 0$ for every t in (a, b), then there is a point μ in (a, b) where

$$\frac{f'(\mu)}{g'(\mu)} = \frac{f(b) - f(a)}{g(b) - g(a)}.$$

Proof. First, we assert that $g(b) \neq g(a)$, for by the Law of the Mean there is a point t^* in (a, b) where $g'(t^*) = [g(b) - g(a)]/[b - a]$, and we are assuming that $g'(t^*) \neq 0$.

Now define the function ϕ on $[a, b]$ by the formula

$$\phi(t) = f(t) - f(a) - \left[\frac{f(b) - f(a)}{g(b) - g(a)}\right][g(t) - g(a)].$$

As in the proof of the Law of the Mean, we verify that ϕ satisfies the hypotheses of Rolle's Theorem, so there is a point μ in (a, b) where $\phi'(\mu) = 0$. The conclusion then follows by routine differentiation of ϕ and substitution of μ for t. (These details are requested in Exercise 6.4.4.)

One of the very strong consequences of continuity is the Intermediate Value Property (Theorem 5.3), according to which, if a function is continuous on an interval, then any number between two values in its range must also be in its range. We have seen examples of derivatives that fail to be continuous, but as the next theorem asserts, these derivatives still possess the Intermediate Value Property. Recall Exercise 6.3.9, in which the given step function—which does not have the Intermediate Value Property—could not be the derivative of any function.

THEOREM 6.5. If f is the derivative of some function F on an interval, then f has the Intermediate Value Property there; that is, if r is a number between $f(a)$ and $f(b)$, then there is some μ between a and b such that $f(\mu) = r$.

Proof. Consider the function ϕ given by $\phi(x) = F(x) - rx$. Then $\phi'(x) = F'(x) - x = f(x) - r$, so we want to show that $\phi'(\mu) = 0$ for some μ between a and b. (We assume for definiteness that $a < b$.) Since ϕ is continuous on the closed interval $[a, b]$, it takes on both maximum and minimum values in $[a, b]$. If ϕ takes on any extreme value in the interior (a, b), then we are done, because Lemma 6.1 (see p. 79) guarantees that ϕ is zero there. Therefore

it suffices for us to eliminate the possibility that both the maximum
and the minimum occur at the endpoints. There are only two ways
that this can happen:

   (i) $\phi(a)$ is a maximum and $\phi(b)$ is a minimum, or
  (ii) $\phi(a)$ is a minimum and $\phi(b)$ is a maximum.

In Case (i),

$$\phi'(a) = \lim_{h \to 0+} \frac{\phi(a + h) - \phi(a)}{h} \leq 0$$

because $\phi(a) \geq \phi(a + h)$, and

$$\phi'(b) = \lim_{h \to 0-} \frac{\phi(b + h) - \phi(b)}{h} \leq 0.$$

But these imply that $f(a) \leq r$ and $f(b) \leq r$, which contradicts our
assumption that r is between $f(a)$ and $f(b)$. Similarly, in Case (ii), we
would have $f(a) \geq r$ and $f(b) \geq r$. Hence, ϕ has an extreme value at
some point μ in (a, b) where $\phi'(\mu) = f(\mu) - r = 0$.

   Note that the numbers a and b in the statement of Theorem
6.5 are not necessarily the endpoints of the interval on which $F' =$
f. Indeed, the significance of the Intermediate Value Property is
that for *any* two points a and b, and *any* r between $f(a)$ and $f(b)$, the
function f takes on the value r at some point *between a and b*. It
should also be recalled that a function with this property cannot
have a "jump" discontinuity; that is, the only way that $\lim_{x \to c} f(x) =$
$f(c)$ can be false is for the limit to fail to exist from one or
both sides.

Exercises 6.4

1. Let G be the graph $\{(x, y): x = t^2 \text{ and } y = t^3, 0 \leq t \leq 1\}$. Find
   a number μ in $(0, 1)$ where the tangent to G is parallel to the
   line through $(0, 0)$ and $(1, 1)$.

2. Prove that $f(x) = [x]$ is not the derivative of any function on
   an interval that contains two integers.

3. Let $g(x) = \begin{cases} \dfrac{x}{|x|}, & \text{if } x \neq 0, \\ 0, & \text{if } x = 0. \end{cases}$

   Prove that $g(x)$ is not the derivative of any function on any
   interval containing zero.

4. Give the details of the proof of Cauchy's Law of the Mean, Theorem 6.4, by following the proof of the Law of the Mean, Theorem 6.3.

6.5. Taylor's Formula with Remainder

In this section we return to the idea, first raised in Section 6.3, that the Law of the Mean gives us a way of calculating $f(x)$ by knowing $f(a)$ and $f'(\mu)$ at some point μ between x and a. Suppose that we want to approximate f with a first-degree polynomial on some interval about a. The function whose graph is the tangent line to the graph of f at a would seem to be a good approximation, and using techniques of elementary calculus we find that that polynomial is given by

$$P_1(x) = f(a) + f'(a)(x - a). \tag{1}$$

Suppose f has a second-order derivative f'' and we wish to choose a second-degree polynomial to approximate f. If we choose

$$P_2(x) = f(a) + f'(a)(x - a) + \frac{f''(a)}{2}(x - a)^2, \tag{2}$$

then

$$P_2(a) = f(a),$$
$$P'_2(a) = f'(a),$$
$$P''_2(a) = f''(a).$$

Therefore P_2 has a graph that is similar to that of f for points near a. Specifically, the graphs of P_2 and f at a have the same height, the same slope, and the same curvature. Now suppose f has an nth-order derivative, and we wish to approximate f with an nth-degree polynomial. Following the preceding pattern, we choose

$$P_n(x) = f(a) + f'(a)(x - a) + \cdots + \frac{f^{(n)}(a)}{n!}(x - a)^n. \tag{3}$$

Then it is easy to check that $P_n(a) = f(a)$, $P'_n(a) = f'(a)$, . . . , and $P_n^{(n)}(a) = f^{(n)}(a)$, so P_n would seem to be a very good approximation of f for points near a. In the first-degree case, the Law of the Mean tells us that we could get exact equality by evaluating the derivative at some point μ instead of using $f'(a)$ as in P_1. Perhaps we can get equality in the nth-degree case by replacing $f^{(n)}(a)$ with $f^{(n)}(\mu)$ for some appropriately chosen μ between a and x. That is precisely what the next theorem assures us.

THEOREM 6.6: TAYLOR'S FORMULA WITH REMAINDER. Suppose that f is a function such that each of $f, f', \ldots, f^{(n-1)}$ is continuous on $[a, b]$ and $f^{(n)}$ exists on (a, b). Then there is a point μ in (a, b) such that

$$f(b) = f(a) + f'(a)(b - a) + \frac{f''(a)}{2!}(b - a)^2 + \cdots$$
$$+ \frac{f^{(n-1)}(a)}{(n - 1)!}(b - a)^{n-1} + \frac{f^{(n)}(\mu)}{n!}(b - a)^n. \tag{4}$$

Proof. Let K be the number that satisfies

$$f(b) = f(a) + f'(a)(b - a) + \cdots$$
$$+ \frac{f^{(n-1)}(a)}{(n - 1)!}(b - a)^{n-1} + \frac{K}{n!}(b - a)^n. \tag{5}$$

We wish to show that $f^{(n)}(\mu) = K$ for some μ in (a, b). Define

$$\phi(x) = f(b) - f(x) - f'(x)(b - x) - \frac{f''(x)}{2!}(b - x)^2 - \cdots$$
$$- \frac{f^{(n-1)}(x)}{(n - 1)!}(b - x)^{n-1} - \frac{K}{n!}(b - x)^n. \tag{6}$$

We can verify that ϕ satisfies the hypotheses of Rolle's Theorem (these details are requested in Exercise 6.5.1), which implies the existence of a point μ in (a, b) where $\phi'(\mu) = 0$. Differentiation of ϕ in formula (6) yields

$$\phi'(x) = -f'(x) + f'(x) - f''(x)(b - x) + f''(x)(b - x) - \cdots$$
$$- \frac{f^{(n)}(x)}{(n - 1)!}(b - x)^{n-1} + \frac{K}{(n - 1)!}(b - x)^{n-1}. \tag{7}$$

Substituting $x = \mu$, we get $f^{(n)}(\mu) = K$.

The proof of Theorem 6.6 is admittedly lacking in details, but the technique is exactly the same as that used in proving Theorems 6.3 and 6.4. It is a good exercise (and a mathematical necessity) to fill in these details (see Exercise 6.5.1).

Exercises 6.5

1. Give the details of the proof of Theorem 6.6.

2. Use Taylor's Formula to get a cubic polynomial that approximates $\sqrt{x + 1}$. (Take $a = 0$ in Equation (4).)

3. Use Taylor's Formula to get a cubic polynomial that approximates $\log(1 + x)$.

4. Use Taylor's Formula to get a 7th-degree polynomial that approximates $\sin x$.

5. Use Taylor's Formula to get an nth-degree polynomial that approximates e^x.

6. Prove: If $x > 0$, then $1 + x + (x^2/2) < e^x < 1 + x + (x^2/2) e^x$.

7. Prove that e is an irrational number. (*Hint:* In Taylor's Formula, take $f(x) = e^x$, $a = 0$, and $b = 1$; assume $e = p/q$ and $n > q$, and multiply through by $(n - 1)!$.)

6.6. L'Hôpital's Rule

In the evaluation of any derivative, we take a limit of a quotient in which both the numerator and the denominator tend to zero. Therefore a limiting expression of the form 0/0 is not unfamiliar, and we know that such quotients may have any value at all for their limits (or the limit may not exist). More simply, consider the quotients $2x/3x$, x^2/x, $2x/x^2$, and $(x \sin 1/x)/x$ as x tends to zero. Although each is of the form 0/0, the limits of the first three are 2/3, 0, and ∞, respectively, whereas the fourth quotient does not approach a limit. The next result offers a simple technique of evaluating the limit of such an indeterminate form.

THEOREM 6.7: L'HÔPITAL'S RULE. Suppose that throughout some interval containing a, each of f and g is a differentiable function and $g'(x) \neq 0$. If

$$\lim_{x \to a} f(x) = \lim_{x \to a} g(x) = 0$$

and

$$\lim_{x \to a} \frac{f'(x)}{g'(x)} = L,$$

then

$$\lim_{x \to a} \frac{f(x)}{g(x)} = L.$$

Note: This result holds in case $\lim_{x \to a}$ is replaced throughout by $\lim_{x \to a+}$, $\lim_{x \to a-}$, $\lim_{x \to +\infty}$, or $\lim_{x \to -\infty}$. The limit value L can also be interpreted in the extended sense; that is, L can be $+\infty$ or $-\infty$, and the assertion is still true.

Proof. Suppose that x is close enough to a so that throughout the interval between a and x, f and g are differentiable with g' non-zero. Then by Cauchy's Law of the Mean, there is a point μ between a and x such that

$$\frac{f(x)}{g(x)} = \frac{f'(\mu)}{g'(\mu)}.$$

(Note that $f(a) = g(a) = 0$.)

Since μ is between a and x, $x \to a$ implies that $\mu \to a$. Hence

$$\lim_{x \to a} \frac{f(x)}{g(x)} = \lim_{\mu \to a} \frac{f'(\mu)}{g'(\mu)} = L.$$

It is clear that the preceding argument applies to left-hand, right-hand, and two-sided limits. The case in which $x \to \infty$ requires a slight twist: we introduce the auxiliary function $u(t) = 1/t$ and make use of the Chain Rule as well as the preceding case:

$$L = \lim_{x \to +\infty} \frac{f'(x)}{g'(x)} = \lim_{t \to 0+} \frac{f'(1/t)}{g'(1/t)}$$

$$= \lim_{t \to 0+} \frac{-f'(1/t)/t^2}{-g'(1/t)/t^2} = \lim_{t \to 0+} \frac{f'(u)u'(t)}{g'(u)u'(t)}$$

$$= \lim_{t \to 0+} \frac{(f \circ u)'(t)}{(g \circ u)'(t)} = \lim_{t \to 0+} \frac{(f \circ u)(t)}{(g \circ u)(t)}$$

$$= \lim_{t \to 0+} \frac{f(1/t)}{g(1/t)} = \lim_{x \to \infty} \frac{f(x)}{g(x)}.$$

The proof for the case in which $x \to -\infty$ is done exactly the same. In Exercise 6.6.1, the details are requested for the first case if L is replaced by ∞; that is,

$$\lim_{x \to a} \frac{f'(x)}{g'(x)} = \infty \quad \text{implies} \quad \lim_{x \to a} \frac{f(x)}{g(x)} = \infty.$$

Applications of L'Hôpital's Rule should be familiar to students from elementary calculus. We illustrate its use with three examples below:

EXAMPLE 6.9. Evaluate $\lim_{x \to 1} (\log x)/(x - 1)$. The expression satisfies the hypotheses of L'Hôpital's Rule for an indeterminate form of the 0/0 type; therefore

$$\lim_{x \to 1} \frac{\log x}{x - 1} = \lim_{x \to 1} \frac{1/x}{1} = 1.$$

EXAMPLE 6.10. Evaluate $\lim_{x \to 0} (1 - \cos x)/x^2$. After one application of L'Hôpital's Rule, we will still be left with an indeterminate form, to which we can reapply L'Hôpital's Rule. Existence of the final limit implies that of the middle limit, which implies the existence of the first limit (as well as the equalities). The computation looks like this:

$$\lim_{x \to 0} \frac{1 - \cos x}{x^2} = \lim_{x \to 0} \frac{\sin x}{2x}$$

$$= \lim_{x \to 0} \frac{\cos x}{2}$$

$$= \frac{1}{2}.$$

In a repeated application of L'Hôpital's Rule, such as in the preceding example, it is necessary to make sure at each step that the expression is still an indeterminate form. Otherwise the hypotheses of the theorem are not satisfied and the resulting computation may be wrong. Consider the following situation:

EXAMPLE 6.11. Evaluate $\lim_{x \to 2} (x^2 - x - 2)/(x - 2)$. We proceed as follows:

$$\lim_{x \to 2} \frac{x^2 - x - 2}{x^2 - 2x} = \lim_{x \to 2} \frac{2x - 1}{2x - 2}$$

$$\stackrel{?}{=} \lim_{x \to 2} \frac{2}{2} = 1.$$

The first equality is correct, but the second limit expression is not indeterminate; it should be evaluated using Theorem 4.6 as $(4 - 1)/(4 - 2) = 3/2$. Thus, a mechanical repetition of L'Hôpital's Rule can lead to an incorrect answer.

Another type of indeterminate form involves a quotient in which both the numerator and denominator increase without bound. For brevity we refer to this as the "∞/∞ form." An occurrence of this form is encountered when one determines the horizontal asymptotes of the graph of a relation such as $y = (2x^2 + 5)/(x^2 - 2)$. As $x \to \infty$, both the numerator and the denominator tend to ∞, but it is easy to see that the quotient tends to 2. In case $y = e^x/x$, we again have an ∞/∞ form, and it is not so easy to evaluate the limit as $x \to \infty$. As we see in the next theorem, we can again use L'Hôpital's Rule and examine the quotient of the respective derivatives of e^x and x, which yields $\lim_{x \to \infty} e^x/1 = \infty$.

THEOREM 6.8: L'HÔPITAL'S THEOREM. Suppose that throughout some interval containing a, each of f and g is a differentiable function and $g'(x) \neq 0$. If

$$\lim_{x \to a} f(x) = \lim_{x \to a} g(x) = \infty \qquad (1)$$

and

$$\lim_{x \to a} \frac{f'(x)}{g'(x)} = L, \qquad (2)$$

then

$$\lim_{x \to a} \frac{f(x)}{g(x)} = L. \qquad (3)$$

Note: The same broad interpretation of $x \to a$ is used here as in Theorem 6.7, namely, the result holds for right-hand, left-hand, and two-sided limits, as well as for $x \to +\infty$ or $x \to -\infty$. Also, L can be replaced by $+\infty$ or $-\infty$.

Proof. We here consider only the case $\lim_{x \to +\infty} f(x)/g(x)$ in detail. By (1) there is an N^* large enough so that on (N^*, ∞) none of the functions f, g, or g' has the value zero. For any x greater than N^*, Theorem 6.4 guarantees that there is a point μ in (N^*, x) satisfying

$$\frac{f'(\mu)}{g'(\mu)} = \frac{f(x) - f(N^*)}{g(x) - g(N^*)} = \frac{f(x)}{g(x)} \left[\frac{1 - \{f(N^*)/f(x)\}}{1 - \{g(N^*)/g(x)\}} \right],$$

or

$$
\begin{aligned}
\frac{f(x)}{g(x)} &= \frac{f'(\mu)}{g'(\mu)} \left[\frac{1 - \{g(N^*)/g(x)\}}{1 - \{f(N^*)/f(x)\}} \right] \\
&= \frac{f'(\mu)}{g'(\mu)} \left[\frac{1 - \{f(N^*)/f(x)\}}{1 - \{f(N^*)/f(x)\}} + \frac{f(N^*)/f(x) - g(N^*)/g(x)}{1 - \{f(N^*)/f(x)} \right] \\
&= \frac{f'(\mu)}{g'(\mu)} \left[1 + \frac{f(N^*)/f(x) - g(N^*)/g(x)}{1 - \{f(N^*)/f(x)\}} \right].
\end{aligned}
\tag{4}
$$

From (2) we know that the first factor of the right-hand member will be close to L so long as N^* is chosen sufficiently large. Therefore we can show that (3) holds by showing that

$$\lim_{x \to \infty} \frac{f(N^*)/f(x) - g(N^*)/g(x)}{1 - \{f(N^*)/f(x)\}} = 0. \tag{5}$$

But this follows immediately from our assumption that $\lim_{x \to \infty} f(x) = \lim_{x \to \infty} g(x) = \infty$. Note that *first* N^* is chosen so that when $\mu > N^*$, $f'(\mu)/g'(\mu)$ is close to L. Then after N^* is fixed, we choose N greater than N^* so that when $x > N$, the fractions in (5) are close to zero.

EXAMPLE 6.12. Evaluate $\lim_{x \to \infty} (\log x)^n/x^\varepsilon$, where $\varepsilon > 0$ and n is a positive integer. Consider these calculations:

$$\lim_{x \to \infty} \frac{(\log x)^n}{x^\varepsilon} = \lim_{x \to \infty} \frac{n(\log x)^{n-1}(1/x)}{\varepsilon x^{\varepsilon-1}}$$

$$= \lim_{x \to \infty} \frac{n(\log x)^{n-1}}{\varepsilon x^\varepsilon}$$

$$= \lim_{x \to \infty} \frac{n(n-1)(\log x)^{n-2}(1/x)}{\varepsilon^2 x^{\varepsilon-1}}$$

$$= \lim_{x \to \infty} \frac{n(n-1)(\log x)^{n-2}}{\varepsilon^2 x^{\varepsilon}}$$

.

.

.

$$= \lim_{x \to \infty} \frac{n!(\log x)^0}{\varepsilon^n x^{\varepsilon}}$$

$$= 0.$$

The first equality is justified by Theorem 6.8, whereas the second is just an algebraic rewriting. This is followed by another application of Theorem 6.8, then another algebraic rewriting, and so on.

EXAMPLE 6.13. Evaluate $\lim_{x \to \infty} x^n/e^x$, where n is a positive integer. Repeated applications (n times) of Theorem 6.8 yield

$$\lim_{x \to \infty} \frac{x^n}{e^x} = \lim_{x \to \infty} \frac{nx^{n-1}}{e^x} = \cdots = \lim_{x \to \infty} \frac{n!}{e^x} = 0.$$

There are other indeterminate forms, such as 0^0, ∞^0, and $0 \cdot \infty$, but these do not require another variation of L'Hôpital's Rule. Instead, they are evaluated by rewriting them into one of the preceding indeterminate forms. This technique is illustrated in the following two examples.

EXAMPLE 6.14. Evaluate $\lim_{x \to 0+} x^x$.

Let $y = x^x$; then $\log y = x \log x = (\log x)/(1/x)$. Applying Theorem 6.7 to the last quotient, we get

$$\lim_{x \to 0+} \log y = \lim_{x \to 0+} \frac{1/x}{-1/x^2} = \lim_{x \to 0+}(-x) = 0.$$

Hence

$$\lim_{x \to 0+} y = \lim_{x \to 0+} e^{\log y} = e^{\lim(\log y)} = e^0 = 1.$$

EXAMPLE 6.15. Show that $\lim_{n \to \infty} (1 + 1/n)^n = e$. By letting n take on all values instead of just integer values, we have a quotient of differentiable functions to which we can apply Theorem 6.8. Let $y = (1 + 1/x)^x$, so

$$\log y = x \log\left(1 + \frac{1}{x}\right) = \frac{\log(1 + 1/x)}{1/x}.$$

By Theorem 6.8,

$$\lim_{x \to \infty} \frac{\log(1 + 1/x)}{1/x} = \lim_{x \to \infty} \frac{\dfrac{-1/x^2}{1 + 1/x}}{-1/x^2} = \lim_{x \to \infty} \frac{1}{1 + 1/x}.$$

Hence

$$\lim_{x \to \infty} \log y = 1, \quad \text{so} \quad \lim_{x \to \infty} y = e^1 = e.$$

Exercises 6.6

1. Give the details of the proof of Theorem 6.7 for the following case:

$$\lim_{x \to a} \frac{f'(x)}{g'(x)} = \infty \quad \text{implies} \quad \lim_{x \to a} \frac{f(x)}{g(x)} = \infty.$$

In Exercises 2–18, evaluate the limits.

2. $\lim_{x \to 0} \dfrac{x^2}{[\log(1 + x)]^2}$

3. $\lim_{x \to \infty} \dfrac{3x^3 - x + 6}{x^3 + x^2 + 5}$

4. $\lim_{x \to \pi} \dfrac{1 + \cos x}{\sin 2x}$

5. $\lim_{x \to 0+} \dfrac{\cot x}{\log x}$

6. $\lim_{x \to 1} \dfrac{x^3 - 3x + 1}{x^4 - x^2 - 2x}$

7. $\lim_{x \to 0+} (x^2 \log x)$

8. $\lim_{x \to 0+} x^{\sin x}$

9. $\lim_{x \to \pi/2} (\sec x - \tan x)$

10. $\lim_{x \to 0+} (\cos \sqrt{x})^{1/x}$

11. $\lim_{x \to 0} \dfrac{e^{-1/x^2}}{x^n}$, where n is any positive integer

12. $\lim_{n} \dfrac{\log n}{n^2}$

13. $\lim_{n} \dfrac{e^n}{\pi^n}$

14. $\lim_{x \to 0} \dfrac{xe^x - \sin x}{\sin^2 x}$

15. $\lim_{x \to 1} \dfrac{\log x}{x^2 - 4x + 3}$

16. $\lim_{x \to 0+} \dfrac{x}{\sqrt{1 + x} - \sqrt{1 - x}}$

17. $\lim_{x \to 0} \dfrac{x^2}{1 - \cos x}$

18. $\lim_{n} \dfrac{e^{2n}}{n^2}$

7

THE
RIEMANN
INTEGRAL

7.1. Riemann Sums and
Integrable Functions

The main topic of this chapter is familiar to students of elementary calculus, where the integral is frequently introduced using upper and lower sums. Here we develop the theory from a different definition, one that employs a more general type of approximating sum. In Section 7.3 the resulting Riemann integral is shown to be equivalent to the Darboux integral, which is defined via upper and lower sums. This allows us to take advantage of the fact that some properties of the integral are easier to prove using Riemann sums, whereas others can be proved more easily using Darboux sums. With this dual approach, we give two proofs of the Fundamental Theorem of Calculus, the result that establishes the connection between the derivative and the integral.

Let f be a function with a domain that includes the interval $[a, b]$. A *partition* $\mathcal{P}$ of $[a, b]$ is a number set $\{x_k\}_{k=0}^n$ satisfying $a = x_0 < x_1 < \cdots < x_n = b$. The *norm* of the partition $\mathcal{P}$, denoted $\|\mathcal{P}\|$, is defined by

$$\|\mathcal{P}\| = \max\{|x_k - x_{k-1}|: k = 1, 2, \ldots, n\}.$$

Thus $\mathcal{P}$ determines n subintervals of $[a, b]$, the largest of which has length $\|\mathcal{P}\|$; the subinterval $[x_{k-1}, x_k]$ is called the kth subinterval (determined by $\mathcal{P}$). In each of the n subintervals, choose a number, say, μ_k in $[x_{k-1}, x_k]$, and form the sum

partition

$$\mathcal{P}(f, \mu) = \sum_{k=1}^{n} f(\mu_k)(x_k - x_{k-1}). = \sum_{k=1}^{n} f(\mu_k) \Delta t_k$$

This is called a *Riemann sum* for the function f on $[a, b]$. With some familiarity with integral calculus, one can immediately recognize that $\mathcal{P}(f, \mu)$ gives an approximation to the area of the region between the graph of f and the horizontal axis. (Of course, if f is not continuous on $[a, b]$, it may be hard to think of the graph of f as serving as a boundary for any such region.) Another interpretation, one that has the advantage of being free of geometric or pictorial dependence, is that $\mathcal{P}(f, \mu)$ represents some sort of "average value" of f on $[a, b]$. This can be seen as follows:

$$\mathcal{P}(f, \mu) = (b - a) \sum_{k=1}^{n} f(\mu_k) \frac{x_k - x_{k-1}}{b - a}.$$

In the sum, each of the images $f(\mu_k)$ is multiplied by the fraction $(x_k - x_{k-1})/(b - a)$ that gives the fractional part of $[a, b]$ from which that number μ_k is chosen. Thus the sum is a weighted average of the values $f(\mu_1), \ldots, f(\mu_n)$ from the range of f. We multiply this weighted average by the total length of $[a, b]$ to get the Riemann sum $\mathcal{P}(f, \mu)$. Therefore it is natural to expect that $\mathcal{P}(f, \mu)/(b - a)$ gives an average of the values of the function f on $[a, b]$. This average is a better indicator of the behavior of f if the partition $\mathcal{P}$ determines only "small" subintervals, so we examine the behavior of $\mathcal{P}(f, \mu)$ as $\|\mathcal{P}\|$ tends to zero.

DEFINITION 7.1. The function f is said to be *Riemann integrable* on $[a, b]$ provided that $\lim_{\|\mathcal{P}\| \to 0} \mathcal{P}(f, \mu)$ exists, and in this case the limit value is denoted by $\int_a^b f$ and is called the *Riemann integral* of f on $[a, b]$.

Before proceeding, we must acknowledge that this definition has no meaning at present because the limit concept $\lim_{\|\mathcal{P}\| \to 0} \mathcal{P}(f, \mu)$ on which the definition is based is completely new to us. It is certainly not a sequential limit, and it cannot be a function limit, because $\mathcal{P}(f, \mu)$ is not a function of the variable $\|\mathcal{P}\|$. To fully appreciate the latter assertion, consider the fact that for a given value of $\|\mathcal{P}\|$, there could be many partitions of $[a, b]$ whose largest subinterval has length $\|\mathcal{P}\|$. For example, the interval $[0, 1]$ may be partitioned by $\mathcal{P}_1 = \{0, 1/3, 2/3, 1\}$, $\mathcal{P}_2 = \{0, 1/3, 1/2, 3/4, 1\}$, or $\mathcal{P}_3 = \{0, 1/5, 2/5, 1/2, 2/3, 1\}$; and then $\|\mathcal{P}_1\| = \|\mathcal{P}_2\| = \|\mathcal{P}_3\| = 1/3$. Furthermore, for each partition of $[a, b]$ having the given norm $\|\mathcal{P}\|$, there are many ways of choosing the points $\{\mu_k\}_{k=1}^{n}$ from the n subintervals. Thus the value of $\|\mathcal{P}\|$ does not determine the value of the sum $\mathcal{P}(f, \mu)$. It is therefore necessary to define the limit concept that was used in the above definition.

DEFINITION 7.2. The statement "$\lim_{\|\mathcal{P}\| \to 0} \mathcal{P}(f, \mu) = I$" means that if $\varepsilon > 0$, then there is a positive number δ such that for every partition $\mathcal{P}$ of $[a, b]$ with $\|\mathcal{P}\| < \delta$ and any choice of the points $\{\mu_k \in [x_{k-1}, x_k], k = 1, \ldots, n\}$, the inequality $|\mathcal{P}(f, \mu) - I| < \varepsilon$ holds.

For a given example, a direct verification of the existence of $\lim_{\|\mathcal{P}\| \to 0} \mathcal{P}(f, \mu)$ is usually too complicated to consider, but this definition of the integral $\int_a^b f$ has certain advantages for developing the theory. In Section 7.3 we prove an equivalent formulation of $\lim_{\|\mathcal{P}\| \to 0} \mathcal{P}(f, \mu)$ which is easier to use in verifying the integrability of particular example functions. In the meantime, however, we can examine the integrability of two simple examples.

EXAMPLE 7.1. If f is a constant function on $[a, b]$, say, $f(x) = C$, then f is integrable on $[a, b]$ and $\int_a^b f = C(b - a)$.

To prove this assertion, we note that for *any* Riemann sum we have

$$\mathcal{P}(f, \mu) = \sum_{k=1}^{n} C(x_k - x_{k-1}) = C \sum_{k=1}^{n} (x_k - x_{k-1})$$
$$= C(x_n - x_0) = C(b - a).$$

Thus, for a given $\varepsilon > 0$, the inequality $|\mathcal{P}(f, \mu) - C(b - a)| < \varepsilon$ is satisfied trivially. Hence $\lim_{\|\mathcal{P}\| \to 0} \mathcal{P}(f, \mu) = C(b - a)$.

EXAMPLE 7.2. If

$$g(x) = \begin{cases} 1, & \text{if } x \in \mathbb{Q}, \\ 0, & \text{if } x \notin \mathbb{Q}, \end{cases}$$

then g is not integrable on any interval $[a, b]$.

To prove that g cannot be integrable, we take $\varepsilon = (b - a)/2$ and use the density of the rational numbers and of the irrational numbers. For any partition $\mathcal{P}$, no matter how small $\|\mathcal{P}\|$ may be, each subinterval determined by $\mathcal{P}$ must contain both rational numbers and irrational numbers, say, $\mu_k' \in [x_{k-1}, x_k]$ and $\mu_k'' \in [x_{k-1}, x_k]$, where μ_k' is in $\mathbb{Q}$ and μ_k'' is in $\sim \mathbb{Q}$. Then

$$\mathcal{P}(g, \mu') = \sum_{k=1}^{n} g(\mu'_k)(x_k - x_{k-1}) = \sum_{k=1}^{n} (x_k - x_{k-1}) = b - a$$

and

$$\mathcal{P}(g, \mu'') = \sum_{k=1}^{n} g(\mu''_k)(x_k - x_{k-1}) = 0.$$

Since the numbers $\mathcal{P}(g, \mu')$ and $\mathcal{P}(g, \mu'')$ are $b - a$ units apart, there is no number I that can be within $(b - a)/2$ of both sums. Hence, $\lim_{\|\mathcal{P}\| \to 0} \mathcal{P}(g, \mu)$ does not exist, so g is not integrable on $[a, b]$.

The notation $\int_a^b f(x)dx$ to denote the value of the integral is familiar from elementary calculus. Although we use the brief notation $\int_a^b f$ in our theoretical discussions, the longer notation has some advantages when working with a specific function. There should be no confusion when this notation is used in stating examples and exercises.

Exercises 7.1

In Exercises 1–5, prove that f is integrable on its domain interval $[a, b]$ and verify the value of $\int_a^b f$.

1. On $[0, 1], f(x) = \begin{cases} 1, & \text{if } x = 1/2, \\ 0, & \text{if } x \neq 1/2; \end{cases}$ $\int_0^1 f = 0.$

2. On $[0, 1], f(x) = \begin{cases} 1, & \text{if } x = 0 \text{ or } 1, \\ 0, & \text{otherwise}; \end{cases}$ $\int_0^1 f = 0.$

3. On $[0, 1], f(x) = \begin{cases} 1, & \text{if } x = 1, 1/2, 1/3, \ldots, \\ 0, & \text{otherwise}; \end{cases}$ $\int_0^1 f = 0.$

4. On $[0, 2], f(x) = \begin{cases} 1, & \text{if } 0 \leq x \leq 1, \\ -2, & \text{if } 1 < x \leq 2; \end{cases}$ $\int_0^2 f = -1.$

5. On $[0, 2], f(x) = \begin{cases} 2, & \text{if } 0 \leq x < 1, \\ 5, & \text{if } x = 1, \\ 1, & \text{if } 1 < x \leq 2; \end{cases}$ $\int_0^2 f = 3.$

6. On $[0, 1], f(x) = \begin{cases} 1, & \text{if } x \in (3/4, 1] \cup (3/8, 1/2] \\ & \quad \cup (3/16, 1/4] \cup \ldots, \\ -1, & \text{if } x \in (1/2, 3/4] \cup (1/4, 3/8] \\ & \quad \cup (1/8, 3/16] \cup \ldots; \end{cases}$

$\int_0^1 f = 0.$

7. Prove: If f is integrable on $[0, 1]$, then

$$\lim_n \frac{1}{n} \sum_{k=1}^{n} f\left(\frac{k}{n}\right) = \int_0^1 f.$$

8. Given that $f(x) = \sqrt{1 - x^2}$ and f is integrable on $[0, 1]$, prove that

$$\lim_n \frac{1}{n^2} \sum_{k=1}^{n} \sqrt{n^2 - k^2} = \int_0^1 f.$$

9. Given that $f(x) = \dfrac{1}{x - 3}$ and f is integrable on $[0, 1]$, prove that

$$\lim_n \sum_{k=1}^{n} \frac{1}{k - 3n} = \int_0^1 f.$$

(handwritten margin note: "same formula as 7")

7.2. Basic Properties

Since we are working with an entirely new limit concept, we must begin by proving such basic results as the uniqueness of the limit value and algebraic closure. Throughout this chapter, we shorten the phrase "f is Riemann integrable" to "f is integrable." In later chapters we study other types of integrals, but until that time there should be no confusion.

THEOREM 7.1. If the function f is integrable on $[a, b]$, then the value $\int_a^b f$ is unique.

Proof. Suppose that $\lim_{\|\mathscr{P}\| \to 0} \mathscr{P}(f, \mu) = I$, and let J be a number not equal to I. We show that J cannot be the limit of $\mathscr{P}(f, \mu)$ by using $\varepsilon = |I - J|/2$. Since ε is half the distance between I and J, it follows that the intervals $(I - \varepsilon, I + \varepsilon)$ and $(J - \varepsilon, J + \varepsilon)$ do not intersect. When $\mathscr{P}$ is a partition of $[a, b]$ with $\|\mathscr{P}\|$ sufficiently small, $\mathscr{P}(f, \mu)$ is in the interval $(I - \varepsilon, I + \varepsilon)$ for any choice of $\{\mu_k\}_{k=1}^{n}$. Then $\mathscr{P}(f, \mu)$ is not in $(J - \varepsilon, J + \varepsilon)$, so $|\mathscr{P}(f, \mu) - J| \geq \varepsilon$ whenever $\|\mathscr{P}\|$ is sufficiently small. Hence $\lim_{\|\mathscr{P}\| \to 0} \mathscr{P}(f, \mu)$ cannot equal J.

PROPOSITION 7.1. If the function f is integrable on $[a, b]$, then f is bounded there.

Proof. Suppose f is defined but not bounded on $[a, b]$, and let I be any number. We assert that no matter how small the norm

$\|\mathcal{P}\|$ may be, it is possible to select $\{\mu_k\}_{k=1}^n$ so that $|\mathcal{P}(f, \mu)| > |I| + 1$, which implies that $|\mathcal{P}(f, \mu) - I| > 1$. Let $\mathcal{P}$ be any partition of $[a, b]$. Then f must be unbounded on at least one of the subintervals determined by $\mathcal{P}$, say, $[x_{m-1}, x_m]$. For $k \neq m$, choose the numbers $\{\mu_k\}_{k=1}^n$ satisfying $x_{k-1} \leq \mu_k \leq x_k$ in any way at all. Thus we have determined the number

$$|I| + \Sigma_{k \neq m} |f(\mu_k)|(x_k - x_{k-1}).$$

Now we use the unboundedness of f on $[x_{m-1}, x_m]$ to choose μ_m so that

$$|f(\mu_m)(x_m - x_{m-1})| > 1 + |I| + \Sigma_{k \neq m} |f(\mu_k)|(x_k - x_{k-1}).$$

Thus the mth term dominates the sum $\mathcal{P}(f, \mu)$, and we have $|\mathcal{P}(f, \mu)| > |I| + 1$. Hence $\lim_{\|\mathcal{P}\| \to 0} \mathcal{P}(f, \mu)$ cannot exist, so f is not integrable on $[a, b]$.

THEOREM 7.2. If each of f and g is an integrable function on $[a, b]$, and c is a number, then $f + g$ and cf are also integrable on $[a, b]$. Moreover,

$$\int_a^b (f \pm g) = \int_a^b f \pm \int_a^b g,$$

and

$$\int_a^b cf = c \int_a^b f.$$

Proof. Suppose $\varepsilon > 0$ and choose δ_f and δ_g to be positive numbers such that

$$\|\mathcal{P}\| < \delta_f \quad \text{implies} \quad \left| \mathcal{P}(f, \mu) - \int_a^b f \right| < \varepsilon/2$$

and

$$\|\mathcal{P}\| < \delta_g \quad \text{implies} \quad \left| \mathcal{P}(g, \mu) - \int_a^b g \right| < \varepsilon/2.$$

Now define $\delta = \min(\delta_f, \delta_g)$. This guarantees that if $\|\mathcal{P}\| < \delta$, then

$$\left| \mathcal{P}(f + g, \mu) - \left(\int_a^b f + \int_a^b g \right) \right| = \left| \Sigma_{k=1}^n [f(\mu_k) + g(\mu_k)](x_k - x_{k-1}) \right.$$
$$\left. - \left(\int_a^b f + \int_a^b g \right) \right|$$
$$= \left| \Sigma_{k=1}^n f(\mu_k)(x_k - x_{k-1}) - \int_a^b f \right.$$
$$\left. + \Sigma_{k=1}^n g(\mu_k)(x_k - x_{k-1}) - \int_a^b g \right|$$

$$\leq \left| \mathcal{P}(f, \mu) - \int_a^b f \right| + \left| \mathcal{P}(g, \mu) - \int_a^b g \right|$$

$$< \varepsilon/2 + \varepsilon/2$$

$$= \varepsilon.$$

Hence $f + g$ is integrable, and $\int_a^b (f + g) = \int_a^b f + \int_a^b g$.

The integrability of cf is obvious if $c = 0$, so assume that $c \neq 0$ and choose δ so that $\|\mathcal{P}\| < \delta$ implies $|\mathcal{P}(f, \mu) - \int_a^b f| < \varepsilon/|c|$. Then $\|\mathcal{P}\| < \delta$ also implies that

$$\left| \mathcal{P}(cf, \mu) - c\int_a^b f \right| = \left| \sum_{k=1}^n cf(\mu_k)(x_k - x_{k-1}) - c\int_a^b f \right|$$

$$\leq |c| \left| \sum_{k=1}^n f(\mu_k)(x_k - x_{k-1}) - c\int_a^b f \right|$$

$$< |c| \, (\varepsilon/|c|)$$

$$= \varepsilon.$$

Hence, $\int_a^b cf = c\int_a^b f$. The integrability of $f - g$ now follows from the preceding cases with a consideration of $f + cg$, with $c = -1$.

Although Theorem 7.2 is stated and proved for the sum of two functions, it can be extended to the sum of m functions by using an induction argument (see Exercise 7.2.5). There is another type of additivity property that can be proved for the Riemann integral. In the next theorem we consider just one function, but its integral is evaluated on two adjacent intervals, and the resulting values are added.

THEOREM 7.3. If the function f is integrable on $[a, b]$ and on $[b, c]$, then f is integrable on $[a, c]$ and

$$\int_a^c f = \int_a^b f + \int_b^c f.$$

Proof. Any partition $\mathcal{P}$ on $[a, c]$ determines partitions $\mathcal{P}_1$ and $\mathcal{P}_2$ on $[a, b]$ and $[b, c]$, respectively, for if b is not one of the partition points of $\mathcal{P}$, it can be inserted between x_{j-1} and x_j, where j is the least integer such that $b \leq x_j$. In the Riemann sum $\mathcal{P}(f, \mu)$, the number μ_j is selected from $[x_{j-1}, x_j]$, and we may have either $\mu_j \in [x_{j-1}, b]$ or $\mu_j \in [b, x_j]$. In the former case, we can write

$$\mathcal{P}(f, \mu) = \sum_{k=1}^{j-1} f(\mu_k)(x_k - x_{k-1}) + f(\mu_j)(b - x_{j-1})$$

$$- f(b)(x_j - b) + f(\mu_j)(x_j - b)$$

$$+ f(b)(x_j - b) + \sum_{k=j}^n f(\mu_k)(x_k - x_{k-1})$$

$$= \mathcal{P}_1(f, \mu) + [f(\mu_j) - f(b)](x_j - b) + \mathcal{P}_2(f, \mu).$$

Similarly, if μ_j is in $[b, x_j]$, we can write

$$\mathcal{P}(f, \mu) = \mathcal{P}_1(f, \mu) + [f(\mu_j) - f(b)](b - x_{j-1}) + \mathcal{P}_2(f, \mu).$$

If $f(x) < K$ on $[a, b]$, then in either case we have

$$|\mathcal{P}(f, \mu) - \mathcal{P}_1(f, \mu) - \mathcal{P}_2(f, \mu)| \le 2K \|\mathcal{P}\|,$$

because $b - x_{j-1} \le x_j - x_{j-1} \le \|\mathcal{P}\|$ and $x_j - b \le x_j - x_{j-1} \le \|\mathcal{P}\|$. If $\varepsilon > 0$, the integrability of f on $[a, b]$ and on $[b, c]$ allows us to choose a positive number δ so that $\|\mathcal{P}_1\| < \delta$ and $\|\mathcal{P}_2\| < \delta$ imply that

$$\left| \mathcal{P}_1(f, \mu) - \int_a^b f \right| < \varepsilon/3 \quad \text{and} \quad \left| \mathcal{P}_2(f, \mu) - \int_b^c f \right| < \varepsilon/3.$$

We can also choose δ even smaller, if necessary, so that $\delta < \varepsilon/(6K)$. Now since $\|\mathcal{P}_1\| \le \|\mathcal{P}\|$ and $\|\mathcal{P}_2\| \le \|\mathcal{P}\|$, we see that $\|\mathcal{P}\| < \delta$ implies

$$\begin{aligned}
\left| \mathcal{P}(f, \mu) - \left(\int_a^b f + \int_b^c f \right) \right| &\le |\mathcal{P}(f, \mu) - \mathcal{P}_1(f, \mu) - \mathcal{P}_2(f, \mu)| \\
&\quad + \left| \mathcal{P}_1(f, \mu) - \int_a^b f \right| + \left| \mathcal{P}_2(f, \mu) - \int_b^c f \right| \\
&< 2K\|\mathcal{P}\| + \varepsilon/3 + \varepsilon/3 \\
&< \varepsilon.
\end{aligned}$$

Hence $\lim_{\|\mathcal{P}\| \to 0} \mathcal{P}(f, \mu) = \int_a^b f + \int_b^c f$.

The next property is analogous to Propositions 2.1 and 2.2. There is a similar result for function limits, too. The general idea is that if we evaluate a limit of some expression that is bounded (above or below), then the same bound applies to the limit value (see also Exercise 7.2.4).

THEOREM 7.4. If each of f and g is an integrable function on $[a, b]$ and $f(x) \le g(x)$ for every x in $[a, b]$, then

$$\int_a^b f \le \int_a^b g.$$

Proof. First consider the case in which $f(x)$ is identically zero. Then $g(x) \ge 0$ on $[a, b]$, so for every Riemann sum, $\mathcal{P}(g, \mu) \ge 0$. If $I < 0$ and $\varepsilon = -I$, then it is not possible to have $|\mathcal{P}(g, \mu) - I| < \varepsilon$, because the distance between $\mathcal{P}(g, \mu)$ and I is at least equal to $|I|$ (see Figure 7.1).

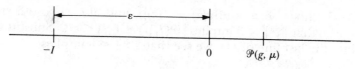

Figure 7.1

Now consider the general case in which f is an arbitrary integrable function and $f(x) \leq g(x)$. Define $h(x) = g(x) - f(x)$. Clearly, $h(x) \geq 0$, and, by Theorem 7.2, h is integrable on $[a, b]$. So the first case of this proof ensures that

$$0 \leq \int_a^b h = \int_a^b (g - f) = \int_a^b g - \int_a^b f.$$

One of the motivating interpretations of the integral was that of a limit of weighted averages of values from the range of f. In averaging a large set of numbers, one should be able to change a few values without altering the average very much. This is true for $\mathcal{P}(f, \mu)$ and the resulting integral value; that is, $f(x)$ may be changed for a finite number of x's in $[a, b]$ without changing either the integrability of f or the value of its integral. This property was illustrated in Exercises 7.1.1–7.1.5.

THEOREM 7.5. Suppose that f is an integrable function on $[a, b]$ and $g(x) = f(x)$ for all but a finite number of points in $[a, b]$; then g is also integrable on $[a, b]$ and $\int_a^b g = \int_a^b f$.

Proof. It is sufficient to prove the assertion in the case in which g differs from f at exactly one point in $[a, b]$, because we can produce n changes in f by changing the value at one point and repeating the procedure n times. So assume that $g(x) = f(x)$ for all x in $[a, b] \sim \{z\}$. For any partition $\mathcal{P}$, z can be in at most two of the subintervals (z may be a partition point), say, $[x_{m-1}, x_m]$ and $[x_m, x_{m+1}]$. Then

$$\left| \mathcal{P}(g, \mu) - \int_a^b f \right| = \left| \mathcal{P}(g, \mu) - \mathcal{P}(f, \mu) + \mathcal{P}(f, \mu) - \int_a^b f \right|$$

$$\leq \left| \sum_{k=1}^n [g(\mu_k) - f(\mu_k)](x_k - x_{k-1}) \right| + \left| \mathcal{P}(f, \mu) - \int_a^b f \right|$$

$$\leq |g(z) - f(z)| (x_m - x_{m-1}) + |g(z) - f(z)| (x_{m+1} - x_m)$$

$$\quad + \left| \mathcal{P}(f, \mu) - \int_a^b f \right|$$

$$\leq |g(z) - f(z)| (2\|\mathcal{P}\|) + \left| \mathcal{P}(f, \mu) - \int_a^b f \right|. \tag{1}$$

If $\varepsilon > 0$, choose $\delta \leq \varepsilon/(4|g(z) - f(z)|)$, and also choose δ small enough so that $\|\mathcal{P}\| < \delta$ implies $|\mathcal{P}(f, \mu) - \int_a^b f| < \varepsilon/2$. Substituting this into the last line of (1), we see that $\|\mathcal{P}\| < \delta$ implies

$$\left|\mathcal{P}(g, \mu) - \int_a^b f\right| < |g(z) - f(z)|(2\delta) + \varepsilon/2$$
$$< \varepsilon/2 + \varepsilon/2$$
$$= \varepsilon.$$

Exercises 7.2

1. Given $f(x) = \begin{cases} 2, & \text{if } x \leq 0, \\ 1/x, & \text{if } x > 0, \end{cases}$ determine whether f is integrable on each of the intervals $[0, 1]$, $[-1, 0]$, $[-1, 1]$.

2. Given that $\int_0^{\pi/4} \sec^2 x \, dx = 1$, prove that $\int_0^{\pi/4} \tan^2 x \, dx = 1 - (\pi/4)$.

3. Assuming that $f(x) = \sqrt{1 + \sin^4 x}$ defines an integrable function, show that

$$\int_0^\pi f = \int_\pi^{2\pi} f \quad \text{and} \quad \int_0^{n\pi} f = n \int_0^\pi f.$$

4. Prove: If f is integrable on $[a, b]$ and $m \leq f(x) \leq M$ for every x in $[a, b]$, then $m(b - a) \leq \int_a^b f \leq M(b - a)$.

5. Prove the extension of Theorem 7.2 to the sum of m integrable functions:

$$\int_a^b \left(\sum_{i=1}^m f_i\right) = \sum_{i=1}^m \int_a^b f_i.$$

6. The function σ is a *step function* on $[a, b]$ if there is a partition $\mathcal{P}$ of $[a, b]$ such that σ is constant on each open subinterval (x_{k-1}, x_k) determined by $\mathcal{P}$. Use Theorems 7.3 and 7.4 to prove that a step function is integrable, and if $\sigma(x) = y_k$ on (x_{k-1}, x_k), then

$$\int_a^b \sigma = \sum_{k=1}^n y_k(x_k - x_{k-1}).$$

7. Prove the Mean Value Theorem for Integrals (MVTI): If f is continuous and integrable[†] on $[a, b]$, then there is a number μ in $[a, b]$ such that

$$f(\mu)(b - a) = \int_a^b f.$$

[†]We prove later that all continuous functions are integrable, so this part of the hypothesis becomes redundant.

7.3. The Darboux Criterion
for Integrability

In this section we develop a characterization of Riemann integrable functions that does not rely on predicting the value of the limit $\lim_{\|\mathscr{P}\| \to 0} \mathscr{P}(f, \mu)$. In this sense, it is analogous to the Cauchy Criterion for sequential convergence. This characterization is extremely useful in showing the integrability of certain classes of functions. In particular, we use it to prove that continuous functions and monotonic functions are integrable.

In order to lay the foundation for this work, we must introduce some notation and terminology. Let f be a bounded function on $[a, b]$ and let $\mathscr{P}$ be a partition of $[a, b]$. Since f is bounded on each subinterval $[x_{k-1}, x_k]$, we can define

$$m_k = \text{glb } \{f(x): x_{k-1} \leq x \leq x_k\}$$
$$M_k = \text{lub } \{f(x): x_{k-1} \leq x \leq x_k\}.$$

Now we form the lower sum and upper sum for f with respect to $\mathscr{P}$:

$$\text{L}(f, \mathscr{P}) = \sum_{k=1}^{n} m_k(x_k - x_{k-1});$$
$$\text{U}(f, \mathscr{P}) = \sum_{k=1}^{n} M_k(x_k - x_{k-1}).$$

We observe that for any Riemann sum $\mathscr{P}(f, \mu)$, we have

$$\text{L}(f, \mathscr{P}) \leq \mathscr{P}(f, \mu) \leq \text{U}(f, \mathscr{P}),$$

because $m_k \leq f(\mu_k) \leq M_k$ for $k = 1, \ldots, n$. Also note that the lower and upper sums need not be Riemann sums, because the numbers m_k and M_k may not be in the range of f.

If each of $\mathscr{P}$ and $\mathscr{P}'$ is a partition of $[a, b]$ and $\mathscr{P} \subseteq \mathscr{P}'$, then $\mathscr{P}'$ is called a *refinement* of $\mathscr{P}$. Thus, one can produce a refinement of $\mathscr{P}$ by inserting additional partition points between those of $\mathscr{P}$. It is clear that if $\mathscr{P}'$ is a refinement of $\mathscr{P}$, then $\|\mathscr{P}'\| \leq \|\mathscr{P}\|$. Also, if $\mathscr{P}$ and $\mathscr{P}*$ are partitions of $[a, b]$, then they have a common refinement; for example, $\mathscr{P} \cup \mathscr{P}*$ is a refinement of both $\mathscr{P}$ and $\mathscr{P}*$, since it contains both sets of partition points.

LEMMA 7.1. If each of $\mathscr{P}$ and $\mathscr{P}*$ is a partition of $[a, b]$, then

$$\text{L}(f, \mathscr{P}) \leq \text{U}(f, \mathscr{P}*);$$

that is, no lower sum can exceed any upper sum.

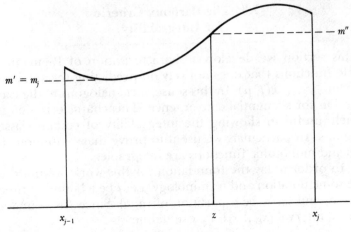

Figure 7.2

Proof. First note that as partition points are added to a partition $\mathscr{P}$, the lower sums increase and the upper sums decrease. For if

$$L(f, \mathscr{P}) = \sum_{k=1}^{n} m_k(x_k - x_{k-1})$$

and we add a point z in (x_{j-1}, x_j) to form $\mathscr{P}' = \mathscr{P} \cup \{z\}$, then

$$L(f, \mathscr{P}') = \sum_{k \neq j} m_k(x_k - x_{k-1}) + m'(z - x_{j-1}) + m''(x_j - z),$$

where m' and m'' are the greatest lower bounds of $f(x)$ on $[x_{j-1}, z]$ and $[z, x_j]$, respectively. Thus $m' \geq m_j$ and $m'' \geq m_j$ (see Figure 7.2), so

$$m'(z - x_{j-1}) + m''(x_j - z) \geq m[(z - x_{j-1}) + (x_j - z)] = m(x_j - x_{j-1}).$$

Hence $L(f, \mathscr{P}') \geq L(f, \mathscr{P})$. Similarly, $U(f, \mathscr{P}') \leq U(f, \mathscr{P})$.

Now suppose $\mathscr{P}$ and $\mathscr{P}*$ are any two partitions on $[a, b]$, and let $\mathscr{P}' = \mathscr{P} \cup \mathscr{P}*$. Since $\mathscr{P}'$ is a refinement of both $\mathscr{P}$ and $\mathscr{P}*$, the preceding observation gives us

$$L(f, \mathscr{P}) \leq L(f, \mathscr{P}') \leq U(f, \mathscr{P}') \leq U(f, \mathscr{P}*),$$

and Lemma 7.1 is proved.

Lemma 7.1 assures us that the set of all lower bounds of f is bounded above; any upper sum will do as an upper bound. Therefore, by the LUB Axiom, there is a least upper bound of this set, say,

$$\lambda(f) = \text{lub}_{\mathscr{P}} \{L(f, \mathscr{P})\}. \tag{1}$$

Furthermore, for any upper sum, we must have

$$\lambda(f) \le U(f, \mathscr{P}). \tag{1a}$$

Similarly, the set of all upper sums is bounded below by every lower sum, so we can define $\Lambda(f)$ by

$$\Lambda(f) = \text{glb}_{\mathscr{P}}\{U(f, \mathscr{P})\}. \tag{2}$$

And for any lower sum we have

$$L(f, \mathscr{P}) \le \Lambda(f). \tag{2a}$$

It follows from the definitions (1) and (2) and from Lemma 7.1 that

$$\lambda(f) \le \Lambda(f);$$

for if $\Lambda(f) < \lambda(f)$ and $\mu = [\Lambda(f) + \lambda(f)]/2$, then we could find an upper sum $U(f, \mathscr{P})$ in the interval $(\Lambda(f), \mu)$ and a lower sum $L(f, \mathscr{P})$ in $(\mu, \lambda(f))$, which would contradict Lemma 7.1.

Since the Riemann sums lie between the upper and lower sums, it is natural to ask what happens when there are upper and lower sums that are arbitrarily close together, that is, when $\lambda(f) = \Lambda(f)$. One would expect that this forces the Riemann sums to converge, with their limit being equal to the common value of $\lambda(f)$ and $\Lambda(f)$. That is precisely what does happen, and it provides the characterization of integrability that we seek. But before attempting this theorem, we must prove a lemma that gives the necessary connection between the limit concept as $\|\mathscr{P}\|$ tends to zero and the lub and glb concepts of (1) and (2).

LEMMA 7.2. If f is a bounded function on $[a, b]$, then

$$\lim_{\|\mathscr{P}\| \to 0} L(f, \mathscr{P}) = \lambda(f) \quad \text{and} \quad \lim_{\|\mathscr{P}\| \to 0} U(f, \mathscr{P}) = \Lambda(f);$$

that is, if $\varepsilon > 0$, there is a positive number δ such that $\|\mathscr{P}\| < \delta$ implies

$$L(f, \mathscr{P}) > \lambda(f) - \varepsilon \quad \text{and} \quad U(f, \mathscr{P}) < \Lambda(f) + \varepsilon.$$

Proof. Suppose $|f(x)| < K$ for every x in $[a, b]$ and $\varepsilon > 0$. By (2), there exists a partition $\mathscr{P}^* = \{z_i\}_{i=0}^q$ of $[a, b]$ such that $U(f, \mathscr{P}^*)$

$< \Lambda(f) + \varepsilon/2$. Define $\delta = \varepsilon/(4qK)$, and let $\mathcal{P}$ be a partition satisfying $\|\mathcal{P}\| < \delta$. In order to show that $U(f, \mathcal{P}) < \Lambda(f) + \varepsilon$, let us consider the common refinement $\mathcal{P}' = \mathcal{P} \cup \mathcal{P}*$ of $\mathcal{P}$ and $\mathcal{P}*$. Then some of the partition points $\{x_k\}_{k=0}^n$ of $\mathcal{P}$ are z_i's from $\mathcal{P}*$, and some are not. In the sum $U(f, \mathcal{P}')$, we can separate the terms into two groups:

  (i) $\Sigma M_k(x_k - x_{k-1})$, where the interval $[x_{k-1}, x_k]$ contains no z_i;
  (ii) $\Sigma M_k(x_k - x_{k-1})$, where either x_{k-1} or x_k (or both) is a z_i.

Since the terms of type (i) are also terms in the sum $U(f, \mathcal{P})$, it follows that $U(f, \mathcal{P}) - U(f, \mathcal{P}')$ consists of precisely the terms of type (ii). The sum of all the type (ii) terms consists of at most $2q - 1$ terms, each of which does not exceed $K\|\mathcal{P}'\| < K\|\mathcal{P}\| < K\delta = (K\varepsilon)/(4qK) = \varepsilon/(4q)$. Therefore

$$U(f, \mathcal{P}) - U(f, \mathcal{P}') < (2q - 1)[\varepsilon/(4q)] < \varepsilon/2.$$

Hence

$$U(f, \mathcal{P}) < U(f, \mathcal{P}') + \varepsilon/2$$
$$\leq U(f, \mathcal{P}*) + \varepsilon/2$$
$$< \Lambda(f) + \varepsilon/2 + \varepsilon/2$$
$$= \Lambda(f) + \varepsilon,$$

and we have proved that $\lim_{\|\mathcal{P}\| \to 0} U(f, \mathcal{P}) = \Lambda(f)$. The dual assertion that $\lim_{\|\mathcal{P}\| \to 0} L(f, \mathcal{P}) = \lambda(f)$ can be proved by a similar argument.

THEOREM 7.6: DARBOUX INTEGRABILITY THEOREM. The bounded function f is integrable on $[a, b]$ if and only if $\lambda(f) = \Lambda(f)$, and in this case $\int_a^b f = \lambda(f) = \Lambda(f)$.

Proof. First assume that $\lambda(f) = I$, and suppose $\varepsilon > 0$. Using Lemma 7.2, we can choose δ so that $\|\mathcal{P}\| < \delta$ implies

$$L(f, \mathcal{P}) > I - \varepsilon \quad \text{and} \quad U(f, \mathcal{P}) < I + \varepsilon.$$

Then for any choice of $\{\mu_k\}_{k=0}^n$, we have

$$I - \varepsilon < L(f, \mathcal{P}) \leq \mathcal{P}(f, \mu) \leq U(f, \mathcal{P}) < I + \varepsilon,$$

so $|\mathcal{P}(f, \mu) - I| < \varepsilon$ whenever $\|\mathcal{P}\| < \delta$. Hence f is integrable on $[a, b]$, and $\int_a^b f = I$.

Now assume that f is integrable on $[a, b]$ and $\int_a^b f = I$, and suppose that $\varepsilon > 0$. If we can show that for some partition $\mathcal{P}$, $U(f, \mathcal{P}) - L(f, \mathcal{P}) < \varepsilon$, then we can conclude that $\lambda(f) = \Lambda(f)$; for $\Lambda(f) - \lambda(f) \le U(f, \mathcal{P}) - L(f, \mathcal{P})$, so we will have $\Lambda(f) \le \lambda(f) + \varepsilon$. Since ε is an arbitrary positive number, this implies that $\Lambda(f) \le \lambda(f)$. But $\Lambda(f) \ge \lambda(f)$ is always true, so we can conclude that $\Lambda(f) = \lambda(f)$.

Choose $\delta > 0$ so that $\|\mathcal{P}\| < \delta$ implies $|\mathcal{P}(f, \mu) - I| < \varepsilon/4$. Since $M_k = \text{lub}\{f(x): x \in [x_{k-1}, x_k]\}$ and $m_k = \text{glb}\{f(x): x \in [x_{k-1}, x_k]\}$, there exist points μ_k' and μ_k'' in $[x_{k-1}, x_k]$ satisfying

$$f(\mu_k') > M_k - \frac{\varepsilon}{4(b-a)} \quad \text{and} \quad f(\mu_k'') < m_k + \frac{\varepsilon}{4(b-a)}.$$

Then

$$U(f, \mathcal{P}) = \sum_{k=1}^n M_k(x_k - x_{k-1})$$

$$< \sum_{k=1}^n \left[f(\mu_k') + \frac{\varepsilon}{4(b-a)} \right](x_k - x_{k-1})$$

$$= \mathcal{P}(f, \mu') + \frac{\varepsilon}{4(b-a)} \sum_{k=1}^n (x_k - x_{k-1})$$

$$< I + \frac{\varepsilon}{4} + \frac{\varepsilon}{4}$$

$$= I + \frac{\varepsilon}{2}.$$

Similarly,

$$L(f, \mathcal{P}) > \sum_{k=1}^n \left[f(\mu_k'') - \frac{\varepsilon}{4(b-a)} \right](x_k - x_{k-1})$$

$$= \mathcal{P}(f, \mu'') - \frac{\varepsilon}{4}$$

$$< I - \frac{\varepsilon}{4} - \frac{\varepsilon}{4}$$

$$= I - \frac{\varepsilon}{2}.$$

Hence, $U(f, \mathcal{P}) - L(f, \mathcal{P}) < \varepsilon$, and the proof is complete.

In using Darboux's Theorem to show that a function is integrable, we frequently use it in the form that was developed in the proof of Theorem 7.6. In order to facilitate this use of it, we isolate the part we want by stating it here as a lemma.

LEMMA 7.3: DARBOUX'S INTEGRABILITY CRITERION (DIC). If
f is a function on $[a, b]$ such that for any positive number ε there is
a partition $\mathcal{P}$ satisfying

$$U(f, \mathcal{P}) - L(f, \mathcal{P}) < \varepsilon,$$

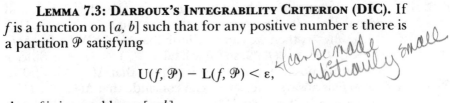

(can be made arbitrarily small)

then f is integrable on $[a, b]$.

Since this assertion is proved as part of Theorem 7.6, it is not
necessary to prove it here.

do some problems

Exercises 7.3

1. Given $f(x) = x$ and $\mathcal{P}_n = \{k/n\}_{k=0}^{n}$, find $L(f, \mathcal{P}_n)$ and
   $U(f, \mathcal{P}_n)$. Then show that f is integrable on $[0, 1]$ by using
   Lemma 7.3. Finally, find $\int_0^1 f$.

2. Given $f(x) = x^2$ and $\mathcal{P}_n = \{k/n\}_{k=0}^{n}$, do the same as in
   Exercise 1.

3. Given $f(x) = 1/(x + 1)$ and $\mathcal{P}_n = \{k/n\}_{k=0}^{n}$, show that
   $\lim_n \{U(f, \mathcal{P}_n) - L(f, \mathcal{P}_n)\} = 0$, thus proving that f is integrable on $[0, 1]$. (*Hint:* Do not try to simplify $U(f, \mathcal{P}_n)$ or
   $L(f, \mathcal{P}_n)$; just work with their difference.)

4. Given $f(x) = \sin x$ and $\mathcal{P}_n = \{k\pi/2n\}_{k=0}^{n}$, show that f is integrable on $[0, \pi/2]$ by proceeding as in Exercise 3.

Exercises 5–9 are concerned with the absolute value $|f|$ of a func-
tion f. For a given f we introduce the functions f^+ and f^-:

$$f^+(x) = \max\{f(x), 0\} \quad \text{and} \quad f^-(x) = \max\{-f(x), 0\}.$$

5. Prove that for any function f,

$$f = f^+ - f^- \quad \text{and} \quad |f| = f^+ + f^-.$$

6. Prove: If f is integrable on $[a, b]$, then f^+ is also integrable
   there. (*Hint:* Compare $M_k - m_k$ for f with the corresponding
   differences for f^+.)

7. Prove: If f is integrable on $[a, b]$, then $|f|$ is integrable there.

8. Show by example that $|f|$ can be integrable even though f is
   not integrable.

9. Prove: If f is integrable on $[a, b]$, then $|\int_a^b f| \leq \int_a^b |f|$.

7.4. Integrability of Continuous Functions

Before resuming the development of our theory, it would be helpful to gain a little more familiarity with Darboux's upper and lower sums and the roles they play in determining the integrability of a function. With this in mind, consider an example that we saw in Chapter 4:

$$f(x) = \begin{cases} 1/q, & \text{if } x = p/q \in \mathbb{Q} \text{ and } p/q \text{ is in lowest terms,} \\ 0, & \text{if } x \in \mathbb{R} \sim \mathbb{Q}. \end{cases}$$

Recall that f is continuous at each irrational number and discontinuous at each rational number. We now assert that f is integrable on any interval $[a, b]$ and $\int_a^b f = 0$. To prove this we use Lemma 7.3 (DIC). First, the density of the irrational numbers ensures that $L(f, \mathcal{P}) = 0$ for every partition $\mathcal{P}$. Thus, for an arbitrary positive number ε, we seek a partition $\mathcal{P}$ such that $U(f, \mathcal{P}) < \varepsilon$. In the interval $[a, b]$ there are only a finite number of points p/q such that $1/q > \varepsilon/2(b - a)$. Let m be the number of such points in $[a, b]$. Let $\mathcal{P}$ be a partition such that $\|\mathcal{P}\| < \varepsilon/(4m)$. In the sum $U(f, \mathcal{P}) = \sum_{k=1}^{n} M_k(x_k - x_{k-1})$ there are at most $2m$ terms where $M_k > \varepsilon/2(b - a)$; for each of these terms we have $M_k(x_k - x_{k-1}) \le 1 \cdot \|\mathcal{P}\| < \varepsilon/(4m)$. Therefore the total of these terms is less than $(2m)[\varepsilon/(4m)] = \varepsilon/2$. In each of the remaining terms, $M_k \le \varepsilon/[2(b - a)]$, so their total is less than $\{\varepsilon/[4(b - a)]\} \sum_{k=1}^{n}(x_k - x_{k-1}) = \varepsilon/2$. Hence, $U(f, \mathcal{P}) < \varepsilon$, so the DIC implies that f is integrable and $\int_a^b f = 0$.

Up to this point, we have seen very few examples that we could verify as being integrable functions. The next two theorems provide us with a very large set of such examples.

THEOREM 7.7. If the function f is continuous on $[a, b]$, then f is integrable there.

Proof. We use the DIC by showing that for an arbitrary positive number ε, there is a partition $\mathcal{P}$ such that $U(f, \mathcal{P}) - L(f, \mathcal{P}) < \varepsilon$. By Theorem 5.5, the continuous function f is uniformly continuous on $[a, b]$. Let δ be a positive number such that for x' and x'' in $[a, b]$,

$$|x' - x''| < \delta \quad \text{implies} \quad |f(x') - f(x'')| < \varepsilon/(b - a).$$

Now choose a partition $\mathcal{P}$ such that $\|\mathcal{P}\| < \delta$, and consider the kth subinterval $[x_{k-1}, x_k]$. Choose points μ_k' and μ_k'' in $[x_{k-1}, x_k]$ such that

$f(\mu_k') = M_k$ and $f(\mu_k'') = m_k$. (Recall that by Theorem 5.2 a continuous function *assumes* its lub and glb.) Because $|\mu_k' - \mu_k''| \le x_k - x_{k-1}$ $\le \|\mathscr{P}\| < \delta$, it follows that

$$M_k - m_k = f(\mu_k') - f(\mu_k'') < \frac{\varepsilon}{(b-a)}.$$

Therefore

$$U(f, \mathscr{P}) - L(f, \mathscr{P}) = \sum_{k=1}^{n} (M_k - m_k)(x_k - x_{k-1})$$

$$< \frac{\varepsilon}{b-a} \sum_{k=1}^{n} (x_k - x_{k-1})$$

$$= \varepsilon;$$

hence, by the DIC, f is integrable.

THEOREM 7.8. If the function f is monotonic on $[a, b]$, then f is integrable there.

Proof. Since f is monotonic, either f or $-f$ is nondecreasing, and by Theorem 7.2, if one of these is integrable, then so is the other one. Therefore we may assume, for the sake of definiteness, that f is nondecreasing. Then for any subinterval $[x_{k-1}, x_k]$ of $[a, b]$, M_k occurs at the right endpoint; that is, $M_k = f(x_k)$. Similarly, $m_k = f(x_{k-1})$. Therefore if $\mathscr{P}$ is any partition of $[a, b]$, then

$$U(f, \mathscr{P}) - L(f, \mathscr{P}) = \sum_{k=1}^{n} (M_k - m_k)(x_k - x_{k-1})$$

$$= \sum_{k=1}^{n} [f(x_k) - f(x_{k-1})](x_k - x_{k-1})$$

$$\le \|\mathscr{P}\| \sum_{k=1}^{n} [f(x_k) - f(x_{k-1})]$$

$$= \|\mathscr{P}\| [f(x_n) - f(x_0)]$$

$$= \|\mathscr{P}\| [f(b) - f(a)].$$

Now if $\varepsilon > 0$, we simply select a partition $\mathscr{P}$ such that $\|\mathscr{P}\| < \varepsilon/[f(b) - f(a)]$. (We may assume that $f(a) \ne f(b)$, for otherwise f is constant and the conclusion is trivial.) Then $U(f, \mathscr{P}) - L(f, \mathscr{P}) < \varepsilon$, so by the DIC, f is integrable.

The collection of functions we can now prove to be integrable can be enlarged even more by combining Theorems 7.7 and 7.8 with Theorem 7.3. For example, a function may have several jump discontinuities, which produce breaks in its graph. But so long as it is continuous on the intervals between these discontinuities, the function is integrable on each of these abutting intervals. Then Theorem 7.3 allows us to infer that the function is integrable on the union of the intervals. Such a function is said to be *piecewise continuous*. The concept of *piecewise monotonic* can be defined similarly. In order to define a function that is neither piecewise continuous nor piecewise monotonic, we must use a rather involved construction (for example, a function like the one above that relies on the density of the sets of rational numbers and irrational numbers). This should serve to create the correct impression that the collection of integrable functions is really quite large.

Exercises 7.4

1. Prove: If f is continuous on $[a, b] \sim \{c_i\}_{i=1}^{m}$, then f is integrable on $[a, b]$.

2. If $f(x) = \begin{cases} (\sin x)/x, & \text{if } x \neq 0, \\ 3, & \text{if } x = 0, \end{cases}$ then f is integrable on $[0, 5]$.

3. If $f(x) = m_n x + b_n$ on $[n - 1, n)$ for $n = 1, 2, \ldots, k$, then f is integrable on $[0, k - (1/2)]$.

4. If f is continuous and nonnegative but not identically zero on $[a, b]$, then $\int_a^b f > 0$.

7.5. Products of Integrable Functions

In Theorem 7.2 we proved that the set of integrable functions is closed under certain algebraic operations, namely, those of addition and constant multiples. In order to completely establish the algebraic closure of the integrable functions, we now prove the result for the multiplication of integrable functions.

THEOREM 7.9. If each of f and g is an integrable function on $[a, b]$, then fg is integrable there.

Proof. We first prove the assertion for the special case in which f and g are nonnegative functions. Suppose $\mathcal{P}$ is any partition of $[a, b]$, and let M_f, M_g, and M_{fg} denote the least upper bounds of f, g, and fg, respectively, on the kth subinterval $[x_{k-1}, x_k]$. It is not hard to see that for nonnegative functions $M_{fg} \leq M_f M_g$. Similarly, if m_f, m_g, and m_{fg} are the corresponding greatest lower bounds, then $m_{fg} \geq m_f m_g$. Therefore

$$M_{fg} - m_{fg} \leq M_f M_g - m_f m_g. \tag{1}$$

Now let B_f and B_g be upper bounds of $f(x)$ and $g(x)$ for x in $[a, b]$, and rewrite (1) as follows:

$$M_{fg} - m_{fg} \leq M_f M_g - m_f M_g + m_f M_g - m_f m_g$$
$$= (M_f - m_f)M_g + m_f(M_g - m_g) \qquad (2)$$
$$\leq (M_f - m_f)B_g + B_f(M_g - m_g).$$

The inequality (2) holds for each term in the sum

$$U(fg, \mathcal{P}) - L(fg, \mathcal{P}) = \sum_{k=1}^{n} (M_{fg,\,k} - m_{fg,\,k})(x_k - x_{k-1}),$$

where we have written $M_{fg,\,k} - m_{fg,\,k}$ with the index k to indicate the term corresponding to the kth subinterval. Therefore, from (2) we get

$$U(fg, \mathcal{P}) - L(fg, \mathcal{P}) \leq B_g \sum_{k=1}^{n} (M_{f,\,k} - m_{f,\,k})(x_k - x_{k-1})$$
$$+ B_f \sum_{k=1}^{n} (M_{g,\,k} - m_{g,\,k})(x_k - x_{k-1})$$
$$= B_g[U(f, \mathcal{P}) - L(f, \mathcal{P})]$$
$$+ B_f[U(g, \mathcal{P}) - L(g, \mathcal{P})].$$

We may assume that neither B_f nor B_g is zero, and by the DIC, if $\varepsilon > 0$ the integrability of f and g allows us to choose $\mathcal{P}$ so that

$$U(f, \mathcal{P}) - L(f, \mathcal{P}) < \frac{\varepsilon}{2B_g} \quad \text{and} \quad U(g, \mathcal{P}) - L(g, \mathcal{P}) < \frac{\varepsilon}{2B_f}.$$

This yields $U(fg, \mathcal{P}) - L(fg, \mathcal{P}) < \varepsilon$, so by the DIC, we conclude that fg is integrable.

To prove the general case in which f and g need not be non-negative, we use Proposition 7.1, which assures us that the integrable functions f and g are bounded below, say, $f(x) \geq K$ and $g(x) \geq L$ for every x in $[a, b]$. Then $f - K$ and $g - L$ are nonnegative integrable functions, so the case just proved can be applied to conclude that $(f - K)(g - L)$ is integrable. Since

$$fg = (f - K)(g - L) + Kg + Lf + KL,$$

and each term of the right-hand member is integrable, it follows from Theorem 7.2 that fg is integrable.

In combining this last result with Theorem 7.2 to get the algebraic closure of the set of integrable functions, it may be noted that something is missing here that was part of the conclusion of Theorem 7.2. In the earlier result, we were able to give explicit formulas for the integral of the sum and the integral of the constant multiple, namely,

$$\int_a^b (f + g) = \int_a^b f + \int_a^b g \quad \text{and} \quad \int_a^b cf = c\int_a^b f.$$

In the case of the product, there is no such formula for the value of the integral $\int_a^b fg$, so in Theorem 7.9 we proved only the existence of the integral. There is a relationship that gives a rough estimate of the value of $\int_a^b fg$, but it is an inequality, so it provides only a bound on the possible values. This result appears in various forms throughout the area of analysis, so it is not surprising that there are three mathematicians' names associated with it.

THEOREM 7.10: CAUCHY-BUNYAKOVSKY-SCHWARZ INEQUALITY. If each of f and g is integrable on $[a, b]$, then

$$\left| \int_a^b fg \right| \le \left[\left(\int_a^b f^2 \right)\left(\int_a^b g^2 \right) \right]^{1/2}.$$

Proof. It follows immediately from Theorem 7.9 that each of fg, f^2, and g^2 is integrable. Also, $\int_a^b f^2 \ge 0$ and $\int_a^b g^2 \ge 0$ by Exercise 7.2.4, so there is no question about the existence of the square root. Define the numbers A, B, and C by

$$A = \int_a^b f^2, \quad B = 2 \int_a^b fg, \quad \text{and} \quad C = \int_a^b g^2,$$

and let q be the quadratic function defined by

$$q(x) = Ax^2 + Bx + C.$$

We assert that q is nonnegative, because

$$q(x) = x^2 \int_a^b f^2 + 2x \int_a^b fg + \int_a^b g^2$$
$$= \int_a^b (x^2 f^2 + 2xfg + g^2)$$
$$= \int_a^b (xf + g)^2,$$

and the nonnegative function $(xf + g)^2$ has a nonnegative integral.  From elementary algebra we recall that $q(x) = 0$ when

$$x = \frac{-B \pm \sqrt{B^2 - 4AC}}{2A}.$$

Since $q(x)$ is never negative, the quadratic equation $q(x) = 0$ can not have two real roots, which means that $B^2 - 4AC \leq 0$. But this says that

$$4\left(\int_a^b fg\right)^2 - 4\left(\int_a^b f^2\right)\left(\int_a^b g^2\right) \leq 0,$$

or

$$\left|\int_a^b fg\right| \leq \left[\left(\int_a^b f^2\right)\left(\int_a^b g^2\right)\right]^{1/2}.$$

COROLLARY 7.10: MINKOWSKI INEQUALITY. If each of f and g is an integrable function on $[a, b]$, then

$$\left[\int_a^b (f + g)^2\right]^{1/2} \leq \left[\int_a^b f^2\right]^{1/2} + \left[\int_a^b g^2\right]^{1/2}.$$

Proof. This is left as Exercise 7.5.3.

In Theorem 7.3 it was shown that a function that is integrable on two intervals is integrable on their union. Now we reverse this, in a sense, by showing that integrability on an interval implies integrability on any subinterval:

PROPOSITION 7.2. If the function f is integrable on $[a, b]$ and $[c, d] \subset [a, b]$, then f is integrable on $[c, d]$.

Proof. This is left as Exercise 7.5.6.

The preceding implication has a partial converse, which can be used to show the integrability of functions that behave awkwardly at the endpoints of an interval:

THEOREM 7.11. If the function f is bounded on $[a, b]$ and integrable on every closed subinterval of (a, b), then f is integrable on $[a, b]$.

Proof. Suppose $\varepsilon > 0$ and $|f(x)| \leq B$ for every x in $[a, b]$. Let $[c, d]$ be a subinterval of (a, b) such that

$$a < c \leq a + \varepsilon/(6B) \quad \text{and} \quad b - \varepsilon/(6B) \leq d < b.$$

Since f is integrable on $[c, d]$, the DIC allows us to choose a partition $\mathcal{P}^*$ of $[c, d]$ such that $U(f, \mathcal{P}^*) - L(f, \mathcal{P}^*) < \varepsilon/3$. Now let $\mathcal{P} = \{a\} \cup \mathcal{P}^* \cup \{b\}$. Then $\mathcal{P}$ is a partition of $[a, b]$, and

$$U(f, \mathcal{P}) - L(f, \mathcal{P}) = (M_1 - m_1)(c - a) + (U(f, \mathcal{P}^*) - L(f, \mathcal{P}^*))$$
$$+ (M_n - m_n)(b - d)$$
$$\leq 2B(c - a) + \{U(f, \mathcal{P}^*) - L(f, \mathcal{P}^*)\}$$
$$+ 2B(b - d)$$
$$< 2B(\varepsilon/6B) + \{\varepsilon/3\} + 2B(\varepsilon/6B)$$
$$= \varepsilon.$$

Hence, by the DIC, f is integrable on $[a, b]$.

As an application of Theorem 7.11, consider the function defined by $f(x) = \sin(1/x)$ when $x \neq 0$. This function is not piecewise monotonic or continuous on any interval that contains zero, but we can still show very easily that f is integrable on $[0, 1]$. Regardless of how f may be defined at $x = 0$, f is still bounded on $[0, 1]$ and continuous on $(c, 1)$ for every positive number c. Therefore by Theorem 7.7, f is integrable on every interval (c, d) in $[0, 1]$, so by Theorem 7.11, f is integrable on $[0, 1]$.

Exercises 7.5

1. Prove: $\int_0^\pi \sqrt{x} \sin x \, dx \leq \pi$.

2. Prove: $\int_0^{\pi/4} (1 + \tan x) \sqrt{x} \sec x \, dx \leq \sqrt{\dfrac{7}{96}} \, \pi$.

3. Prove the Minkowski Inequality (Corollary 7.10).

4. Prove: $\int_0^{\pi/2} (\sqrt{\cos x} + x)^2 dx \leq \left(1 + \sqrt{\dfrac{\pi^3}{24}}\right)^2$.

5. Prove: If f is integrable and $|f(x)| \geq \delta > 0$ for every x in $[a, b]$, then $1/f$ is integrable on $[a, b]$. (*Hint:* First assume f is nonnegative, then extend to the general case as in the proof of Theorem 7.9.)

6. Prove Proposition 7.2.

7. Prove Bliss's Theorem: If each of f and g is continuous on $[a, b]$, then
$$\int_a^b fg = \lim_{\|\mathcal{P}\| \to 0} \sum_{k=1}^n f(\mu_k')g(\mu_k'')(x_k - x_{k-1}),$$

where μ_k' and μ_k'' are arbitrary points in the kth subinterval $[x_{k-1}, x_k]$. (*Hint:* Show that for $\|\mathcal{P}\|$ sufficiently small, the right-hand sum is within $\varepsilon/2$ of the Riemann sum $\mathcal{P}(fg, \mu')$.)

7.6. The Fundamental Theorem of Calculus

In several areas of mathematics, one can find a result that is labeled as the fundamental theorem of that area. In the area of calculus, there can be little argument that the result that is most deserving of that designation is the one that provides the connection between the concepts of the integral and the derivative. Its importance can hardly be overstated. Without such a connection, the Riemann integral would have been too cumbersome for the wide range of applications that it has found. But the problem-solving aspects of calculus should already be familiar, so we proceed immediately to the terminology and proof of this very basic result.

If f and F are functions such that f is the derivative of F, then F is called a *primitive* of f. Of course, F can have at most one derivative function, but if f has a primitive, then it has many primitives; however, by Corollary 6.3b to the Law of the Mean, any two primitives of f must differ by a constant.

THEOREM 7.12: FUNDAMENTAL THEOREM OF CALCULUS. If the function f is integrable on $[a, b]$ and F is a primitive of f on $[a, b]$, then

$$\int_a^b f = F(b) - F(a).$$

Proof. Let $\mathcal{P}$ be a partition of $[a, b]$, and consider the following collapsing sum:

$$F(b) - F(a) = \sum_{k=1}^n \{F(x_k) - F(x_{k-1})\}. \tag{1}$$

Since F is differentiable on each subinterval $[x_{k-1}, x_k]$, we can apply the Law of the Mean to get a number μ_k in $[x_{k-1}, x_k]$ such that

$$F'(\mu_k) = \frac{F(x_k) - F(x_{k-1})}{x_k - x_{k-1}},$$

or

$$F(x_k) - F(x_{k-1}) = F'(\mu_k)(x_k - x_{k-1}) = f(\mu_k)(x_k - x_{k-1}). \tag{2}$$

Substituting the right-hand member of (2) into (1), we get

$$F(b) - F(a) = \sum_{k=1}^{n} f(\mu_k)(x_k - x_{k-1})$$
$$= \mathcal{P}(f, \mu),$$

where $\mathcal{P}(f, \mu)$ is a Riemann sum for f with respect to $\mathcal{P}$. Thus we have shown that for *any* partition $\mathcal{P}$, the points $\{\mu_k\}_{k=1}^{n}$ can be chosen so that the value of $\mathcal{P}(f, \mu)$ is $F(b) - F(a)$. Therefore $F(b) - F(a)$ is the only possible value of the limit $\lim_{\|\mathcal{P}\| \to 0} \mathcal{P}(f, \mu)$. Since we are assuming that f is integrable, this limit must exist and its value is denoted by $\int_a^b f$. Hence,

$$\int_a^b f = F(b) - F(a).$$

It should be emphasized that the hypothesis of Theorem 7.12 includes the *assumption* that f is integrable. The statement that f is the derivative of some F does not imply *a priori* that f must be integrable. A problem that is sometimes used to trick students in elementary calculus is to ask them to evaluate some integral such as $\int_{-1}^{1} x^{-2} dx$. Of course, the integral does not exist, because the integrand is unbounded on $[-1, 1]$. But many students blindly find a primitive $F(x) = -x^{-1}$ and "apply" the Fundamental Theorem of Calculus to claim that the integral has the value -2. It could be pointed out that $-x^{-1}$ is not a primitive of x^{-2} throughout the interval $[-1, 1]$ because these functions are not defined at $x = 0$. The next example is not vulnerable to this criticism, and yet it provides a function that is a primitive of a nonintegrable derivative:

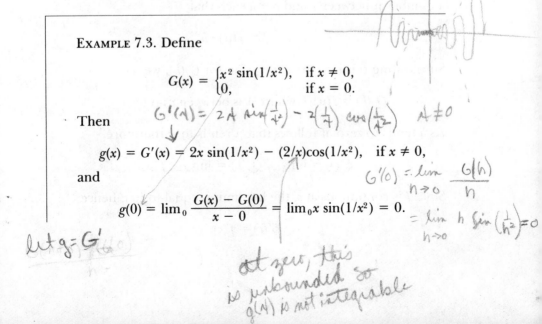

EXAMPLE 7.3. Define

$$G(x) = \begin{cases} x^2 \sin(1/x^2), & \text{if } x \neq 0, \\ 0, & \text{if } x = 0. \end{cases}$$

Then $\qquad G'(x) = 2x \sin\left(\frac{1}{x^2}\right) - 2\left(\frac{1}{x}\right) \cos\left(\frac{1}{x^2}\right) \quad x \neq 0$

$$g(x) = G'(x) = 2x \sin(1/x^2) - (2/x)\cos(1/x^2), \quad \text{if } x \neq 0,$$

and

$$g(0) = \lim_0 \frac{G(x) - G(0)}{x - 0} = \lim_0 x \sin(1/x^2) = 0.$$

$G'(0) = \lim_{h \to 0} \frac{G(h)}{h}$

$= \lim_{h \to 0} h \sin\left(\frac{1}{h^2}\right) = 0$

let $g = G'$

at zero, this is unbounded so $g(x)$ is not integrable

Thus G is a primitive of g throughout $\mathbb{R}$, but the amplitude factor $2/x$ in the second term of $g(x)$ tells us that g is unbounded in any interval that contains zero. Hence g is not integrable on $[-1, 1]$, for example.

In most elementary textbooks, the Fundamental Theorem of Calculus is stated in a somewhat weaker form; namely, it is assumed that f is continuous—rather than merely integrable as in Theorem 7.12. The stronger hypothesis is needed so that a proof can be obtained from the following theorem:

THEOREM 7.13. If f is a continuous function on $[a, b]$, let ϕ be defined for x in $[a, b]$ by $\phi(x) = \int_a^x f$. Then ϕ is a primitive of f on $[a, b]$.

Proof. Let c be a number in $[a, b]$, and consider the difference quotient

$$Q_c(h) = \frac{\phi(c + h) - \phi(c)}{h}$$

$$= \frac{1}{h} \left(\int_a^{c+h} f - \int_a^c f \right)$$

$$= \frac{1}{h} \left(\int_a^c f + \int_c^{c+h} f - \int_a^c f \right)$$

$$= \frac{1}{h} \int_c^{c+h} f.$$

By the Mean Value Theorem for Integrals (Exercise 7.2.7), there is a number μ between c and $c + h$ such that

$$\int_c^{c+h} f = f(\mu) \cdot h.$$

Substituting this into the expression for $Q_c(h)$, we have

$$Q_c(h) = f(\mu), \quad \text{where } \mu \text{ is between } c \text{ and } c + h.$$

As h tends to zero, it follows that μ tends to c; therefore

$$\phi'(c) = \lim_{h \to 0} Q_c(h) = \lim_{\mu \to c} f(\mu).$$

Since f is continuous at c, the last limit is equal to $f(c)$; hence

$$\phi'(c) = f(c).$$

The Fundamental Theorem of Calculus (for continuous functions) can now be deduced as follows. The function ϕ of the preceding paragraph is a primitive of the continuous function f, so if F is an arbitrary primitive of f, then $F' = f = \phi'$. Therefore by Corollary 6.3b following the Law of the Mean, F and ϕ differ by a constant, say,

$$F(x) - \phi(x) = C \quad \text{for every } x \text{ in } [a, b].$$

Substituting a for x, we get

$$F(a) - \int_a^a f = C,$$

so $C = F(a)$. Now substitute b for x:

$$F(b) - \phi(b) = C = F(a),$$

which is the same as

$$F(b) - F(a) = \int_a^b f.$$

Exercises 7.6

1. Given that $f(x) = \begin{cases} 0, & \text{if } -0 \leq x < 1, \\ 1, & \text{if } 1 \leq x \leq 2, \end{cases}$ prove that f is integrable on $[0, 2]$ but has no primitive there. (*Hint:* See Theorem 6.5.)

2. Define $F(x) = \int_0^x f$, where f is given in Exercise 1. Is F continuous on $[0, 2]$? Is F differentiable on $[0, 2]$? Explain your answers.

3. Given that $g(x) = \begin{cases} \sin 1/x, & \text{if } x \neq 0, \\ 0, & \text{if } x = 0, \end{cases}$ is g continuous on $[0, 1/\pi]$? Is g integrable on $[0, 1/\pi]$? Explain your answers.

4. Define $G(x) = \int_0^x g$, where g is given in Exercise 3. Is G continuous on $[0, 1/\pi]$? Is G differentiable there? Explain your answers.

5. Define $h(x) = \begin{cases} 1, & \text{if } x = 1, 1/2, 1/3, \ldots, \\ 0, & \text{otherwise}, \end{cases}$ and $H(x) = \int_0^x h$.

Is H continuous on $[0, 1]$? Is H differentiable there? Explain your answers.

6. Prove: If f is integrable on $[a, b]$ and $F(x) = \int_a^x f$, then F is continuous on $[a, b]$.

7. Prove: If f is continuous and strictly positive on $[a, b]$ and $F(x) = \int_a^x f$, then F is strictly increasing on $[a, b]$. (The phrase "strictly increasing" means that $x_1 < x_2$ implies $F(x_1) < F(x_2)$.)

8. Prove the Integration by Parts Theorem: If f and g are differentiable functions such that f' and g' are integrable on $[a, b]$, then

$$\int_a^b fg' = f(b)g(b) - f(a)g(a) - \int_a^b f'g.$$

8

IMPROPER
INTEGRALS

8.1. Types of Improper Integrals

In Chapter 7 the Riemann integral was defined and studied for certain functions whose domains are closed intervals. Recall that Proposition 7.1 asserts that a function must be bounded if it is to be integrable. Yet there are many useful, commonly encountered functions whose graphs have vertical asymptotes, which means the functions are unbounded and therefore not integrable. In this chapter we develop a theory of integration for some of these unbounded functions. Another restriction of the Riemann integral is that the domain of integration must be a (bounded) closed interval. The necessity of this is somewhat more subtle than that of the boundedness of the integrand function, but a moment's reflection should make it clear that we cannot form a Riemann sum—or even define a partition—over an unbounded set. Yet many of the most common functions are defined on all of $\mathbb{R}$, or at least on $(0, \infty)$, so it is desirable to have a theory of integration over such extended intervals. Thus the theory of the Riemann integral is here extended to cover two new situations: in the first, the domain is unbounded; in the second, the integrand function is unbounded. We also consider combinations and variations of these two cases. In both cases, we define the new integral to be the limit of previously defined Riemann integrals.

8.2. Integrals over Unbounded Domains

DEFINITION 8.1. Let *a* be a real number and let *f* be a function that is (Riemann) integrable on $[a, t]$ for every $t \geq a$. Then the *improper integral* of *f* on $[a, \infty)$ is the expression

$$\lim_{t \to \infty} \int_a^t f.$$

If the limit exists, the improper integral is said to be *convergent,* and its limit value is denoted by $\int_a^\infty f$. If the limit does not exist, the improper integral is *nonconvergent.* If $\lim_{t \to \infty} \int_a^t f = \infty$ or $\lim_{t \to \infty} \int_a^t f = -\infty$, then the improper integral is *divergent.*

The improper integral of a function on $(-\infty, b]$ is defined similarly:

$$\int_{-\infty}^b f = \lim_{t \to -\infty} \int_t^b f.$$

The following examples should be recognized as those seen in an elementary calculus course. They are stated here for the purpose of illustrating our current notation and terminology.

EXAMPLE 8.1.

$$\int_1^\infty \frac{1}{x^2}\, dx = \lim_{t \to \infty} \int_1^t \frac{1}{x^2}\, dx$$

$$= \lim_{t \to \infty} (1 - 1/t)$$

$$= 1.$$

Hence the improper integral is convergent.

EXAMPLE 8.2.

$$\int_0^\infty \cos x\, dx = \lim_{t \to \infty} \int_0^t \cos x\, dx$$

$$= \lim_{t \to \infty} (\cos t - 1).$$

The improper integral is nonconvergent, since cos *t* oscillates between 1 and −1 as $t \to \infty$.

EXAMPLE 8.3.

$$\int_1^\infty \frac{1}{\sqrt{x}}\,dx = \lim_{t\to\infty} \int_1^t \frac{1}{\sqrt{x}}\,dx$$

$$= \lim_{t\to\infty} (2\sqrt{t} - 1)$$

$$= \infty.$$

Hence the improper integral is divergent.

To extend the domain of integration to the entire real line, we use a combination of the improper integrals $\int_a^\infty f$ and $\int_{-\infty}^b f$.

DEFINITION 8.2. If the function f is integrable on $[-t, t]$ for every number t, then the improper integral $\int_{-\infty}^\infty f$ is the expression

$$\lim_{t\to -\infty} \int_t^c f + \lim_{t\to\infty} \int_c^t f,$$

where c is any real number. It is *convergent* if both $\int_{-\infty}^c f$ and $\int_c^\infty f$ are convergent, and in that case

$$\int_{-\infty}^\infty f = \int_{-\infty}^c f + \int_c^\infty f.$$

Note that the convergence or nonconvergence of $\int_{-\infty}^\infty f$ is independent of the number c in the preceding formula (see Exercise 8.2.2).

EXAMPLE 8.4.

$$\int_{-\infty}^\infty \frac{1}{x^2 + 1}\,dx = \int_{-\infty}^0 \frac{1}{x^2 + 1}\,dx + \int_0^\infty \frac{1}{x^2 + 1}\,dx$$

$$= \lim_{t\to -\infty} \arctan x\big|_t^0 + \lim_{t\to\infty} \arctan x\big|_0^t$$

$$= -\left(-\frac{\pi}{2}\right) + \frac{\pi}{2}$$

$$= \pi.$$

Hence the improper integral is convergent.

EXAMPLE 8.5.

$$\int_{-\infty}^{\infty} e^{-x}dx = \int_{-\infty}^{0} e^{-x}dx + \int_{0}^{\infty} e^{-x}dx$$
$$= \lim_{t \to -\infty} -e^{-x}\big|_{t}^{0} + \lim_{t \to \infty} -e^{-x}\big|_{0}^{t}$$
$$= \lim_{t \to -\infty}(-1 + e^{-t}) + \lim_{t \to -\infty}(-e^{-t} + 1)$$
$$= \infty.$$

Hence the improper integral is divergent.

Defining the improper integral in terms of the definite Riemann integral has both advantages and disadvantages. On the positive side, there is no need to develop a separate limit theory for the new type of integral. But, on the other hand, we are required to evaluate $\int_{a}^{t} f$ as a function of t in order to determine if $\int_{a}^{\infty} f$ is convergent, and the difficulty of finding the primitive of a given function is well known to the calculus student. Since it is frequently sufficient to determine convergence of $\int_{a}^{\infty} f$ without finding its value, it will be very helpful to have a test that says that under the right conditions the convergence of a simple integral implies that of a complicated one. Such a result is called a "comparison test." That is the nature of the next theorem, but first we prove a preliminary result.

LEMMA 8.1. Let f be a nonnegative function that is integrable *(bound* on $[a, t]$ for every $t \geq a$. If there is a number B such that $\int_{a}^{t} f \leq B$ for every $t \geq a$, then $\int_{a}^{\infty} f$ is convergent.

Proof. Define $F(t) = \int_{a}^{t} f$ for $t \geq a$. Since f is nonnegative and $F(t + h) - F(t) = \int_{t}^{t+h} f$, it follows from Theorem 7.4 that F is nondecreasing on $[a, \infty)$. Therefore $F(t) \leq B$ ensures that $\lim_{t \to \infty} F(t) = \text{lub}\{F(t): t \geq a\}$; that is, $\lim_{t \to \infty} \int_{a}^{t} f$ exists.

THEOREM 8.1. Let f and g be nonnegative functions such that for every $t \geq a$, f and g are integrable on $[a, t]$ and $f(t) \leq g(t)$. If $\int_{a}^{\infty} g$ is convergent, then $\int_{a}^{\infty} f$ is convergent also.

Proof. Define $h(x) = g(x) - f(x)$ for $x \geq a$. Then h is nonnegative and integrable on $[a, t]$ for every $t \geq a$, and

$$\int_{a}^{t} h \leq \int_{a}^{t} g \leq \int_{a}^{\infty} g.$$

Thus $\int_a^\infty g$ is an upper bound for $\int_a^t h$ for $t > a$, so by Lemma 8.1, $\int_a^\infty h$ is convergent. Now

$$\lim_{t \to \infty} \int_a^t f = \lim_{t \to \infty} \int_a^t g - h$$
$$= \lim_{t \to \infty} \int_a^t g - \lim_{t \to \infty} \int_a^t h,$$

and the last two limits exist because both $\int_a^\infty g$ and $\int_a^\infty h$ are convergent.

The utility of this result is illustrated in the following calculation.

EXAMPLE 8.6. Show that $\int_1^\infty (x^3 + 1)^{-1}\, dx$ is convergent.
A primitive for $1/(x^3 + 1)$ could be obtained by a partial fraction decomposition, but it is much easier to note that

$$\frac{1}{x^3 + 1} < \frac{1}{x^2} \quad \text{for } x \geq 1,$$

and by Example 8.1, $\int_1^\infty x^{-2}\, dx$ is convergent. Hence, by Theorem 8.1, $\int_1^\infty (x^3 + 1)^{-1}\, dx$ is also convergent.

Exercises 8.2

1. Prove: If $\int_a^\infty f$ is convergent and $\alpha > a$, then $\int_\alpha^\infty f$ is also convergent.

2. Prove that in Definition 8.2 both the convergence of $\int_{-\infty}^\infty f$ and its value are independent of the choice of the number c.

3. Prove the following "extension" of Theorem 8.1: Let f and g be nonnegative functions such that f is integrable on $[a, t]$ for every $t \geq a$ and g is integrable on $[b, t]$ for every $b \geq a$. If $\int_b^\infty g$ is convergent and there is a number c such that $f(t) \leq g(t)$ for every $t \geq c$, then $\int_a^\infty f$ is also convergent.

4. For what values of p is $\int_1^\infty x^p\, dx$ convergent?

In Exercises 5–16, determine whether the improper integral is convergent.

5. $\int_1^\infty \dfrac{1}{x\sqrt{x + 1}}\, dx$

6. $\int_2^\infty \dfrac{1}{(x-1)^{2/3}}\,dx$

7. $\int_2^\infty \dfrac{1}{x \log x}\,dx$

standard integral but split it; no comparison

8. $\int_0^\infty e^{-x}\,dx$

9. $\int_0^\infty x\,e^{-x}\,dx$

10. $\int_{-\infty}^\infty x\,e^{-x}\,dx$ *Hint:*

11. $\int_{-\infty}^\infty x\,e^{-x^2}\,dx$

12. $\int_1^\infty \dfrac{1}{\sqrt{x^4+1}}\,dx$

13. $\int_{-\infty}^\infty \dfrac{x}{1+x^2}\,dx$ *split, standard integrals*

14. $\int_4^\infty \dfrac{1}{x^2-x}\,dx$

15. $\int_0^\infty f$, where $f(x) = \begin{cases} 1, & \text{if } 2n-2 \le x < 2n-1, \\ -1, & \text{if } 2n-1 \le x < 2n, \end{cases}$
    for $n = 1, 2, \ldots$

    graph helpful of function integrate

16. $\int_0^\infty g$, where $g(x) = \begin{cases} 1/n, & \text{if } 2n-2 \le x < 2n-1, \\ -1/n, & \text{if } 2n-1 \le x < 2n, \end{cases}$
    for $n = 1, 2, \ldots$

8.3. Integrals of Unbounded Functions

DEFINITION 8.3. Let f be a function that is Riemann integrable on every closed subinterval of $[a, b)$ but not integrable on $[a, b]$. Then the *improper integral* of f on $[a, b]$ is the expression

$$\lim_{t \to b-} \int_a^t f.$$

If the limit exists, the improper integral is said to be *convergent* and its limit value is denoted by $\int_a^{b-} f$; otherwise the improper integral is *nonconvergent*. If $\lim_{t \to b-} \int_a^t f = \infty$ or $\lim_{t \to b-} \int_a^t f = -\infty$, then the improper integral is *divergent*. Similarly, we define the analogous improper integral

$$\int_{a+}^b f = \lim_{t \to a+} \int_t^b f.$$

Note that the symbols $\int_a^{b-} f$ and $\int_{a+}^{b} f$ that are used for the value of the improper integral are very similar to the symbol for the value of the (proper) Riemann integral. Many authors use $\int_a^b f$ for both proper and improper integrals. There is little chance of confusion though, because in the case of an improper integral on $[a, b]$, f must be unbounded on $[a, b]$ (see Exercise 8.3.1), and this is usually obvious. For example,

$$\int_{1+}^{2} \frac{1}{x^2 - 1} \, dx \quad \text{is an improper integral, and}$$

$$\int_{2}^{3} \frac{1}{x^2 - 1} \, dx \quad \text{is an ordinary Riemann integral.}$$

This definition parallels so closely the definitions of Section 8.2 that we can omit the examples and leave them for the exercises. As before, there are situations not covered by Definition 8.3 that we can handle by a simple combination of the integrals already defined. For example, if f is unbounded as $x \to c$, where $a < c < b$, then we write

$$\int_a^b f = \int_a^{c-} f + \int_{c+}^{b} f$$

$$= \lim_{t \to c-} \int_a^t f + \lim_{t \to c+} \int_t^b f. \tag{1}$$

If f is integrable on every closed subinterval of $[a, c)$ and of $(c, b]$, then both $\int_a^c f$ and $\int_c^b f$ are improper integrals as defined above. As before, the left-hand improper integral in (1) is convergent if and only if both of the right-hand improper integrals are convergent.

Another variation of this type of improper integral is one in which f is unbounded at both endpoints of $[a, b]$. Then we choose a number c in (a, b) and write

$$\int_{a+}^{b-} f = \int_{a+}^{c} f + \int_{c}^{b-} f$$

$$= \lim_{t \to a+} \int_t^c f + \lim_{t \to b-} \int_c^t f,$$

where each expression on the right-hand side is an improper integral of the previous type (Definition 8.3) in which f is unbounded at *one* of the endpoints of the interval. It can be shown that the choice of c in (a, b) does not affect the convergence or the value of the improper integral $\int_a^b f$ (see Exercise 8.3.2).

It is now plain that the preceding types of improper integrals can be used in combination to form the improper integral of any function with a graph that has at most a finite number of asymptotes. For example, if $f(x) = 1/(x - a)(x - b)$, where $a < b$, then we choose an arbitrary number c in (a, b) and write

$$\int_{-\infty}^{\infty} f = \int_{-\infty}^{a-1} f + \int_{a-1}^{a-} f + \int_{a+}^{c} f + \int_{c}^{b-} f + \int_{b+}^{b+1} f + \int_{b+1}^{\infty} f.$$

Although the calculation of such an integral may be cumbersome—we have just decomposed one improper integral into *six* limits—the major achievement here is that we do not need a separate theory for each situation. There are just two basic types of improper integrals, and finite combinations of these two types suffice for most functions.

This section closes with a comparison test for improper integrals of unbounded functions. It is analogous to Theorem 8.1, and its proof, which is requested in Exercise 8.3.4, can be achieved by an argument similar to the proof of Theorem 8.1.

Corresponds to th. 8.1

THEOREM 8.2. Let f and g be nonnegative functions such that for every t in $[a, b)$, f and g are integrable on $[a, t]$ and $f(t) \le g(t)$. If $\int_{a}^{b-} g$ is convergent, then $\int_{a}^{b-} f$ is convergent also.

This theorem has an obvious dual for $\int_{a+}^{b} f$, which we illustrate by an example.

EXAMPLE 8.7. Show that $\int_{0+}^{1} [x(x + 1)]^{-1} dx$ is divergent.
For $0 < x \le 1$,

+ indicates pole

$$\frac{1}{x(x + 1)} \ge \frac{1}{x(1 + 1)} = \frac{1}{2x};$$

use comparison test

and since

$$\int_{0+}^{1} \frac{1}{x} dx = \lim_{a \to 0+} \log x \big|_{a}^{1} = \infty,$$

we can conclude that the larger function also yields a divergent integral.

Exercises 8.3

1. Prove: If $\int_{a}^{b+} f$ is convergent but f is not Riemann integrable on $[a, b]$, then f is unbounded on $[a, b]$.

2. Suppose f is integrable on every closed subinterval of $[a, b]$ and f is unbounded as $x \to a+$ and $x \to b-$. If $a < c < c' < b$, prove that

$$\int_{a+}^{c} f + \int_{c}^{b-} f = \int_{a+}^{c'} f + \int_{c'}^{b-} f,$$

so that any choice of c (or c') in (a, b) can be used to determine $\int_{a+}^{b-} f$.

3. Suppose that f is integrable on every closed subinterval of (a, ∞) and f is unbounded as $x \to a+$. Prove that the definition

$$\int_{a+}^{\infty} f = \int_{a+}^{c} f + \int_{c}^{\infty} f$$

is independent of the choice of c in (a, ∞). (Compare Exercise 8.3.2.)

4. Prove Theorem 8.2 using an argument similar to the proof of Theorem 8.1. *(in notes, similar to proof of Th.8.1)*

5. For what values of $p < 0$ does the improper integral $\int_{0+}^{1} x^p\, dx$ converge?

In Exercises 6–13, determine whether the improper integral is convergent.

improper integral ½ to zero makes it a pole

6. $\int_{0}^{1-} \dfrac{1}{\sqrt{1 - x^2}}\, dx$ *(1-) pole*

7. $\int_{0}^{\pi/2-} \tan x\, dx$ *pole*

8. $\int_{0+}^{1-} \dfrac{1}{x^2 - x}\, dx$

9. $\int_{0}^{2} \dfrac{1}{(x - 1)^{2/3}}\, dx$ *split*

10. $\int_{0+}^{\infty} \dfrac{\log x}{\sqrt{x}}\, dx$

11. $\int_{0+}^{\infty} \dfrac{1}{x\sqrt{x} + 1}\, dx$ *split*

12. $\int_{0+}^{1} \log x\, dx$

13. $\int_{1+}^{\infty} \dfrac{1}{x(\log x)^2}\, dx$

Handwritten marginalia:

$\int_{0+}^{1} x^p\, dx \qquad p \geq 0 \; R\text{-integral}$
$= \lim_{t \to 0+} \int_{t}^{1} x^p\, dt$

$p < 0 \quad p = -1$
$\lim_{t \to 0+} \int_{t}^{1} x^{-1}\, dx = \ln x \Big|_{t}^{1} = (\ln 1 - \ln t)$
$= 0 -$

$\lim_{t \to 0+} \int_{t}^{1} x^p\, dt = \lim_{t \to 0+} \dfrac{x^{p+1}}{p+1}\Big|_{t}^{1} = \lim_{t \to 0+}\left[\dfrac{1}{p+1} - \dfrac{t^{p+1}}{p+1}\right]$
$= \dfrac{1}{p+1} - \lim_{t \to 0+}\dfrac{t^{p+1}}{p+1}$

$p+1 > 0 \text{ makes this convergent} \quad 0 > p > -1$
$\lim t^{p+1} \to 0 \text{ or } \dfrac{1}{t^{p+1}} \text{ if } p+1 \text{ appears}$

$\int_{0}^{1} x^p\, dt = \dfrac{1}{p+1} \quad \text{(defined as an improper integral)}$

8.4. The Gamma Function

In this section we study a function that is defined by an improper integral. It is called the *gamma function*, and it is very useful in analysis and applications because its values at the positive integers form the sequence of factorials: $n! = 1 \cdot 2 \cdot 3 \cdot \ldots \cdot n$. Therefore the gamma function can be thought of as an extension of the factorial sequence to a function that is defined for noninteger values as well as for the positive integers.

DEFINITION 8.4. For each $x > 0$, let

$$\Gamma(x) = \int_{0+}^{\infty} e^{-t} t^{x-1} dt.$$

It is necessary to show that this improper integral is convergent so that $\Gamma(x)$ is well defined.

PROPOSITION 8.1. If $x > 0$, then the improper integral $\int_{0+}^{\infty} e^{-t} t^{x-1} dt$ is convergent.

Proof. We write

$$\int_{0+}^{\infty} e^{-t} t^{x-1} dt = \int_{0+}^{1} e^{-t} t^{x-1} dt + \int_{1}^{\infty} e^{-t} t^{x-1} dt.$$

First consider $\int_{0+}^{1} e^{-t} t^{x-1} dt$ and note that if $x \geq 1$, then $e^{-t} t^{x-1}$ is bounded for $0 \leq t \leq 1$. Thus we have a (proper) Riemann integral and there is nothing to prove. In case $0 \leq x < 1$, we note that $e^{-t} t^{x-1} < t^{x-1}$ for $t > 0$, and $\int_{0+}^{1} t^{x-1} dt$ is convergent (Exercise 8.2.5). Therefore, by Theorem 8.2, we infer that $\int_{0+}^{1} e^{-t} t^{x-1} dt$ is also convergent.

To deal with the integral on $[1, \infty)$ we recall that $\lim_{t \to \infty} e^{-t} t^n = 0$ for every positive integer n (see Example 6.12). Applying this fact with $n \geq x + 1$, we see that for t sufficiently large,

$$e^{-t} t^{x+1} \leq e^{-t} t^n < 1;$$

thus

$$e^{-t} t^{x-1} < t^{-2}$$

for t greater than some number c. Therefore, by Theorem 8.1 and Exercises 8.1.3 and 8.1.4, the convergence of $\int_{1}^{\infty} t^{-2} dt$ implies the convergence of $\int_{1}^{\infty} e^{-t} t^{x-1} dt$.

Now that we are assured that $\Gamma(x)$ is well defined, our next task is to prove a functional equation for Γ that establishes its relationship with the factorials. A *functional equation* is an equation involving the values of a function at two or more points in its domain. For example, the function f given by

$$f(x) = 2x + 1$$

satisfies the functional equation

$$f(x + 3) = f(x) + 6$$

for every real number x.

THEOREM 8.3. If $x > 0$, then $\Gamma(x + 1) = x\,\Gamma(x)$.

Proof. Suppose $0 < \varepsilon < 1 < b$ and use integration by parts (Exercise 7.6.8) to get

$$\int_{\varepsilon}^{1} e^{-t}t^{x-1}dt = e^{-1}/x - e^{-\varepsilon}\varepsilon^{x}/x + \frac{1}{x}\int_{\varepsilon}^{1} e^{-t}t^{x}dt \tag{1}$$

and

$$\int_{1}^{b} e^{-t}t^{x-1}dt = e^{-b}b^{x}/x - e^{-1}/x + \frac{1}{x}\int_{1}^{b} e^{-t}t^{x}dt. \tag{2}$$

Letting $\varepsilon \to 0+$ in (1), we get

$$\int_{0+}^{1} e^{-t}t^{x-1}dt = \frac{1}{ex} + \frac{1}{x}\int_{0+}^{1} e^{-t}t^{x}dt; \tag{3}$$

letting $b \to \infty$ in (2), we get

$$\int_{1}^{\infty} e^{-t}t^{x-1}dt = \frac{-1}{ex} + \frac{1}{x}\int_{1}^{\infty} e^{-t}t^{x}dt. \tag{4}$$

Now add equations (3) and (4) to obtain

$$\begin{aligned}
\int_{0+}^{\infty} e^{-t}t^{x-1}dx &= \int_{0+}^{1} e^{-t}t^{x-1}dt + \int_{1}^{\infty} e^{-t}t^{x-1}dt \\
&= \frac{1}{x}\int_{0+}^{1} e^{-t}t^{x}dt + \int_{1}^{\infty} e^{-t}t^{x}dt \\
&= \frac{1}{x}\int_{0+}^{\infty} e^{-t}t^{x}dt.
\end{aligned} \tag{5}$$

Applying Definition 8.4 to the first and last integrals of (5), we conclude that

$$\Gamma(x) = \frac{1}{x}\Gamma(x + 1),$$

which is obviously equivalent to the asserted functional equation.

The next result is now only a corollary of Theorem 8.3, but it is the best-known property of Γ, so we give it the title "theorem."

THEOREM 8.4. If n is a positive integer, then

$$\Gamma(n) = (n - 1)!. \tag{6}$$

Proof. We prove this using the Principle of Mathematical Induction (see Appendix A). First, we show that (6) is true for the case $n = 1$ by computing

$$\Gamma(1) = \int_0^\infty e^{-t}dt = 1$$

(see Exercise 8.2.8). And since 0! is defined to have the value 1, we have

$$\Gamma(1) = 0!.$$

Next we assume that (6) is true for some arbitrary value of n, say, $n = k$:

$$\Gamma(k) = (k - 1)!. \tag{7}$$

We must now show that it follows that (6) is true for the case $n = k + 1$. From Theorem 8.3 we have

$$\Gamma(k + 1) = k\Gamma(k),$$

and (7) allows us to replace $\Gamma(k)$ with $(k - 1)!$. Hence

$$\Gamma(k + 1) = k(k - 1)! = k!,$$

so (6) is true for $n = k + 1$ whenever it is true for $n = k$. Therefore the Principle of Mathematical Induction allows us to conclude that (6) is true for every positive integer n.

With the aid of Theorem 8.4 it is possible to extend the domain of Γ to include negative numbers that are not integers. The essential idea is to define Γ so that (6) continues to hold throughout the extended domain. For example, we define

$$\Gamma(x) = \frac{1}{x}\Gamma(x + 1), \quad \text{if } -1 < x < 0; \tag{8}$$

since $x + 1$ is in $(0, 1)$ when x is in $(-1, 0)$, the right-hand member of (8) is already defined by Definition 8.4. Also, it is obvious that

(8) is equivalent to (6). Now that Γ is defined on $(-1, 0)$, we can define

$$\Gamma(x) = \frac{1}{x}\Gamma(x + 1), \quad \text{if } -2 < x < -1.$$

This is continued inductively so that for any positive integer n,

$$\Gamma(x) = \frac{1}{x}\Gamma(x + 1), \quad \text{if } -n < x < -n + 1. \tag{9}$$

Therefore the domain of Γ now consists of all real numbers except $0, -1, -2, \ldots$. We make no attempt to define $\Gamma(x)$ when x is a nonpositive integer, for not only is it not possible to use $x = 0$ in (8), but also Γ is unbounded as $x \to 0+$. This property is contained in the next result (compare Exercise 8.4.6).

PROPOSITION 8.2. $\lim_{x \to 0+} \Gamma(x) = \infty$.

Proof. If $x > 0$, then

$$\begin{aligned}
\Gamma(x) &> \int_{0+}^{1} e^{-t}t^{x-1}dt \\
&\geq \int_{0+}^{1} e^{-1}t^{x-1}dt \\
&= e^{-1} \lim_{a \to 0+} \int_{a}^{1} t^{x-1}dt \\
&= e^{-1} \lim_{a \to 0+} \frac{1}{x}(1 - a^x) \\
&= \frac{1}{ex}.
\end{aligned}$$

Therefore

$$\lim_{x \to 0+} \Gamma(x) \geq \lim_{x \to 0+} \frac{1}{ex} = \infty.$$

The left-hand limit of $\Gamma(x)$ at $x = 0$ is easily found using either (8) or (9):

$$\lim_{x \to 0-} \Gamma(x) = \lim_{x \to 0-} \frac{1}{x}\Gamma(x + 1) = -\infty.$$

From these limits we can conclude that the graph of $y = \Gamma(x)$ has vertical asymptotes at $x = 0, -1, -2, \ldots$. The details of this deduction are requested in Exercise 8.4.8. Because of its many appli-

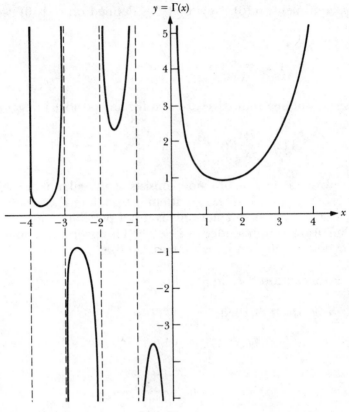

Figure 8.1

cations, values of the gamma function have been tabulated. We calculate one nonintegral value of $\Gamma(x)$ in the next proposition. The
graph of $y = \Gamma(x)$ is shown in Figure 8.1.

PROPOSITION 8.3. $\Gamma(1/2) = \sqrt{\pi}$.

Proof. By Definition 8.4 we have

$$\Gamma(1/2) = \int_{0+}^{\infty} e^{-t}t^{-1/2}dt.$$

Substituting $x = \sqrt{t}$, we see that this is equivalent to

$$\Gamma(1/2) = 2\int_{0}^{\infty} e^{-x^2}dx,$$

so our task is to evaluate a nonelementary integral ("nonelementary" in the sense that e^{-x^2} does not have a primitive function that

we have encountered before). The technique of this evaluation employs an iterated double integral, and although this is seen in elementary calculus courses, we do not develop the theory of double integrals until Chapter 16. So with the understanding that this is not in logical sequence, we use this double integral for this calculation.

We have

$$\int_0^\infty e^{-x^2}dx = \lim_{b \to \infty} \int_0^b e^{-x^2}dx,$$

so let

$$I = \int_0^b e^{-x^2}dx.$$

Then

$$I^2 = \left(\int_0^b e^{-x^2}dx\right)\left(\int_0^b e^{-y^2}dy\right) = \int_0^b \int_0^b e^{-(x^2 + y^2)}dxdy.$$

The last integral is the double integral over the square

$$\{(x, y): 0 \le x \le b \quad \text{and} \quad 0 \le y \le b\}.$$

Compare this integral to the integrals over the two quarter-circle regions R_1 and R_2 in the first quadrant that are bounded by $x^2 + y^2 = b^2$ and $x^2 + y^2 = 2b^2$, respectively (see Figure 8.2). Since the integrand is positive, we have

$$\int_{R_1} \int e^{-(x^2+y^2)}dxdy \le \int_0^b \int_0^b e^{-(x^2+y^2)}dxdy \le \int_{R_2} \int e^{-(x^2+y^2)}dxdy.$$

Converting two of these integrals to polar coordinates, we have

$$\int_{\theta=0}^{\pi/2} \int_{r=0}^{b} e^{-r^2} rdrd\theta \le I^2 \le \int_{\theta=0}^{\pi/2} \int_{r=0}^{b\sqrt{2}} e^{-r^2} rdrd\theta,$$

so

$$\frac{\pi}{4}(1 - e^{-b^2}) \le I^2 \le \frac{\pi}{4}(1 - e^{-2b^2}).$$

Therefore it is clear that

$$\lim_{b \to \infty} I^2 = \frac{\pi}{4},$$

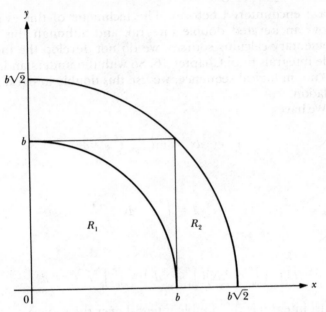

Figure 8.2

and hence

$$\Gamma(1/2) = 2 \int_0^\infty e^{-x^2} dx$$
$$= 2 \lim_{b \to \infty} I$$
$$= 2(\sqrt{\pi}/2)$$
$$= \sqrt{\pi}.$$

Exercises 8.4

1. Find the values: *use recursive relationship of Γ function*
   (a) $\Gamma(1/2)$ (b) $\Gamma(-1/2)$ (c) $\Gamma(5/2)$ (d) $\Gamma(-7/2)$

2. Evaluate $\int_0^\infty e^{-t} \sqrt{t^3}\, dt$.

3. Evaluate $\int_0^\infty e^{-3t} t^{1/2} dt$.

4. Evaluate $\int_{0+}^1 \left(\log \frac{1}{u}\right)^{1/2} du$.   *improper at lower limit*

5. Evaluate $\int_{0+}^1 \left(\log \frac{1}{u}\right)^{-1/2} du$.

6. Use Theorem 8.3 to prove that $\lim_{x \to 0+} x\Gamma(x) = 1$ and therefore deduce that $\lim_{x \to 0+} \Gamma(x) = \infty$.

7. Prove: $\lim_{x \to 0-} \Gamma(x) = -\infty$.

8. Prove that the graph of $y = \Gamma(x)$ has vertical asymptotes at $x = 0, -1, -2, \ldots$.

9. Prove: If $\alpha > 0$ and $x > 0$, then $\int_{0+}^{\infty} e^{-\alpha t} t^{x-1}\, dt = \alpha^{-x}\Gamma(x)$.

10. Prove: If $\alpha > 0$, then $\int_{0+}^{\infty} e^{-t^\alpha}\, dt = \dfrac{1}{\alpha}\Gamma\left(\dfrac{1}{\alpha}\right)$.

11. Use the Principle of Mathematical Induction to prove that

$$\Gamma\left(n + \frac{1}{2}\right) = \frac{(2n)!\,\sqrt{\pi}}{4^n\, n!} \quad \text{for } n = 0, 1, 2, \ldots.$$

8.5. The Laplace Transform

In this section we see how the improper integral can be used to transform a given function into another function. The motive for such a transformation is the hope that the new function will be simpler than the original, thus allowing an easy solution to a problem in the new system, and then a transformation of that solution back into the original system. Probably the best-known application of the Laplace transform is in the solution of initial value problems in differential equations.

DEFINITION 8.5. If the domain of the function f includes $[0, \infty)$, then the *Laplace transform* of f is the function $\mathcal{L}\{f\}$ defined by

$$\mathcal{L}\{f(x)\} = \int_{0+}^{\infty} e^{-xt} f(t)\,dt.$$

The domain of $\mathcal{L}\{f\}$ consists of all numbers x for which the improper integral is convergent.

Note that if $f(t)$ is bounded as $t \to 0+$, then the integral in Definition 8.5 is of the type $\int_0^\infty g$, and if $f(t)$ is unbounded as $t \to 0+$, then we write

$$\mathcal{L}\{f(x)\} = \int_{0+}^{b} e^{-xt} f(t)\,dt + \int_{b}^{\infty} e^{-xt} f(t)\,dt.$$

EXAMPLE 8.8. If $f(t) = e^{-t}$, then its Laplace transform is

$$\mathcal{L}\{e^{-x}\} = (x + 1)^{-1} \quad \text{for } x > -1.$$

This is verified by evaluating the improper integral:

$f(x) = e^{-x}$

$$\mathcal{L}\{e^{-x}\} = \int_0^\infty e^{-xt}e^{-t}dt$$

$$= \int_0^\infty e^{-t(x+1)}dt$$

$$= \lim_{b \to \infty} \left[\frac{-e^{-t(x+1)}}{x+1} \right]\Big|_{t=0}^{b}$$

$$= \frac{1}{x+1}.$$

$= \frac{-1}{x+1} \left[\lim_{b \to \infty} e^{-(x+1)b} - 1 \right]$

$x + 1 > 0$

$x > -1$

Note that the integral is convergent if and only if $\lim_{b \to \infty} e^{-b(x+1)} = 0$, which shows that the domain of $\mathcal{L}\{e^{-x}\}$ is $(-1, \infty)$.

EXAMPLE 8.9. If $f(t) = t$, then its Laplace transform is

$$\mathcal{L}\{x\} = x^{-2} \quad \text{for} \quad x > 0.$$

For

$$\mathcal{L}\{x\} = \int_{0+}^\infty e^{-xt}t\, dt = \lim_{b \to \infty} \left[\frac{-te^{-xt}}{x} - \frac{e^{-xt}}{x^2} \right]\Big|_{t=0}^{b} = x^{-2},$$

at upper limit goes to zero

$0 - \left(-\frac{1}{x^2}\right) = \frac{1}{x^2}, \quad x > 0$

and the limit exists if and only if $x > 0$.

In Exercises 8.5, the calculation of $\mathcal{L}\{f(x)\}$ is requested for several elementary functions. They provide additional experience in evaluating improper integrals, so we do not calculate them here. But the value of the modest list of examples in those exercises is enhanced by the fact that they can be combined to yield the transforms of many more functions. That is the main thrust of the next two theorems.

THEOREM 8.5. If a and b are real numbers, then

linearity

$$\mathcal{L}\{af(x) + bg(x)\} = a\mathcal{L}\{f(x)\} + b\mathcal{L}\{g(x)\},$$

where the domain of the left-hand function contains the domains of $\mathcal{L}\{f\}$ and $\mathcal{L}\{g\}$.

Proof. This is an immediate consequence of Definition 8.5 and Theorem 7.2.

A transformation that has the property expressed in Theorem 8.5 is called a *linear transformation*. The name is taken from the fact that this property is exhibited by a function $f(x) = mx$ whose graph is a line through the origin; for

$$f(ax + by) = m(ax + by)$$
$$= a(mx) + b(my)$$
$$= af(x) + bf(y).$$

The next result concerns the transform of another type of combination of functions. It is called a *shifting property* and is reminiscent of translation of axes in analytic geometry.

THEOREM 8.6. If a is a real number and $\mathcal{L}\{f(x)\}$ is defined for $x > b$, then

$$\mathcal{L}\{e^{ax}f(x)\} = \mathcal{L}\{f(x - a)\} \quad \text{for} \quad x > a + b.$$

Proof. By Definition 8.5, we have

$$\mathcal{L}\{e^{ax}f(x)\} = \int_{0+}^{\infty} e^{-xt} e^{at} f(t)\,dt$$
$$= \int_{0+}^{\infty} e^{-(x-a)t} f(t)\,dt$$
$$= \mathcal{L}\{f(x - a)\}.$$

The last equality is justified by observing that the last integral is precisely the defining integral for $\mathcal{L}\{f(x)\}$ with x replaced by $x - a$.

The observation that "x is replaced by $x - a$" expresses the way that this theorem is applied to yield other Laplace transforms.

EXAMPLE 8.10. If $f(t) = te^{at}$, then its Laplace transform is

$$\mathcal{L}\{xe^{ax}\} = (x - a)^{-2}.$$

We simply apply Theorem 8.6 to the identity function of Example 8.9.

In the next theorem it is shown that the Laplace transforms of a function and its derivative are connected by a very simple relationship. It is this property that makes the Laplace transform so useful in solving differential equations.

THEOREM 8.7. Suppose the function f has a continuous derivative on $[0, \infty)$. If $\lim_{t \to \infty} e^{-xt} f(t) = 0$ and $\mathcal{L}\{f(x)\}$ exists when $x > a$, then

$$\mathcal{L}\{f'(x)\} = x\, \mathcal{L}\{f(x)\} - f(0) \quad \text{for} \quad x > a. \tag{1}$$

Proof. Integration by parts yields

$$\int_0^b e^{-xt} f'(t)dt = e^{-xb} f(b) - f(0) + \int_0^b xe^{-xt} f(t)dt. \tag{2}$$

Since $\lim_{b \to \infty} e^{-xb} f(b) = 0$, as $b \to \infty$, Equation (2) becomes

$$\mathcal{L}\{f'(x)\} = \int_{0+}^{\infty} e^{-xt} f'(t)dt$$
$$= -f(0) + \int_{0+}^{\infty} xe^{-xt} f(t)dt$$
$$= -f(0) + x\, \mathcal{L}\{f(x)\}.$$

According to (1) the Laplace transform of f' is obtained simply by multiplying the transform of f by the independent variable and then subtracting the initial value of f.

EXAMPLE 8.11. Let $f(t) = te^{at}$; then $f'(t) = e^{at} + ate^{at}$, and the Laplace transform of f' is

$$\mathcal{L}\{a^{ax} + axe^{ax}\} = x\, \mathcal{L}\{xe^{ax}\} - f(0)$$
$$= \frac{x}{(x-a)^2}$$

by (1) and Example 8.10.

Exercises 8.5

1. Prove: If f is a nonnegative function such that the integral for $\mathcal{L}\{f(x_0)\}$ is convergent, then $\mathcal{L}\{f(x)\}$ exists for every $x > x_0$.

2. Prove: For any nonnegative function f, the domain of $\mathcal{L}\{f(x)\}$ is one of the following types of sets: $\emptyset$, (a, ∞), $[a, \infty)$, or $(-\infty, \infty)$.

3. Prove: Suppose the function f has a continuous second-order derivative on $[0, \infty)$, $\lim_{t \to \infty} e^{-xt} f(t) = 0$,

$\lim_{t \to \infty} e^{-xt} f'(t) = 0$, and $\mathscr{L}\{f(x)\}$ and $\mathscr{L}\{f'(x)\}$ exist when $x > a$; then

$$\mathscr{L}\{f''(x)\} = x^2 \, \mathscr{L}\{f(x)\} - xf(0) - f'(0).$$

Verify the following formulas. The symbols a and b denote constants in each case, and n is a positive integer.

4. $\mathscr{L}\{a\} = \dfrac{a}{x}$, $x > 0$.

5. $\mathscr{L}\{x^n\} = \dfrac{n!}{x^{n+1}}$, $x > 0$.

6. $\mathscr{L}\{\sin ax\} = \dfrac{a}{x^2 + a^2}$, $x > 0$.

7. $\mathscr{L}\{\cos ax\} = \dfrac{x}{x^2 + a^2}$, $x > 0$.

8. $\mathscr{L}\{e^{ax}\} = \dfrac{1}{x - a}$, $x > 0$.

9. $\mathscr{L}\{x^n \, e^{ax}\} = \dfrac{n!}{(x - a)^{n+1}}$, $x > 0$.

10. $\mathscr{L}\{e^{ax} \sin bx\} = \dfrac{b}{(x - a)^2 + b^2}$, $x > a$.

11. $\mathscr{L}\{e^{ax} \cos bx\} = \dfrac{x - a}{(x - a)^2 + b^2}$, $x > a$.

12. $\mathscr{L}\{\sinh ax\} = \dfrac{a}{x^2 - a^2}$, $x > a$. (Recall that $\sinh z = (e^z - e^{-z})/2$.)

13. $\mathscr{L}\{\cosh ax\} = \dfrac{x}{x^2 - a^2}$, $x > a$. (Recall that $\cosh z = (e^z + e^{-z})/2$.)

14. $\mathscr{L}\{ax^2 + bx + c\} = \dfrac{2a}{x^3} + \dfrac{b}{x^2} + \dfrac{c}{x}$, $x > 0$.

15. $\mathscr{L}\{\sin^2 ax\} = \dfrac{2a^2}{x(x^2 + 4a^2)}$, $x > 0$. (*Hint:* Let $f(x) = \sin^2 ax$, so $f'(x) = a \sin 2ax$; then use Theorem 8.7.)

16. $\mathscr{L}\{\cos^2 ax\} = \dfrac{x^2 + 2a^2}{x(x^2 + 4a^2)}$, $x > 0$.

17. $y'' - y' - 2y = 3e^t$

9

INFINITE SERIES

9.1. Convergent and Divergent Series

Let $\{a_k\}$ be a number sequence, and define the related sequence $\{s_n\}$ by the formula

$$s_n = \sum_{k=1}^{n} a_k.$$

Then we say that $\{s_n\}$ is the *sequence of partial sums* of the *infinite series* Σa_k. The number s_n is called the *n*th partial sum, and a_k is called the *k*th term of the series. The discerning reader may note that this statement does not define the term *infinite series*. Indeed, there is no essential difference between $\{s_n\}$ and Σa_k; both represent a function (the same function) from $\mathbb{N}$ into $\mathbb{R}$. The only difference between the study of sequences and the study of series is one of viewpoint. Historically, infinite series were investigated first, perhaps because it was natural to ask whether addition could be extended to give a value to the sum of an infinite set of terms. For example,

$$1 + \frac{1}{2} + \frac{1}{4} + \cdots + \left(\frac{1}{2}\right)^{n-1} + \cdots = \; ?$$

$$1 + \frac{1}{2} + \frac{1}{3} + \cdots + \frac{1}{n} + \cdots = \; ?$$

or

$$1 - 1 + 1 - 1 + \cdots + (-1)^{n+1} + \cdots = \; ?$$

Once one realizes that a series is identical to a sequence, it is possible to draw on the theory of sequences that was developed in Chapter 2. For example, the following definition is not needed to introduce the concept of convergence; rather, it is more a matter of introducing the terminology and notation that is used in the setting of infinite series:

DEFINITION 9.1. The series Σa_k is *convergent* provided that its sequence of partial sums is convergent, in which case we write

$$\sum_{k=1}^{\infty} a_k = \lim_n \{\sum_{k=1}^{n} a_k\},$$

and this limit value is called the *sum* of the series. In case $\lim_n \{\sum_{k=1}^{n} a_k\} = \infty$, the series Σa_k is said to be *divergent*, and we write

$$\sum_{k=1}^{\infty} a_k = \infty.$$

We have just introduced some peculiarities of notation and terminology that will probably be new to most readers. In fact, they are not used universally, but they are convenient. First, note that the symbol Σa_k denotes a *function* (from $\mathbb{N}$ into $\mathbb{R}$), whereas $\Sigma_{k=1}^{\infty} a_k$ denotes a *number*. This distinction is made in the same way that we use f to denote a function and $f(x)$ to denote a number in the range of that function. The other peculiarity is the special meaning of the word *divergent*. In standard usage, the word divergent is used to describe any series that is not convergent; here we simply call such a series *nonconvergent*. When we say that Σa_k is divergent, we mean that it is nonconvergent in a particular way, namely, its partial sums tend to infinity.

EXAMPLE 9.1: GEOMETRIC SERIES. The series Σr^{k-1} is convergent if and only if $|r| < 1$. If $r \neq 1$, then the partial sums are given by the elementary formula

$$\sum_{k=1}^{n} r^{k-1} = \frac{1 - r^n}{1 - r},$$

and we have seen the sequence $\{r^n\}$ in Chapter 2. If $|r| < 1$, then $\lim_n r^n = 0$, so we have

$$\sum_{k=1}^{\infty} r^{k-1} = \frac{1}{1 - r} \quad \text{(if } |r| < 1).$$

If $r < -1$, then $\{r^n\}$ is unbounded, so Σr^k is nonconvergent. If $r > 1$, then $\lim_n r^n = \infty$, so Σr^k is divergent. If $r = -1$, then the sequence of partial sums is $\{(-1)^{n+1}\}$, so Σr^k is nonconvergent. Finally, if $r = 1$, then the nth partial sum is given by $s_n = n$, so Σr^k is divergent.

EXAMPLE 9.2. If

$$a_k = \begin{cases} \dfrac{1}{j}, & \text{if } k = 2j - 1, \\[2mm] -\dfrac{1}{j}, & \text{if } k = 2j, \end{cases}$$

then Σa_k is convergent and $\Sigma_{k=1}^{\infty} a_k = 0$. In this case, it is much simpler to write the series in expanded form:

$$\Sigma a_k = 1 - 1 + \frac{1}{2} - \frac{1}{2} + \frac{1}{3} - \frac{1}{3} + \cdots .$$

Now it is clear that the partial sums converge to zero, because

$$\Sigma_{k=1}^{2j} a_k = 0 \quad \text{and} \quad \Sigma_{k=1}^{2j-1} a_k = \frac{1}{j}.$$

EXAMPLE 9.3: HARMONIC SERIES. The series $\Sigma 1/k$ is divergent. To prove this assertion, consider the 2^nth partial sum and group the terms as follows:

$$\Sigma_{k=1}^{2^n} \frac{1}{k} = 1 + \frac{1}{2} + \left(\frac{1}{3} + \frac{1}{4}\right) + \left(\frac{1}{5} + \cdots + \frac{1}{8}\right)$$

$$+ \cdots + \left(\frac{1}{1 + 2^{n-1}} + \cdots + \frac{1}{2^n}\right)$$

$$> 1 + \frac{1}{2} + \left(\frac{1}{4} + \frac{1}{4}\right) + \left(\frac{1}{8} + \cdots + \frac{1}{8}\right)$$

$$+ \cdots + \left(\frac{1}{2^n} + \cdots + \frac{1}{2^n}\right)$$

$$= 1 + \frac{1}{2} + 2\left(\frac{1}{4}\right) + 4\left(\frac{1}{8}\right) + \cdots + 2^{n-1}\left(\frac{1}{2^n}\right)$$

$$= 1 + \frac{1}{2} + \frac{1}{2} + \frac{1}{2} + \cdots + \frac{1}{2}$$

$$= 1 + \frac{n}{2}.$$

If $m > 2^n$, then

$$\sum_{k=1}^{m} \frac{1}{k} > \sum_{k=1}^{2^n} \frac{1}{k} > 1 + \frac{n}{2},$$

and therefore it follows that $\Sigma 1/k$ is divergent.

The first result that we prove about series is a simple consequence of convergence:

PROPOSITION 9.1. If Σa_k is convergent, then $\lim_k a_k = 0$.

Proof. If $s_n = \sum_{k=1}^{n} a_k$ and $\sum_{k=1}^{\infty} a_k = A$, then

$$\lim_n a_n = \lim_n (s_n - s_{n-1})$$

$$= \lim_n s_n - \lim_n s_{n-1}$$

$$= A - A$$

$$= 0.$$

Note that according to this proposition the condition $\lim_k a_k = 0$ is only *necessary* for the convergence of Σa_k. This condition is not *sufficient* to ensure convergence, because the harmonic series in Example 9.3 shows that the converse implication is false.

Suppose that Σa_k is an infinite series and we change a finite number of terms to produce the series Σb_k. If N is the greatest integer such that $b_k \neq a_k$, and $\sum_{k=1}^{N} b_k = C + \sum_{k=1}^{N} a_k$, then $n \geq N$ implies $\sum_{k=1}^{n} b_k = C + \sum_{k=1}^{n} a_k$. Therefore Σb_k is convergent if and only if Σa_k is convergent, and we have proved the following result:

LEMMA 9.1. If Σa_k and Σb_k are series for which there is a number N such that $k > N$ implies $a_k = b_k$, then either both series converge or both fail to converge.

In the same way that Lemma 9.1 was proved, it can easily be shown that a finite number of terms can be deleted (or replaced by zeros) without changing the convergence or divergence of a series.

3 types of non-convergent series:
(1) $\lim S_n = +\infty$
(2) $\lim S_n = -\infty$ *Divergent*
(3) $\lim S_n$ does not exist, oscillates between at least 2 different numbers

Exercises 9.1

1. Prove that the deletion of any finite set of terms from an infinite series Σa_k does not change its convergence or non-convergence. (*Note:* Deleting terms is similar to but not the same as replacing them by zeros.) $K = \sum_{k=1}^{\infty} a_k < \infty$ *use def of series to prove*

2. Prove that the convergence or nonconvergence of Σa_k is not altered by the insertion of finitely many new terms in the series.   *prove*

3. Give an example of a nonconvergent series Σa_k such that $\lim_k a_k = 0$ and s_n takes on infinitely many positive values and infinitely many negative values.
   alternates
   $s_1 = 1$
   $s_2 = 1 - \frac{1}{2}$
   $s_3 = 1 - \frac{1}{2} - \frac{1}{2}$
   $s_4 = 1 - \frac{1}{2} - \frac{1}{2} - \frac{1}{2}$
   $s_n = 1, \frac{1}{2}, 0, -\frac{1}{2}, -1,$
   $-\frac{2}{3}, -\frac{1}{3}, 0, \frac{1}{3}, \frac{2}{3}, 1, \cdots$

4. Give an example of a divergent series Σa_k that has infinitely many negative terms as well as infinitely many positive terms. $1, 1, -1, \frac{1}{2}, \frac{1}{2}, -\frac{1}{2}, \frac{1}{3}, \frac{1}{3}, -\frac{1}{3},$
   $s_1 = 1$ $s_2 = 2$ $s_3 = 1\frac{1}{2}$
   $s_3 = 1$ $\frac{1}{3} + \frac{1}{4}$

5. Give an example of a divergent series Σa_k such that $\lim_n (s_n/n) = 0$; that is, s_n tends to ∞ more slowly than n tends to ∞.   *See pg 27 of notes where A copied by Class*
   $s_1 = 1\frac{1}{2}$
   $s_5 = 2$
   $s_6 = 1\frac{1}{2}$
   $s_7 = 1 + \frac{1}{2} + \frac{1}{3}$
   $s_8 = 1 + \frac{1}{2} + \frac{1}{3} + \frac{1}{3}$
   $s_9 = 1 + \frac{1}{2} + \frac{1}{3}$
   $\lim s_{3n} =$
   $\lim \sum_{k=1}^{n} \frac{1}{k} = 0$

6. Determine convergence or divergence of $\Sigma \dfrac{1}{k(k+1)}$.

   $\left(Hint: \dfrac{1}{k(k+1)} = \dfrac{1}{k} - \dfrac{1}{k+1}.\right)$

7. Determine convergence or divergence of Σa_k, where
   $$a_k = \begin{cases} 1/m, & \text{if } k = m^2 \text{ for } m = 1, 2, \ldots, \\ 0, & \text{otherwise.} \end{cases}$$

8. Determine convergence or divergence of
   $$\Sigma \dfrac{1}{\left[\dfrac{k+3}{4}\right]} \sin\left(\dfrac{\pi k}{2}\right),$$
   where $[x]$ denotes the greatest integer function. (*Hint:* Expand s_n.)

9. Prove the Cauchy Criterion for Series: The series Σa_k is convergent if and only if for each $\varepsilon > 0$ there is a number N such that $n > m > N$ implies $\left|\Sigma_{k=m}^{n} a_k\right| < \varepsilon$. $\sum_{k=1}^{\infty} \left(\frac{1}{2}\right)^{k-1} < \infty$

10. Show by an example that the Cauchy Criterion of Exercise 9 is *not* equivalent to the following property: for each m in $\mathbb{N}$, $\lim_n \{\Sigma_{k=n}^{n+m} a_k\} = 0$.

2/19

9.2. Comparison Tests

In this section we study series whose terms satisfy $a_k \geq 0$ for every k; such series are called *nonnegative series*. Similarly, if $a_k > 0$ for every k, then Σa_k is called a *positive series*. One obvious motive for studying these series is that they are easier to work with than series that have terms of both signs. Another reason is that much of the early work on infinite series was restricted to nonnegative series. In the eighteenth and early nineteenth centuries, some of the greatest mathematicians of the period proved theorems that give criteria to determine whether a nonnegative series converges or diverges. These results are known as "tests" for the convergence of series. First we give a general and very useful result:

LEMMA 9.2. The nonnegative series Σa_k is convergent if and only if its sequence of partial sums is bounded above.

Proof. It is obvious that the sequence of partial sums is non-decreasing, so by the Monotonic Sequence Theorem (Theorem 2.5), $\{\Sigma_{k=1}^n a_k\}$ is convergent if and only if there is a number B such that for each n, $\Sigma_{k=1}^n a_k \leq B$.

It is clear that if a nonnegative series is nonconvergent, then it is divergent (to ∞). The first convergence test that we give here is a very simple one to prove. It uses the concept of *dominance*: the nonnegative series Σb_k is said to *dominate* the nonnegative series Σa_k provided that there are numbers N and B such that $k \geq N$ implies $a_k \leq Bb_k$.

THEOREM 9.1: COMPARISON TEST. Suppose Σa_k and Σb_k are nonnegative series such that Σb_k dominates Σa_k. If Σb_k is convergent, then Σa_k is also convergent, and if Σa_k is divergent, then Σb_k is also divergent.

Proof. Assume that Σb_k is convergent and Σa_k is dominated by Σb_k. Then by Lemma 9.1, we can assume that $a_k \leq Bb_k$ for *all k*, which gives

$$\Sigma_{k=1}^n a_k \leq B \, \Sigma_{k=1}^n b_k \leq B \, \Sigma_{k=1}^\infty b_k < \infty.$$

Hence, the partial sums of Σa_k are bounded above, so by Lemma 9.2, Σa_k is convergent. The second part of the conclusion is merely the contrapositive of the implication we have just proved.

EXAMPLE 9.4. The series $\Sigma 1/(2k - 1)$ is divergent because

$$\frac{1}{2k - 1} > \frac{1}{2k} = \frac{1}{2} \cdot \frac{1}{k},$$

and $\Sigma 1/k$ is divergent.

In many cases it is difficult to exhibit an exact coefficient B to establish the inequality $a_k \leq B b_k$ for dominance. In such cases it is helpful to have another form of the comparison test that does not require a term-by-term inequality.

THEOREM 9.2: ASYMPTOTIC COMPARISON TEST. If Σa_k and Σb_k are nonnegative series such that $\lim_k (a_k/b_k) = L > 0$, then either both series converge or both series diverge. _Limit must be positive_

Proof. Since $\lim_k (a_k/b_k) = L$, there is a number N such that $k > N$ implies _we know the bk series used_

$$\frac{L}{2} < \frac{a_k}{b_k} < 2L, \quad \text{or} \quad \left(\frac{L}{2}\right) b_k < a_k < 2L b_k.$$

Thus Σa_k dominates Σb_k, and Σa_k is dominated by Σb_k; the conclusion follows from Theorem 9.1.

EXAMPLE 9.5. The series $\Sigma (k - 1)/(2k^2 + k + 7)$ is divergent; for,

$$\lim_k \frac{k - 1}{2k^2 + k + 7} \cdot \frac{1}{1/k} = \lim_k \frac{k - 1}{2k + 1 + (7/k)} = 1/2.$$

Since $\Sigma 1/k$ is divergent, we conclude that both series are divergent.

9.3. The Cauchy Condensation Test

In Example 9.3, the harmonic series was investigated by grouping the terms into a more condensed series that could be tested more easily. This technique can be used in more general situations, and it is the content of the next theorem.

THEOREM 9.3: CAUCHY CONDENSATION TEST. If Σa_k is a series of positive terms such that $\{a_k\}$ is nonincreasing, then Σa_k is convergent if and only if the "condensed" series $\Sigma 2^k a_{2^k}$ is convergent.

Proof. Consider the set consisting of the 2^n numbers

$$\{a_{2^n}, a_{1+2^n}, \ldots, a_{-1+2^{n+1}}\}.$$

Since a_{2^n} is the largest member of this set and $a_{2^{n+1}}$ is less than or equal to any member, it follows that

$$2^n a_{2^n} \geq \sum_{k=2^n}^{-1+2^{n+1}} a_k \geq 2^n a_{2^{n+1}} = \frac{1}{2} 2^{n+1} a_{2^{n+1}}.$$

Therefore

$$\sum_{n=0}^{m} 2^n a_{2^n} \geq \sum_{k=1}^{-1+2^{m+1}} a_k \geq \frac{1}{2} \sum_{n=1}^{m} 2^{n+1} a_{2^{n+1}}.$$

Hence, the partial sums of Σa_k are bounded if and only if those of $\Sigma 2^n a_{2^n}$ are bounded.

EXAMPLE 9.6. The series $\Sigma 1/k^p$ is convergent if and only if $p > 1$. Consider the "condensed" series

$$\sum 2^k \frac{1}{(2^k)^p} = \sum 2^{k-kp} = \sum (2^{1-p})^k.$$

This is a geometric series of common ratio $r = 2^{1-p}$. Since $2^{1-p} < 1$ if and only if $p > 1$, the assertion follows from Theorem 9.3 and our knowledge of the geometric series in Example 9.1.

An examination of the comparison tests and the related examples may lead one to seek a "universal comparison test series," that is, a series Σu_k that diverges so slowly that it is dominated by *every* divergent positive series. To put this more precisely, does there exist a divergent positive series Σu_k such that if Σb_k is a positive series satisfying $\lim_k (b_k/u_k) = 0$, then Σb_k must be convergent? Such a series would be an ideal one to use in a comparison test, and for many years mathematicians searched for such a series. But in 1827, it was established that no such universal comparison series exists. That is the import of the next theorem, which was originally proved by Abel.

THEOREM 9.4. If Σa_k is a divergent positive series, then there is a divergent positive series Σb_k such that $\lim_k(b_k/a_k) = 0$.

one divergent series will dominate another divergent series...

Proof. Let $s_k = \Sigma_{j=1}^k a_j$ and define $b_k = a_k/s_k$. Then $\lim_k(b_k/a_k) = \lim_k(1/s_k) = 0$, because Σa_k is divergent. To show that Σb_k is divergent, we show that its partial sums do not form a Cauchy sequence. For any m in $\mathbb{N}$, the unboundedness of $\{s_n\}$ allows us to choose $n > m$ such that $s_n > 2s_m$. Then

$$\Sigma_{k=1}^n b_k - \Sigma_{k=1}^{m-1} b_k = \Sigma_{k=m}^n b_k$$

$$= \Sigma_{k=m}^n \frac{a_k}{s_k}$$

$$= \Sigma_{k=m}^n \frac{s_k - s_{k-1}}{s_k}$$

$$\geq \frac{1}{s_n} \Sigma_{k=m}^n (s_k - s_{k-1})$$

$$= \frac{1}{s_n} \{s_n - s_{m-1}\}$$

$$> \frac{1}{s_n} \left\{ \frac{s_n}{2} \right\}$$

$$= \frac{1}{2}.$$

Hence $\{\Sigma_{k=1}^n b_k\}_{n=1}^\infty$ is not a Cauchy sequence, which implies that Σb_k is divergent.

In attempting to determine convergence or divergence for a given series, one of the greatest challenges is finding which test(s) will yield an answer for that series. For that reason we defer the exercise set until the end of Section 9.5, thereby including all our tests for nonnegative series.

9.4. Elementary Tests

The next three theorems give convergence tests familiar to students of elementary calculus. They are simple to prove and easy to apply, but they yield answers for only special types of series. Of course, the criticism of limited applicability can be made of any convergence test. The reason there are so many convergence tests (we examine only a few of them) is that there is no single test that will indicate convergence or divergence for every series. It can be pointed out that any series is convergent if and only if its partial sums form a Cauchy sequence (compare Exercise 9.5.1), but this is not a realistic test criterion. Determining whether a sequence satisfies Cauchy's criterion involves the same difficulties as checking to see whether it satisfies the definition of convergence.

THEOREM 9.5: INTEGRAL TEST. Let Σa_k be a positive series for which there is a nondecreasing function f such that for each k, $f(k) = a_k$; then Σa_k is convergent if and only if the improper integral $\int_1^\infty f$ is convergent.

Proof. If $k \le x \le k + 1$, then

$$a_k = f(k) \ge f(x) \ge f(k + 1) = a_{k+1}.$$

Therefore

$$a_k = \int_k^{k+1} a_k \ge \int_k^{k+1} f \ge \int_k^{k+1} a_{k+1} = a_{k+1}.$$

(Note that f is integrable because f is monotonic.) Thus

$$\sum_{k=1}^n a_k \ge \sum_{k=1}^n \int_k^{k+1} f = \int_1^{n+1} f,$$

and

$$\sum_{k=1}^n a_k = a_1 + \sum_{k=1}^{n-1} a_{k+1} \le a_1 + \sum_{k=1}^{n-1} \int_k^{k+1} f = a_1 + \int_1^n f.$$

Hence, $\{\sum_{k=1}^n a_k\}$ is bounded above if and only if $\lim_{b \to \infty} \int_1^b f < \infty$.

EXAMPLE 9.7. The series $\Sigma 1/(k \log k)$ is divergent. Consider the integral

$$\int_2^\infty \frac{dx}{x \log x} = \lim_{b \to \infty} [\log(\log x)] \Big|_2^b = \infty.$$

Hence the integral and the series are divergent. Here we integrated over $[2, \infty)$ instead of $[1, \infty)$ to avoid the difficulties of defining $\log(\log 1)$. This omission of the first few terms does not affect the convergence or divergence of the series.

THEOREM 9.6: RATIO TEST. Suppose that Σa_k is a positive series such that $\lim_k (a_{k+1}/a_k) = L$; then

$L < 1$ implies that Σa_k is convergent,

$L > 1$ implies that Σa_k is divergent,

and if $L = 1$, then Σa_k may be either convergent or divergent.

Proof. First assume that $L < 1$, and choose a number r such that $L < r < 1$. Then there is a number N such that $k \geq N$ implies $a_{k+1}/a_k \leq r$. Therefore

$$a_{N+1} \leq ra_N,$$

$$a_{N+2} \leq ra_{N+1} \leq r(ra_N) = r^2 a_N,$$

$$a_{N+3} \leq ra_{N+2} \leq r(r^2 a_N) = r^3 a_N,$$

$$\cdot$$
$$\cdot$$
$$\cdot$$

$$a_{N+m} \leq ra_{N+m-1} \leq r(r^{m-1} a_N) = r^m a_N.$$

Hence Σa_k is dominated by a multiple of the convergent geometric series Σr^k. Therefore, by the Comparison Test, Σa_k is convergent.

Next consider the case in which $L > 1$. This guarantees that there is a number N' such that $k \geq N'$ implies $a_{k+1}/a_k \geq 1$, which is equivalent to $a_{k+1} \geq a_k$. Thus $k \geq N'$ implies $a_k \geq a_N > 0$, so $\{a_k\}$ does not have limit zero. Therefore, by Proposition 9.1, Σa_k is divergent. Finally, the ambiguity of the case $L = 1$ is established by the series $\Sigma 1/k^p$. For *any* value of p we have

$$\lim_k \frac{1/(k + 1)^p}{1/k^p} = \lim_k \left| \frac{k}{k + 1} \right|^p = 1^p = 1;$$

but in Example 9.6, we saw that this series is convergent if $p > 1$ and divergent if $p \leq 1$.

The Ratio Test is applicable for rapidly converging or rapidly diverging series such as those that contain exponential factors or factorial expressions.

EXAMPLE 9.8. The series $\Sigma r^k/(k!)$ is convergent for every real number r; for, applying the Ratio Test, we see that

$$\lim_k \frac{r^{k+1}/(k + 1)!}{r^k/(k!)} = \lim_k \frac{r}{k+1} = 0.$$

THEOREM 9.7: ROOT TEST. Suppose that Σa_k is a nonnegative series such that $\lim_k (a_k)^{1/k} = L$; then

$L < 1$ implies that Σa_k is convergent,

$L > 1$ implies that Σa_k is divergent,

and if $L = 1$ then Σa_k may be either convergent or divergent.

Proof. See Exercise 9.5.1.

9.5. Delicate Tests

In Theorems 9.6 and 9.7, the ambiguous case in which $\lim_k (a_{k+1}/a_k) = 1$ leads one to ask whether there is a way of refining the Ratio Test to yield an answer even though the ratio limit equals 1. For example, we may ask, "How *fast* does a_{k+1}/a_k approach 1?" We could examine the limit of the indeterminate form $k\{(a_{k+1}/a_k)-1\}$, and its limit value may give an indication of whether the ratio a_{k+1}/a_k has a bias toward being greater than 1 (indicating divergence) or less than 1 (indicating convergence). That is the motivation for the next test.

THEOREM 9.8: KUMMER'S TEST. Suppose that Σa_k is a positive series and $\{p_k\}$ is a positive sequence such that

$$\lim_k \left\{ p_k \frac{a_k}{a_{k+1}} - p_{k+1} \right\} = L. \qquad (1)$$

If $L > 0$, then Σa_k is convergent;

if $L < 0$ and $\Sigma 1/p_k$ is divergent, then Σa_k is divergent.

Proof. First assume that $L > 0$, and choose r satisfying $0 < r < L$. Then for some N, $k \geq N$ implies

$$p_k \frac{a_k}{a_{k+1}} - p_{k+1} > r.$$

Therefore

$$p_N a_N - p_{N+1} a_{N+1} > r a_{N+1},$$

$$p_{N+1} a_{N+1} - p_{N+2} a_{N+2} > r a_{N+2},$$
$$\cdot$$
$$\cdot$$
$$\cdot$$
$$p_{N+m-1} a_{N+m-1} - p_{N+m} a_{N+m} > r a_{N+m}.$$

If we add these inequalities, the terms on the left cancel in pairs, and the resulting inequality is

$$p_N a_N - p_{N+m} a_{N+m} > r \sum_{k=N+1}^{N+m} a_k. \tag{2}$$

If $s_n = \sum_{k=1}^{n} a_k$, (2) becomes

$$p_N a_N - p_{N+m} a_{N+m} > r\{s_{N+m} - s_N\},$$

or

$$r s_{N+m} < r s_N + p_N a_N - a_{N+m} a_{N+m} < r s_N + p_N a_N.$$

Hence, for every m in $\mathbb{N}$,

$$s_{N+m} < s_N + (p_N a_N / r);$$

thus $\{s_n\}$ is bounded above, and therefore Σa_k is convergent.

Now assume that $L < 0$ and choose N so that $k \geq N$ implies

$$p_k \frac{a_k}{a_{k+1}} - p_{k+1} \leq 0.$$

Then $k \geq N$ implies $p_k a_k \leq p_{k+1} a_{k+1}$, and so

$$p_N a_N \leq p_{N+1} a_{N+1} \leq \cdots \leq p_m a_m \quad \text{(whenever } m > N).$$

Thus $p_N a_N / p_m \leq a_m$, so Σa_k dominates $\Sigma 1/p_k$. By the Comparison Test, the divergence of $\Sigma 1/p_k$ implies that of Σa_k, and the proof is complete.

COROLLARY 9.8: RAABE'S TEST. Suppose that Σa_k is a positive series such that

$$\lim_k \left\{ k \left(\frac{a_k}{a_{k+1}} - 1 \right) \right\} = L.$$

If $L > 1$, then Σa_k is convergent;

if $L < 1$, then Σa_k is divergent;

if $L = 1$, then Σa_k may be either convergent or divergent.

Proof. Apply Kummer's Test with $p_k = k$. The details are left as an exercise (Exercise 9.5.3).

EXAMPLE 9.9. The following series is divergent:

$$\frac{1}{2} + \frac{1 \cdot 3}{2 \cdot 4} + \frac{1 \cdot 3 \cdot 5}{2 \cdot 4 \cdot 6} + \cdots$$

$$+ \frac{1 \cdot 3 \cdot 5 \cdots (2k - 1)}{2 \cdot 4 \cdot 6 \cdots (2k)} + \cdots.$$

First, by the Ratio Test we have

$$\lim_k \frac{a_{k+1}}{a_k} = \lim_k \frac{2k + 1}{2k + 2} = \lim_k \frac{2 + (1/k)}{2 + (2/k)} = 1,$$

so the test is inconclusive. Next we try Raabe's Test:

$$k\left\{\frac{a_k}{a_{k+1}} - 1\right\} = k\left\{\frac{2k + 2}{2k + 1} - 1\right\}$$

$$= k\left\{\frac{2k + 2 - (2k + 1)}{2k + 1}\right\}$$

$$= \frac{k}{2k + 1}.$$

Since $\lim_k k/(2k + 1) = 1/2$, Raabe's Test shows that the series is divergent.

Exercises 9.5

1. Prove the Root Test (Theorem 9.7). (*Hint:* See the proof of the Ratio Test, Theorem 9.6.)

2. Prove: If Σa_k is a positive series, then

$$\Sigma_{k=1}^{\infty} a_k = \sup_n \{\Sigma_{k=1}^{n} a_k\}.$$

3. Prove Raabe's Test (Corollary 9.8).

$\sum a_k < \infty$ positive series $\Rightarrow \sum a_k^2 < \infty$ If $\sum a_k < \infty$, then $\lim_{k \to \infty} a_k = 0$

$\lim_{k \to \infty} a_k = 0$

$\forall \epsilon > 0 \; \exists N \ni$

$\forall k > N \; |a_k - 0| < \epsilon$

$|a_k| < \epsilon$

$a_k < \epsilon$

$\left(\sum a_k \quad 1 - \frac{1}{\sqrt{2}} + \frac{1}{\sqrt{3}} - \frac{1}{\sqrt{4}} + \frac{1}{\sqrt{5}} + \cdots \right) \qquad \sum a_k^2 = 1 + \frac{1}{2} + \frac{1}{3} + \frac{1}{4} + \frac{1}{5}$

4. Prove: If Σa_k is a convergent positive series, then Σa_k^2 is also convergent. Show by an example that Σa_k can be convergent (but not positive) while Σa_k^2 is divergent.

Exercises 5 and 6 give extensions of the Asymptotic Comparison Test.

Let $\epsilon = 1$

for $\epsilon = 1$

$\exists N \ni \forall k \geq N$

$a_k < 1$

$a_k^2 < a_k$

$\sum_{k=N}^{\infty} a_k^2 < \sum_{k=N}^{\infty} a_k$

$\sum_{k=N}^{\infty} a_k^2 < \infty$ implies

$\sum_{k=1}^{N-1} a_k^2 + \sum_{k=N}^{\infty} a_k^2 < \infty$

$\sum_{k=1}^{\infty} a_k^2 < \infty$

since

$\sum a_k < \infty$ converges so also

5. Prove: If Σa_k and Σb_k are positive series such that Σa_k is convergent and $\lim_k (a_k / b_k) = 0$, then Σb_k is convergent also.

6. Prove: If Σa_k and Σb_k are positive series such that Σb_k is divergent and $\lim_k (a_k / b_k) = \infty$, then Σa_k is divergent also.

Determine convergence or divergence for each of the following series.

7. $\displaystyle\sum \frac{1}{\sqrt{k(k-1)}}$

8. $\displaystyle\sum \frac{1}{k(\log k)^p}$

9. $\displaystyle\sum \frac{k r^k}{k!}$

10. $\displaystyle\sum \frac{1}{k \log k \{\log(\log k)\}^p}$

11. $\displaystyle\sum \frac{k^3 (k+1)^k}{(2k)^k}$

12. $\displaystyle\sum \frac{1}{\sqrt{k^3 - 4}}$

13. $\displaystyle\sum \frac{k!}{k^k}$ $\qquad \frac{5 \cdot 4 \cdot 3 \cdot 2 \cdot 1}{5 \cdot 5 \cdot 5 \cdot 5 \cdot 5}$

14. $\displaystyle\sum \frac{1}{(\log k)^k}$

15. $\displaystyle\sum \frac{k + 3^k}{5^k}$

16. $\displaystyle\sum k e^{-k^2}$

17. $\dfrac{1}{2} + \dfrac{1}{3} + \dfrac{1}{2^2} + \dfrac{1}{3^2} + \dfrac{1}{2^3} + \dfrac{1}{3^3} + \cdots$

18. $\dfrac{1}{3} + \dfrac{1 \cdot 2}{3 \cdot 5} + \dfrac{1 \cdot 2 \cdot 3}{3 \cdot 5 \cdot 7} + \dfrac{1 \cdot 2 \cdot 3 \cdot 4}{3 \cdot 5 \cdot 7 \cdot 9} + \cdots$

19. $\left(\dfrac{1}{2} \right)^p + \left(\dfrac{1 \cdot 3}{2 \cdot 4} \right)^p + \left(\dfrac{1 \cdot 3 \cdot 5}{2 \cdot 4 \cdot 6} \right)^p + \cdots$

5. (1) $\sum a_k < \infty$

(2) $\sum b_k$

(3) $a_k > 0 \quad b_k > 0$

(4) $\lim_{k \to \infty} \dfrac{b_k}{a_k} = 0$ (means b_k gets bigger faster than or a_k "smaller"

$\forall \epsilon > 0 \; \exists N \ni \forall k > N \; \left| \dfrac{b_k}{a_k} - 0 \right| < \epsilon$ Choose $\epsilon =$

$\exists N \ni \forall k \geq N \; \left| \dfrac{b_k}{a_k} - 0 \right| < 1$

$\left| \dfrac{b_k}{a_k} \right| < 1$

$\left| \dfrac{b_k}{a_k} \right| < 1$

$b_k < a_k$

$\sum_{k=N}^{\infty} b_k < \sum_{k=N}^{\infty} a_k$

$L + \sum_{k=1}^{N-1} b_k <$

$\sum_{k=1}^{\infty} b_k$

6. $\lim_{k} \dfrac{a_k}{b_k} = \infty$

$\sum b_k = \infty$

$\forall R > 0 \; \exists N \ni \forall k \geq N \; \dfrac{a_k}{b_k} > R$

if $R = 1 \quad \dfrac{a_k}{b_k} > 1 \quad a_k > b_k$

ratio test $\dfrac{(k+1)!}{(k+1)^{k+1}} \cdot \dfrac{k^k}{k!}$

$= \dfrac{(k+1) k!}{(k+1)^k (k+1)} \cdot \dfrac{k^k}{k!} = \dfrac{k^k}{(k+1)^k} = \left(\dfrac{k}{k+1} \right)^k = \left(\dfrac{1}{1 + \frac{1}{k}} \right)^k$

$\lim_{k \to \infty} = \dfrac{1}{e} < 1$

because

$e = 1 + \frac{1}{k}$ or $(1 + k)^{1/k}$

(indeterminate form) $\dfrac{\infty}{\infty}$

20. $\dfrac{1}{2} \cdot \dfrac{1}{3} + \dfrac{1 \cdot 3}{2 \cdot 4} \cdot \dfrac{1}{5} + \dfrac{1 \cdot 3 \cdot 5}{2 \cdot 4 \cdot 6} \cdot \dfrac{1}{7} + \cdots$

9.6. Absolute and Conditional Convergence

In this section we study series that have both positive and negative terms, and the first thing we do is designate those series that converge regardless of the mixture of signs.

DEFINITION 9.2. The series Σa_k is said to be *absolutely convergent* provided that $\Sigma |a_k|$ is convergent. If Σa_k is convergent but $\Sigma |a_k|$ is divergent, then Σa_k is said to be *conditionally convergent*.

As we see in the next theorem, absolute convergence of Σa_k implies that Σa_k is convergent, so one advantage of absolute convergence is obvious: it allows us to determine convergence by using the tests of the preceding section for the nonnegative series $\Sigma |a_k|$. But this convenient situation does not always occur. There are (conditionally) convergent series Σa_k for which $\Sigma |a_k|$ is divergent. The series of Example 9.2 is such a case. We saw that the series converges to the sum $\Sigma_{k=1}^{\infty} a_k = 0$. But

$$\Sigma |a_k| = 1 + 1 + (1/2) + (1/2) + (1/3) + (1/3) + \cdots,$$

so for each n,

$$s_{2n} = \Sigma_{k=1}^{2n} |a_k| = 2 \Sigma_{k=1}^{n} 1/k.$$

Since $\Sigma 1/k$ is divergent, it follows that $\Sigma |a_k|$ is divergent.

THEOREM 9.9. If Σa_k is absolutely convergent, then Σa_k is convergent.

Proof. Since $\Sigma |a_k|$ is convergent, its partial sums form a Cauchy sequence. If $n > m$, then we can write

$$\Sigma_{k=1}^{n} |a_k| - \Sigma_{k=1}^{m} |a_k| = \Sigma_{k=m+1}^{n} |a_k|$$
$$\geq |\Sigma_{k=m+1}^{n} a_k| \qquad (1)$$
$$= |\Sigma_{k=1}^{n} a_k - \Sigma_{k=1}^{m} a_k|.$$

The left-hand member is the difference between the nth and mth partial sums of $\Sigma |a_k|$, so by the Cauchy Criterion it approaches zero as m and n tend to ∞. The right-hand member of the last line of (1) is the absolute value of the difference between the nth and mth partial sums of Σa_k. Since it, too, must approach zero as m and n tend to ∞, the partial sums of Σa_k form a Cauchy sequence. Hence Σa_k is convergent.

There are no tests for conditional convergence per se. One first tests $\Sigma |a_k|$ to determine whether Σa_k is absolutely convergent. If $\Sigma |a_k|$ is divergent, then one tries to establish (conditional) convergence or nonconvergence of Σa_k by using properties of that particular series. For example, if $\lim a_k \neq 0$, then by Proposition 9.1, Σa_k cannot be convergent.

In the next two theorems, we develop some techniques that are useful in determining conditional convergence. First we prove a lemma that is sometimes known as *summation by parts*. This name is based on its similarity to the Integration by Parts Theorem.

LEMMA 9.3: SUMMATION BY PARTS. If each of $\{a_k\}$ and $\{b_k\}$ is a number sequence and $s_n = \Sigma_{k=1}^n a_k$, then for each n,

$$\Sigma_{k=1}^n a_k b_k = s_n b_{n+1} - \Sigma_{k=1}^n s_k(b_{k+1} - b_k). \tag{2}$$

Proof. If we define $s_0 = 0$, then for each k in $\mathbb{N}$ we have $a_k = s_k - s_{k-1}$. Consequently,

$$
\begin{aligned}
\Sigma_{k=1}^n a_k b_k &= \Sigma_{k=1}^n (s_k - s_{k-1})b_k \\
&= b_1(s_1 - s_0) + b_2(s_2 - s_1) + \cdots + b_n(s_n - s_{n-1}) \\
&= s_1(b_1 - b_2) + s_2(b_2 - b_3) + \cdots + s_{n-1}(b_{n-1} - b_n) + b_n s_n \\
&= \Sigma_{k=1}^n s_k(b_k - b_{k+1}) + s_n b_n - s_n b_{n+1} + s_n b_{n+1} \\
&= -\Sigma_{k=1}^n s_k(b_k - b_{k+1}) + s_n(b_n - b_{n+1}) + s_n b_{n+1} \\
&= s_n b_{n+1} - \Sigma_{k=1}^n s_k(b_{k+1} - b_k).
\end{aligned}
$$

The summation-by-parts technique was discovered and used extensively by Abel. One example of his use of it is seen in the next theorem.

THEOREM 9.10: ABEL'S TEST. If the series Σa_k has bounded partial sums and $\{b_k\}$ is a nonincreasing null sequence, then $\Sigma a_k b_k$ is convergent.

Proof. We show that the partial sums of $\Sigma a_k b_k$ form a Cauchy sequence. Suppose that $s_n = \Sigma_{k=1}^n a_k$, $M = \text{lub}\{|s_n|: n \in \mathbb{N}\}$, and $n > m$. Now apply Lemma 9.3 to the sequences $\{a_k\}_{k=m}^\infty$ and $\{b_k\}_{k=m}^\infty$:

$$\left|\sum_{k=m}^{n} a_k b_k\right| = \left|s_n b_{n+1} - \sum_{k=m}^{n} s_k(b_{k+1} - b_k)\right|$$
$$\leq |s_n b_{n+1}| + \sum_{k=m}^{n} |s_k||b_{k+1} - b_k|.$$

Since $\{b_k\}$ is nonincreasing, we have $|b_{k+1} - b_k| = b_k - b_{k+1}$, so

$$\left|\sum_{k=1}^{n} a_k b_k - \sum_{k=1}^{m-1} a_k b_k\right| = \left|\sum_{k=m}^{n} a_k b_k\right|$$
$$\leq |s_n b_{n+1}| + \sum_{k=m}^{n} |s_k| (b_k - b_{k+1})$$
$$\leq M b_{n+1} + M \sum_{k=m}^{n} (b_k - b_{k+1})$$
$$= M b_{n+1} + M\{b_m - b_{n+1}\}$$
$$= M b_m.$$

Since $\lim_m b_m = 0$, it follows that $\{\sum_{k=1}^{n} a_k b_k\}_{n=1}^{\infty}$ is a Cauchy sequence; hence $\Sigma a_k b_k$ is convergent.

The next theorem deals with a special class of series having both positive and negative terms. The series in this class have positive and negative terms that alternate in their order of appearance, so these are called *alternating series*.

THEOREM 9.11: ALTERNATING SERIES THEOREM. If $\{a_k\}$ is a nonincreasing null sequence, then the alternating series $\Sigma(-1)^{k+1} a_k$ is convergent.

Proof. This result is an easy corollary of Abel's Test. The series $\Sigma(-1)^{k+1}$ obviously has bounded partial sums, and $\{a_k\}$ is a nonincreasing null sequence by hypothesis. Therefore, by Theorem 9.10, the series $\Sigma(-1)^{k+1} a_k$ is convergent.

EXAMPLE 9.10. The alternating harmonic series $\Sigma(-1)^{k+1}/k$ is conditionally convergent. We know from Example 9.3 that this series cannot be absolutely convergent, and it clearly satisfies the hypotheses of Theorem 9.11; therefore it is a conditionally convergent alternating series.

Exercises 9.6

Determine absolute convergence, conditional convergence, or nonconvergence for each series in Exercises 1–8.

[handwritten note: If conditionally convergent, test for absolute convergence (if +, don't need to test, it is automatically absolutely convergent)]

1. $\sum \dfrac{(-1)^{k+1}}{\log(k+1)}$

2. $\sum \dfrac{1}{k+1}\sin\left(\dfrac{\pi k}{2}\right)$

3. $\sum (-1)^{k+1}\dfrac{1}{k+1}$

4. $\sum (-1)^{k+1}\dfrac{k}{k^2+1}$

5. $\sum (-1)^{k+1}\dfrac{k^2}{k^5+1}$

6. $\sum \dfrac{\log k}{k^2}\cos\left(\dfrac{\pi k}{2}\right)$

7. $\sum \dfrac{(-1)^{k+1}}{(k+1)(\log[k+1])^2}$

8. $\sum (-1)^{k+1}\left(\dfrac{3}{e}\right)^k$

9. Let $\sum(-1)^{k+1}a_k$ be an alternating series as in Theorem 9.11; that is, $\lim_k a_k = 0$ and for each k, $a_{k+1} \geq a_k \geq 0$; let $\{s_n\}$ be the sequence of partial sums. Prove the following assertions (which constitute another proof of Theorem 9.11):

(a) The subsequence $\{s_{2n}\}$ is nondecreasing.

(b) The subsequence $\{s_{2n-1}\}$ is nonincreasing.

(c) For each n, $|s_n - s_{n-1}| \leq a_n$.

(d) $\sum (-1)^k a_k$ is convergent and for each n,

$$|s_n - \textstyle\sum_{k=0}^{\infty} (-1)^{k+1}a_k| < a_n.$$

9.7. Regrouping and Rearranging Series

If one thinks of an infinite series as an extension of the operation of addition, there are some natural questions that immediately come to mind. Properties such as associativity and commutativity can be couched in an infinite series setting in the following ways.

  (i) If the terms of the series $\sum a_k$ are regrouped by inserting parentheses, is the convergence or divergence unchanged?
 (ii) If the terms of $\sum a_k$ are rearranged into a different order, is the convergence or divergence unchanged?

The answer to the first question is easy, and it is given in the next proposition. The rearrangement question takes somewhat more work.

PROPOSITION 9.2. If Σa_k is convergent and Σb_k is a regrouping of Σa_k formed by inserting pairs of parentheses, then Σb_k is also convergent and $\Sigma_{k=1}^{\infty} b_k = \Sigma_{k=1}^{\infty} a_k$.

Proof. The assertion follows immediately from the observation that the partial sums of Σb_k form a subsequence of the sequence of partial sums of Σa_k.

EXAMPLE 9.11. Consider the series

$$S = \Sigma a_k = 1 - \frac{1}{2} + \frac{1}{3} - \frac{1}{4} + \frac{1}{5} - \frac{1}{6} + \cdots,$$

[handwritten: $\leftarrow$ cond. convergent]

[handwritten: $2S = 2 - 1 + \frac{2}{3} - \frac{1}{2} + \frac{2}{3} - \frac{1}{3} + \frac{2}{7} - \frac{1}{4} + \cdots \frac{1}{5}$]

and

$$\Sigma b_k = \left(1 - \frac{1}{2}\right) + \left(\frac{1}{3} - \frac{1}{4}\right) + \left(\frac{1}{5} - \frac{1}{6}\right) + \cdots.$$

[handwritten: infinite positive #'s]

We see that Σb_k is a regrouping of Σa_k, and if $s_n = \Sigma_{k=1}^{n} a_k$ then s_{2n} is the nth partial sum of Σb_k.

In order to investigate the question of rearranging the order of the terms of a series, we first give a precise definition of the concept.

DEFINITION 9.3. The series Σb_k is a *rearrangement* of the series Σa_k if there is a one-to-one function π from $\mathbb{N}$ onto $\mathbb{N}$ such that for each k, $b_k = a_{\pi(k)}$.

The stipulation that π maps $\mathbb{N}$ onto $\mathbb{N}$ guarantees that every term of Σa_k appears (at least once) in Σb_k, and the stipulation that π is one-to-one ensures that no term of Σa_k appears more than once in Σb_k. If Σb_k rearranges only a finite number of the terms of Σa_k, then it is clear from Lemma 9.1 that Σb_k is convergent if and only if Σa_k is convergent. Also, it is easy to see that such a finite rearrangement yields $\Sigma_{k=1}^{\infty} b_k = \Sigma_{k=1}^{\infty} a_k$. This is not true for all rearrangements, however, as we see in the next example.

EXAMPLE 9.12. The conditionally convergent series $\Sigma(-1)^{k+1}/k$ can be rearranged to give another conditionally convergent series whose sum is

$$\left(\frac{1}{2}\right)\Sigma_{k=1}^{\infty}\frac{(-1)^{k+1}}{k}.$$

Consider the following:

$$2\Sigma_{k=1}^{\infty}\frac{(-1)^{k+1}}{k} = 2\left(1 - \frac{1}{2} + \frac{1}{3} - \frac{1}{4} + \frac{1}{5} - \frac{1}{6} + \cdots\right)$$

$$= 2 - 1 + \frac{2}{3} - \frac{2}{4} + \frac{2}{5} - \frac{2}{6} + \cdots$$

$$= 2 - 1 + \frac{2}{3} - \frac{1}{2} + \frac{2}{5} - \frac{1}{3} + \cdots$$

$$\overset{?}{=} 2 - 1 - \frac{1}{2} + \frac{2}{3} - \frac{1}{3} - \frac{1}{4}$$

$$+ \frac{2}{5} - \frac{1}{5} - \cdots$$

$$= (2 - 1) - \frac{1}{2} + \left(\frac{2}{3} - \frac{1}{3}\right) - \frac{1}{4}$$

$$+ \left(\frac{2}{5} - \frac{1}{5}\right) - \cdots$$

$$= 1 - \frac{1}{2} + \frac{1}{3} - \frac{1}{4} + \frac{1}{5} - \cdots$$

$$= \Sigma_{k=1}^{\infty}\frac{(-1)^{k+1}}{k}.$$

The steps of doubling each term, reducing fractions, and re-grouping are all valid, so the equality must fail where the question mark is located. That is the step where the order of the terms was rearranged.

The next theorem shows that Example 9.12 is just one of many possible rearrangements that would produce a different sum of the alternating harmonic series:

THEOREM 9.12. If Σa_k is a conditionally convergent series and L is any real number, then there is a rearrangement of Σa_k whose sum is L.

the rearrangement only holds for infinite # of terms - (if you reuse same finite # of terms doesn't hold)

Proof. Let $\{p_k\}$ be the subsequence of $\{a_k\}$ consisting of all the positive terms, and let $\{-q_k\}$ be the subsequence of $\{a_k\}$ consisting of all the nonpositive terms. We assert that both Σp_k and Σq_k must be divergent. For if both were convergent then $\Sigma_{k=1}^n |a_k| \leq \Sigma_{k=1}^\infty p_k + \Sigma_{k=1}^\infty q_k$, and Σa_k would be absolutely convergent. Also, if one of the series Σp_k and Σq_k were convergent and the other divergent, then Σa_k would not be conditionally convergent. For suppose $\Sigma_{k=1}^\infty p_k = \infty$ and $\Sigma_{k=1}^\infty q_k = Q$. Then for each number B there is an N such that $n > N$ implies $\Sigma_{k=1}^n p_k > B + Q$, which implies

$$\Sigma_{k=1}^m a_k > \Sigma_{k=1}^n p_k - \Sigma_{k=1}^\infty q_k > (B + Q) - Q = B$$

whenever m is sufficiently large. Thus Σa_k is nonconvergent. Similarly, if $\Sigma_{k=1}^\infty p_k = P$ and $\Sigma_{k=1}^\infty q_k = \infty$, then we could get $\Sigma_{k=1}^n a_k$ less than any preassigned negative number.

Now we use the divergence of Σp_k and Σq_k to construct a rearrangement of Σa_k that converges to L. First choose $n(1)$ to be the least integer satisfying

$$p_1 + p_2 + \cdots + p_{n(1)} > L.$$

(If $L \leq 0$, we choose $n(1) = 0$, so there are no p_k's in this group.) Next choose $q_1, \ldots, q_{n(2)}$ so that $n(2)$ is the least integer such that

$$p_1 + \cdots + p_{n(1)} - q_1 - q_2 - \cdots - q_{n(2)} < L.$$

We know that there are enough p_k's and q_k's to achieve both of these inequalities, because $\Sigma_{k=1}^\infty p_k = \infty$ and $\Sigma_{k=1}^\infty q_k = \infty$. The process is then continued inductively: As soon as the partial sum exceeds L, we start inserting the next unused negative terms and do so until the partial sum drops below L. In this fashion we use all the terms of Σp_k and Σq_k, and the partial sums of the rearrangement oscillate about L. Furthermore, once we have progressed beyond $p_{n(j)}$, the partial sums do not differ from L by more than $\max\{p_{n(j)}, q_{n(j-1)}\}$. This is true because $n(j)$ is chosen to be the *least* integer such that the inequality holds. But Σa_k is convergent, so $\lim_k a_k = 0$. Therefore p_k and q_k must both approach zero, and so the partial sums of the rearrangement must converge to L.

In the next theorem, which was first proved by Dirichlet, we see that an *absolutely* convergent series is not affected by rearranging its terms:

THEOREM 9.13. If Σa_k is an absolutely convergent series and $\Sigma a_{\pi(k)}$ is a rearrangement of Σa_k, then $\Sigma a_{\pi(k)}$ is absolutely convergent and $\Sigma_{k=1}^\infty a_{\pi(k)} = \Sigma_{k=1}^\infty a_k$.

Proof. We introduce the notation

$$p_k = \max\{a_k, 0\} \quad \text{and} \quad q_k = \min\{a_k, 0\}.$$

A word of caution is needed here: These definitions of p_k and q_k are *not* the same as those used in the proof of Theorem 9.12. We now have

$$a_k = p_k + q_k \quad \text{and} \quad |a_k| = p_k - q_k.$$

Since $\Sigma |a_k|$ is convergent, it is easily seen that both Σp_k and Σq_k are absolutely convergent. Also, if

$$P = \sum_{k=1}^{\infty} p_k \quad \text{and} \quad Q = \sum_{k=1}^{\infty} q_k,$$

then

$$\sum_{k=1}^{\infty} a_k = P + Q.$$

Since $a_{\pi(k)} = p_{\pi(k)} + q_{\pi(k)}$, we see that $\Sigma p_{\pi(k)}$ is obtained by forming a rearrangement of the nonnegative terms of Σa_k and inserting zeros where $a_{\pi(k)} < 0$. Therefore each partial sum of $\Sigma p_{\pi(k)}$ satisfies

$$\sum_{k=1}^{m} p_{\pi(k)} \leq \sum_{k=1}^{\infty} p_k = P.$$

Moreover, since every p_k appears somewhere as a term in $\Sigma p_{\pi(k)}$, it follows that P equals the least upper bound of these partial sums. Therefore by the Monotonic Sequence Theorem,

$$\sum_{k=1}^{\infty} p_{\pi(k)} = P.$$

Similarly,

$$\sum_{k=1}^{\infty} q_{\pi(k)} = Q,$$

so

$$\sum_{k=1}^{\infty} a_{\pi(k)} = \sum_{k=1}^{\infty} (p_{\pi(k)} + q_{\pi(k)}) = P + Q.$$

Finally, $\Sigma a_{\pi(k)}$ is absolutely convergent because both $\Sigma p_{\pi(k)}$ and $\Sigma q_{\pi(k)}$ are absolutely convergent, and $|a_{\pi(k)}| = p_{\pi(k)} - q_{\pi(k)}$.

9.8. Multiplication of Series

The concept of the sum of two series is so simple that we have taken it for granted in some of the prior reasoning in this chapter. If each of Σa_k and Σb_k is a series, then their sum is the series $\Sigma(a_k + b_k)$, which is formed by adding corresponding terms; that is, the kth term of the sum is the sum of the kth terms of the two series. It is clear that each partial sum satisfies

$$\sum_{k=1}^{n}(a_k + b_k) = \sum_{k=1}^{n}a_k + \sum_{k=1}^{n}b_k.$$

Therefore it follows from Theorem 2.3 on the sum of convergent sequences that the convergence of Σa_k and Σb_k implies the convergence of their sum $\Sigma(a_k + b_k)$, and

$$\sum_{k=1}^{\infty}(a_k + b_k) = \sum_{k=1}^{\infty}a_k + \sum_{k=1}^{\infty}b_k.$$

Once again, it is just a matter of changing our emphasis from the sequence $\{\sum_{k=1}^{n}a_k\}$ to the series Σa_k. As we said at the beginning of this chapter, the theory of sequential convergence eliminates the need for another proof of a series statement that is merely a rewording of a sequential statement.

In formulating the concept of multiplying series, the expected similarities do not work out the same. We could form the term-by-term product of two series, which would be $\Sigma a_k b_k$. This type of product works well enough for sequences, but in dealing with series, the distributive law gets in the way. To illustrate the problem, let Σa_k and Σb_k be two series and consider their *inner product* series $\Sigma a_k b_k$. The convergence of this product series is determined by the convergence of the sequence $\{\sum_{k=1}^{n}a_k b_k\}_{n=1}^{\infty}$, and the nth partial sum $\sum_{k=1}^{n}a_k b_k$ is not, in general, equal to the product of the partial sums $\sum_{k=1}^{n}a_k$ and $\sum_{k=1}^{n}b_k$. Therefore it is not likely that we can establish much connection between the sum $\sum_{k=1}^{\infty}a_k b_k$ and the sums $\sum_{k=1}^{\infty}a_k$ and $\sum_{k=1}^{\infty}b_k$. It is easy to prove that if Σa_k is convergent and Σb_k is absolutely convergent, then $\Sigma a_k b_k$ is absolutely convergent (see Exercise 9.8.6). But if Σa_k and Σb_k are both conditionally convergent, then we cannot infer that $\Sigma a_k b_k$ is convergent (absolutely or conditionally). The next example shows this to be the case:

EXAMPLE 9.13. The series $\Sigma(-1)^{k+1}/\sqrt{k}$ is conditionally convergent, whereas the inner product of this series with itself is $\Sigma 1/k$, which is divergent.

It would be very useful to have multiplication of series defined in such a way that the product of the sums is equal to the sum of the products. This, after all, is one way of expressing the Distributive Law:

$$(a_1 + a_2)(b_1 + b_2) = a_1 b_1 + a_1 b_2 + a_2 b_1 + a_2 b_2.$$

If we are to use this as our model for multiplication, then we must define $(\Sigma a_k)(\Sigma b_k)$ so that each a_k is multiplied by each b_k (exactly once). Thus the product should be a "series" of the form $\Sigma a_k b_j$ in which each ordered pair of subscripts (k, j) appears exactly once. In order to make a well-defined series out of this collection of terms, it is necessary to stipulate the order in which these terms are to appear and what, if any, grouping should be employed. The particular grouping and ordering that is customarily used is the following: Group together all terms $a_k b_j$ in which the subscripts have the same sum, say, $k + j = t$; then arrange these groups by the increasing values of t. Thus the product $(\Sigma a_k)(\Sigma b_k)$ would be the series

$$a_1 b_1 + (a_1 b_2 + a_2 b_1) + (a_1 b_3 + a_2 b_2 + a_3 b_1) + \cdots. \qquad (1)$$

In working with the products of series, it is convenient for notational purposes to have the first term written as a_0 rather than a_1; that is, $\Sigma a_k = a_0 + a_1 + a_2 + \cdots$. With this convention, (1) is replaced by

$$a_0 b_0 + (a_0 b_1 + a_1 b_0) + (a_0 b_2 + a_1 b_1 + a_2 b_0) + \cdots. \qquad (2)$$

In this form, the product follows the same pattern that is used in multiplying polynomials:

$$(a_0 + a_1 x + a_2 x^2 + \cdots + a_m x^m)(b_0 + b_1 x + b_2 x^2 + \cdots + b_n x^n)$$

$$= a_0 b_0 + a_0 b_1 x + a_0 b_2 x^2 + a_0 b_3 x^3 + \cdots + a_0 b_n x^n$$

$$+ a_1 b_0 x + a_1 b_1 x^2 + a_1 b_2 x^3 + \cdots + a_1 b_n x^{n+1} \qquad (3)$$

$$+ a_2 b_0 x^2 + a_2 b_1 x^3 + \cdots + a_2 b_n x^{n+2}$$

$$\cdots + \cdots$$

$$= a_0 b_0 + (a_0 b_1 + a_1 b_0)x + (a_0 b_2 + a_1 b_1 + a_2 b_0)x^2 + \cdots.$$

In the last line of (3), when we combine terms of the same power of x, we see that the coefficient of x^t is equal to the sum of those $a_k b_j$'s such that $k + j = t$. Of course, this pattern breaks down when t is greater than either m or n, but if there is an infinite series of such terms, this does not happen. This notion of series multiplication was introduced by Cauchy; we state the formal definition at this time.

DEFINITION 9.4. The *Cauchy product* of the series Σa_k and Σb_k is the series Σc_k in which the kth term is given by

$$c_k = \sum_{j=0}^{k} a_j b_{k-j}. \tag{4}$$

No matter how natural or well motivated this definition may be, it still has to be proved that the sum of the product is equal to the product of the sums. It can be shown that

$$\sum_{k=0}^{\infty} c_k = \left(\sum_{k=0}^{\infty} a_k\right)\left(\sum_{k=0}^{\infty} b_k\right) \tag{5}$$

whenever all three series are convergent, but it is necessary to assume the convergence of all three series. We later see (Example 9.14) that Σc_k may be divergent even though Σa_k and Σb_k are both conditionally convergent. First, however, we prove that (5) holds whenever Σa_k and Σb_k are absolutely convergent.

THEOREM 9.14. If the series Σa_k and Σb_k are absolutely convergent, then their Cauchy product Σc_k also is absolutely convergent and

(even true if one is CC and one is AC)

$$\sum_{k=0}^{\infty} c_k = \left(\sum_{k=0}^{\infty} a_k\right)\left(\sum_{k=0}^{\infty} b_k\right).$$

Proof. In order to interpret the various partial sums that are involved in this argument, it is helpful to visualize the terms $a_j b_k$ as an infinite matrix with entries that include all the possible combinations of a_j's multiplied by b_k's.

Upon examination of Figure 9.1, we see that the Cauchy product term $c_n = \sum_{j=0}^{n} a_j b_{n-j}$ is the sum of the terms in the diagonal running from $a_0 b_n$ in the top row to $a_n b_0$ in the left-hand column. Thus the partial sum $\sum_{k=0}^{n} c_k$ consists of the sum of all the terms in an upper-left triangular subset of the array. Also, the partial sum $\sum_{k=0}^{n} a_k$ is obtained by summing the first $n + 1$ terms in any column and factoring out the common factor b_k. If this is done for the first $n + 1$ columns, we get all the terms in an upper-left square of the array. The total of these $(n + 1)^2$ terms is

$$b_0\left(\sum_{k=0}^{n} a_k\right) + b_1\left(\sum_{k=0}^{n} a_k\right) + \cdots + b_n\left(\sum_{k=0}^{n} a_k\right)$$
$$= \left(\sum_{k=0}^{n} a_k\right)\left(\sum_{k=0}^{n} b_k\right). \tag{6}$$

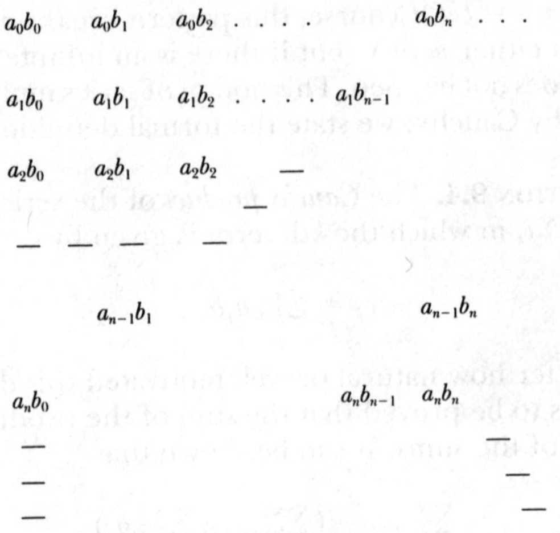

$$a_0b_0 \qquad a_0b_1 \qquad a_0b_2 \; . \; . \; . \qquad\qquad\qquad a_0b_n \; . \; . \; .$$

$$a_1b_0 \qquad a_1b_1 \qquad a_1b_2 \; . \; . \; . \; . \; a_1b_{n-1}$$

$$a_2b_0 \qquad a_2b_1 \qquad a_2b_2 \qquad —$$

$$—\quad\quad—\quad\quad—\quad\quad—$$

$$a_{n-1}b_1 \qquad\qquad\qquad\qquad a_{n-1}b_n$$

$$a_nb_0 \qquad\qquad\qquad\qquad a_nb_{n-1} \quad a_nb_n$$

$$—\qquad\qquad\qquad\qquad —$$

$$—$$

$$—$$

Figure 9.1

From these observations, it is clear that for each n,

$$\textstyle\sum_{k=0}^{n}|c_k| \leq (\sum_{k=0}^{n}|a_k|)(\sum_{k=0}^{n}|b_k|) \leq (\sum_{k=0}^{\infty}|a_k|)(\sum_{k=0}^{\infty}|b_k|).$$

Therefore Σc_k is absolutely convergent.

To show that (5) holds, we first consider the special case in which Σa_k and Σb_k are nonnegative series. Then the above argument shows that for each n,

$$\textstyle\sum_{k=0}^{n}c_k \leq (\sum_{k=0}^{\infty}a_k)(\sum_{k=0}^{\infty}b_k), \tag{7}$$

which implies that

$$\textstyle\sum_{k=0}^{\infty}c_k \leq (\sum_{k=0}^{\infty}a_k)(\sum_{k=0}^{\infty}b_k). \tag{8}$$

Now consider the product $(\Sigma_{k=0}^{n}a_k)(\Sigma_{k=0}^{n}b_k)$. From (6) we know that this is equal to the sum of the $(n + 1)^2$ terms in the upper-left square of Figure 9.1. This square is contained in some upper-left triangle whose total is $\Sigma_{k=0}^{m}c_k$. Thus for m sufficiently large ($m \geq 2n + 2$, to be precise), we have

$$\textstyle\sum_{k=0}^{m}c_k \geq (\sum_{k=0}^{n}a_k)(\sum_{k=0}^{n}b_k). \tag{9}$$

Since (9) holds for all n, we conclude that

$$\textstyle\sum_{k=0}^{\infty}c_k \geq (\sum_{k=0}^{\infty}a_k)(\sum_{k=0}^{\infty}b_k). \tag{10}$$

Combining (8) and (10), we conclude that (5) holds.

To complete the proof, we consider the general case in which Σa_k and Σb_k have terms of arbitrary signs. We have proved above that the terms in the array in Figure 9.1 form an absolutely convergent series; therefore, by Theorem 9.13, their sum is the same for any rearrangement. Consequently, the sum is the same whether we order and group the terms by triangles to get $\Sigma_{k=0}^{\infty} c_k$ or order and group the terms by squares to get $(\Sigma_{k=0}^{\infty} a_k)(\Sigma_{k=0}^{\infty} b_k)$. Hence Equation (5) holds and the proof is complete.

Prior to Theorem 9.14 we said that the convergence of Σa_k and Σb_k was not enough to guarantee the convergence of their Cauchy product. This assertion is now demonstrated in the following example:

EXAMPLE 9.14. If each of Σa_k and Σb_k is the convergent alternating series $\Sigma (-1)^{k+1}/\sqrt{k+1}$, then the Cauchy product of Σa_k and Σb_k is nonconvergent. We show that $\{c_k\}$ cannot have limit 0 by examining the subsequence $\{c_{2k}\}$:

$$c_{2k} = \sum_{j=0}^{2k} \frac{(-1)^{j+1}}{\sqrt{j+1}} \frac{(-1)^{2k-j-1}}{\sqrt{2k-j-1}}$$

$$= (-1)^{2k} \sum_{j=0}^{2k} \frac{1}{\sqrt{j+1}\ \sqrt{2k-j-1}}.$$

(11)

We claim that each of the $2k + 1$ terms in the right-hand sum is equal to or greater than $1/2k$; for,

$$[2k - (j+1)]^2 \ge 0,$$

so

$$4k^2 - 4k(j+1) + (j+1)^2 \ge 0,$$

or

$$4k^2 \ge (j+1)[4k - (j+1)] \ge (j+1)[2k-j-1],$$

which yields

$$\frac{1}{\sqrt{j+1}\ \sqrt{2k-j-1}} \ge \frac{1}{2k}.$$

The right-hand sum in (11) is equal to at least the number of terms times the smallest term, so

$$c_{2k} \geq (2k + 1) \left(\frac{1}{2k} \right) > 1. \qquad (12)$$

Since (12) holds for every k in $\mathbb{N}$, c_k does not approach zero, and therefore, by Proposition 9.1, Σc_k is nonconvergent.

Theorem 9.14 and Example 9.14 do not give the final word on the convergence of the Cauchy product. Although we do not prove it here, there is a theorem (proved by Mertens) that states that the Cauchy product of Σa_k and Σb_k is convergent so long as *one* of the series Σa_k or Σb_k is absolutely convergent and the other is convergent. And, as we stated above, Equation (5) holds whenever all three series are convergent.

Exercises 9.8

1. Prove: If Σa_k is convergent and Σb_k is absolutely convergent, then $\Sigma a_k b_k$ is absolutely convergent.

2. Discuss commutativity of the Cauchy product: $(\Sigma a_k)(\Sigma b_k) \stackrel{?}{=} (\Sigma b_k)(\Sigma a_k)$.

3. Prove: If Σa_k is conditionally convergent, then there exists a rearrangement of Σa_k that diverges (to ∞).

4. Prove: If Σa_k is conditionally convergent, then there exists a rearrangement of Σa_k whose partial sums oscillate between two preassigned numbers L_1 and L_2.

5. Prove: If the series Σa_k is convergent and x is in $[0, 1]$, then the series $\Sigma a_k x^k$ is convergent.

6. Prove that the series

$$\frac{1}{2} - \frac{1}{3} + \frac{1}{4} - \frac{1}{9} + \frac{1}{8} - \frac{1}{27} + \cdots$$

is convergent and that its sum is $\frac{1}{2}$.

7. Prove: If $\sum_{k=1}^{\infty} \frac{(-1)^{k+1}}{k} = \log 2$, then

$$\sum_{k=1}^{\infty} \frac{1}{2k(2k - 1)} = \log 2.$$

8. Prove: If $\sum_{k=1}^{\infty} \frac{1}{k^2} = \frac{\pi^2}{6}$, then

(a) $\frac{\pi^2}{24} = \frac{1}{2^2} + \frac{1}{4^2} + \frac{1}{6^2} + \cdots,$

(b) $\frac{\pi^2}{8} = \frac{1}{1^2} + \frac{1}{3^2} + \frac{1}{5^2} + \cdots,$

(c) $\frac{\pi^2}{12} = 1 - \frac{1}{2^2} + \frac{1}{3^2} - \frac{1}{4^2} + \cdots.$

10

THE RIEMANN-STIELTJES INTEGRAL

10.1. Functions of Bounded Variation

Before beginning a discussion of the topics of primary interest in this chapter, it is helpful to introduce an extension of the least upper bound concept. When we write lub $S = b$, we are stating implicitly that the set S is bounded above. It is often convenient to make a similar statement in situations in which we are not sure that S is bounded above. For this purpose we introduce the *supremum* of the (nonempty) set S (in $\mathbb{R}$):

$$\sup S = \begin{cases} \text{lub } S, & \text{if } S \text{ is bounded above,} \\ \infty, & \text{if } S \text{ is not bounded above.} \end{cases}$$

Thus the statement "$\sup S = \infty$" means simply that S is not bounded above, and "$\sup S = b$" means that S has the least upper bound b. Similarly, we define the *infimum* of a nonempty set S:

$$\inf S = \begin{cases} \text{glb } S, & \text{if } S \text{ is bounded below,} \\ -\infty, & \text{if } S \text{ is not bounded below.} \end{cases}$$

Recall from Chapter 7 that a partition $\mathscr{P}$ of the interval $[a, b]$ is a finite, increasing, number sequence $\{x_k\}_{k=0}^n$ such that $x_0 = a$ and $x_n = b$. Let f be a function on $[a, b]$ and consider the sum

176

$$\mathcal{P}(f) = \sum_{k=1}^{n} |f(x_k) - f(x_{k-1})|. \tag{1}$$

We wish to consider the set of all values of $\mathcal{P}(f)$, where $\mathcal{P}$ ranges over all partitions of $[a, b]$. This set of values may be bounded above or unbounded above, and we write the supremum of this set of values as $\sup_{\mathcal{P}} \mathcal{P}(f)$ to denote that the supremum is taken over all partitions $\mathcal{P}$ of $[a, b]$.

DEFINITION 10.1. The function f is said to have *bounded variation* on $[a, b]$ provided that $\sup_{\mathcal{P}} \mathcal{P}(f)$ is finite. In this case we write

$$V_a^b f = \sup_{\mathcal{P}} \mathcal{P}(f).$$

The number $V_a^b f$ is called the (total) *variation* of f on $[a, b]$.

EXAMPLE 10.1. Any monotonic function has bounded variation, because if $\mathcal{P}$ is *any* partition, then $\mathcal{P}(f) = |f(b) - f(a)|$.

EXAMPLE 10.2. The following step functions have bounded variation on $[0, 2]$:

$$f(x) = \begin{cases} 0, & \text{if } x \le 1, \\ 1, & \text{if } x > 1, \end{cases} \quad \text{and} \quad g(x) = \begin{cases} 0, & \text{if } x \ne 1, \\ 1, & \text{if } x = 1; \end{cases}$$

then $V_0^2 f = 1$ and $V_0^2 g = 2$.

The first theorem on this topic is a simple, sufficient condition to guarantee that a function has bounded variation.

THEOREM 10.1. If f has a bounded derivative on $[a, b]$, then f has bounded variation on $[a, b]$.

Proof. For any partition $\mathcal{P}$, f is differentiable on the kth subinterval $[x_{k-1}, x_k]$, so by the Law of the Mean there exists a number μ_k in that subinterval such that

$$f'(\mu_k) = \frac{f(x_k) - f(x_{k-1})}{x_k - x_{k-1}}.$$

Using this to substitute in (1), we get

$$\mathscr{P}(f) = \sum_{k=1}^{n} |f'(\mu_k)|(x_k - x_{k-1})$$

$$\leq \{\sup_{[a,\,b]} |f'(x)|\} \sum_{k=1}^{n} (x_k - x_{k-1})$$

$$= \{\sup_{[a,\,b]} |f'(x)|\}(b - a).$$

Since the preceding inequality holds for every $\mathscr{P}$, the number on the right side is greater than or equal to V_a^b; hence the variation is finite.

It is obvious that the condition of Theorem 10.1 is not a necessary one for bounded variation. The functions in Example 10.2 are not differentiable on $[0, 2]$ and therefore do not have bounded derivatives. But even if it is assumed that f is differentiable and has bounded variation, it still is not necessary for f' to be bounded (see Example 10.5).

Let us now compare the concept of bounded variation to the previously studied function properties of boundedness, continuity, and differentiability.

PROPOSITION 10.1. If f has bounded variation on $[a, b]$, then f is bounded there.

Proof. If x is in $[a, b]$, let $\mathscr{P}$ be the simple partition consisting of $\{a, x, b\}$. Then

$$\mathscr{P}(f) = |f(x) - f(a)| + |f(b) - f(x)| \leq V_a^b.$$

Therefore

$$|f(x)| - |f(a)| \leq V_a^b,$$

so

$$|f(x)| \leq |f(a)| + V_a^b.$$

Hence, if V_a^b is finite, then f is bounded.

Next we see that continuity does *not* imply bounded variation.

EXAMPLE 10.3. If

$$f(x) = \begin{cases} x \sin \dfrac{\pi}{2x}, & \text{if } x \neq 0, \\ 0, & \text{if } x = 0, \end{cases}$$

then $V_0^1 = \infty$. This is proved by showing that for any number B, there is a partition $\mathscr{P}$ such that $\mathscr{P}(f) > B$. To produce such a $\mathscr{P}$, we choose

$$x_n = 1, x_{n-1} = \frac{1}{3}, x_{n-2} = \frac{1}{5}, \ldots, x_{n-k} = \frac{1}{2k+1}, \ldots ; \quad (2)$$

when $k > 1$, this gives

$$|f(x_{n-k}) - f(x_{n-k+1})| = \left| \frac{1}{(2k+1)} - \frac{-1}{(2k-1)} \right| > \frac{2}{(2k+1)}.$$

Since we know that the series $\Sigma_k 2/(2k+1)$ is divergent, it follows that by choosing n large enough, there will be enough terms in $\mathscr{P}(f)$ so that the corresponding partial sum exceeds B.

Example 10.3 may lead one to conjecture that a function that "oscillates infinitely many times" cannot have bounded variation. This, however, is not the case, as we can establish with an example similar to the preceding one.

EXAMPLE 10.4. If

$$f(x) = \begin{cases} x^2 \sin(1/x), & \text{if } x \neq 0, \\ 0, & \text{if } x = 0, \end{cases}$$

then f has bounded variation on $[0, 1]$. This can be proved by using Theorem 10.1. The following computation shows that f' is bounded on $[0, 1]$:

$$|f'(x)| = |2x \sin(1/x) - \cos(1/x)| \leq 2 + 1 \quad (\text{for } 0 < x \leq 1)$$

and

$$f'(0) = 0.$$

The next example shows that differentiability does not imply bounded variation:

EXAMPLE 10.5. If

$$f(x) = \begin{cases} x^2 \sin \left(\dfrac{\pi}{x}\right)^2, & \text{if } x \neq 0, \\ 0, & \text{if } x = 0, \end{cases}$$

then f does not have bounded variation on $[0, 1]$ although f is differentiable there (see Exercise 10.1.6). The only point where the differentiability of f is not obvious is $x = 0$, and there we have

$$f'(0) = \lim_{x \to 0} \frac{f(x) - f(0)}{x}$$

$$= \lim_{x \to 0} x \sin\left(\frac{\pi}{x}\right)^2$$

$$= 0.$$

The last equality holds because $\left|x \sin(\pi/x)^2\right| \leq |x|$.

Exercises 10.1

In Exercises 1–5, find the total variation V_a^b of the given function on the indicated interval.

1. $f(x) = x \sin x$ on $[0, 3\pi]$.

2. $f(x) = 2x^3 - 3x^2$ on $[-1, 2]$.

3. $f(x) = \begin{cases} \tan x, & \text{if } 0 \leq x < \pi/2, \\ 0, & \text{if } x = \pi/2, \end{cases}$ on $\left[0, \dfrac{\pi}{2}\right]$.

4. $f(x) = \begin{cases} \sin \dfrac{\pi}{x}, & \text{if } x \neq 0, \\ 0, & \text{if } x = 0, \end{cases}$ on $[0, 1]$.

5. $\{a_k\}$ is an infinite sequence of numbers and

$$f(x) = \begin{cases} a_k, & \text{if } x = \dfrac{1}{k + 1}, \\ 0, & \text{otherwise}, \end{cases}$$ on $[0, 1]$.

6. $f(x) = \begin{cases} x^2 \sin\left(\dfrac{\pi}{x}\right)^2, & \text{if } x \neq 0, \\ 0, & \text{if } x = 0, \end{cases}$ on $[0, 1]$.

(*Hint:* Take $x_{n-k} = 1/\sqrt{2k+1}$ and argue similarly to Example 10.3.)

7. Prove that any step function has bounded variation on $[a, b]$.

8. Prove: If the functions f and g have bounded variation on $[a, b]$, then so does $f \pm g$.

9. Prove: If f has bounded variation on $[a, b]$ and $g(x) = f(x)$ for $x \neq c$, where $c \in [a, b]$, then g has bounded variation on $[a, b]$.

10.2. The Total Variation Function

Suppose that f has bounded variation on $[a, b]$ and $a \leq x \leq b$. It is easy to show that f also has bounded variation on the subinterval $[a, x]$, because any partition $\mathcal{P}$ of the subinterval $[a, x]$ can be extended to a partition $\mathcal{P}^*$ of $[a, b]$ by adjoining the one additional point $x_{n+1} = b$. Then clearly

$$\mathcal{P}(f) \leq \mathcal{P}^*(f) \leq V_a^b.$$

Thus we can define the variation function v_f by

$$v_f(x) = V_a^x f \quad \text{when} \quad a \leq x \leq b. \tag{1}$$

LEMMA 10.1. If f has bounded variation on $[a, b]$, then v_f is a nondecreasing function there.

Proof. The argument here is similar to that of the preceding paragraph except that we replace b by y. If $a \leq x \leq y \leq b$, then for any $\mathcal{P}$ on $[a, x]$ there is a corresponding $\mathcal{P}^*$ on $[a, y]$ satisfying $\mathcal{P}(f) \leq \mathcal{P}^*(f)$. Therefore the supremum of the $\mathcal{P}(f)$'s cannot exceed the supremum of the $\mathcal{P}^*(f)$'s; thus $V_a^x f \leq V_a^y$. Hence $v_f(x) \leq v_f(y)$, which shows that v_f is nondecreasing on $[a, b]$.

LEMMA 10.2. If f has bounded variation on $[a, b]$, then $v_f - f$ is nondecreasing there.

Proof. If $a \leq x \leq y \leq b$, then we can show that

$$V_a^y f - f(y) - \{V_a^x f - f(x)\} \geq V_x^y f - \{f(y) - f(x)\}. \tag{2}$$

This is sufficient to prove the assertion because the right-hand member is obviously nonnegative (the subtracted term $|f(y) - f(x)|$ is precisely $\mathcal{P}(f)$ when $\mathcal{P}$ is the trivial partition $\{x, y\}$). Therefore our task is to show that

$$V_a^y f - V_a^x f \geq V_x^y f$$

or

$$V_a^y f \geq V_a^x f + V_x^y f. \tag{3}$$

Inequality (3) is verified as follows: If $\mathcal{P}_1$ and $\mathcal{P}_2$ are partitions of $[a, x]$ and $[x, y]$, respectively, then $\mathcal{P}_1 \cup \mathcal{P}_2$ is a partition of $[a, y]$, and

$$\begin{aligned} \mathcal{P}_1(f) + \mathcal{P}_2(f) &= \Sigma_{x_k \leq x}|f(x_k) - f(x_{k-1})| + \Sigma_{x_k > x}|f(x_k) - f(x_{k-1})| \\ &= \Sigma_{x_k \in \mathcal{P}_1 \cup \mathcal{P}_2}|f(x_k) - f(x_{k-1})| \\ &= (\mathcal{P}_1 \cup \mathcal{P}_2)(f). \end{aligned}$$

Therefore it is not possible that $\sup \mathcal{P}_1(f) + \sup \mathcal{P}_2(f)$ could be greater than $\sup(\mathcal{P}_1 \cup \mathcal{P}_2)(f)$; hence $V_a^x f + V_x^y f \leq V_a^y$. We conclude that $v_f(x) - f(x) \leq v_f(y) - f(y)$, which completes the proof.

THEOREM 10.2. The function f has bounded variation if and only if there exist nondecreasing functions g and h such that $f = g - h$.

Proof. Suppose that $f = g - h$, where g and h are monotonic. Then g and h both have bounded variation, so their difference does also (see Exercise 10.2.5).

Conversely, if f has bounded variation, then Lemmas 10.1 and 10.2 show that v_f and $v_f - f$ are nondecreasing functions, and it is obvious that $f = v_f - (v_f - f)$.

Exercises 10.2

1. Prove: If f is a monotonic function on $[a, b]$, then
   $v_f(x) = |f(x) - f(a)|$ for x in $[a, b]$.

2. Given $f(x) = \begin{cases} x + 1, & \text{if } -1 \leq x < 0, \\ x, & \text{if } 0 \leq x < 1, \\ 1 - x, & \text{if } 1 \leq x \leq 2, \end{cases}$ find $v_f(x)$ for x

   in $[-1, 2]$.

3. Given $f(x) = \begin{cases} 0, & \text{if } x = 0, \\ 1, & \text{if } 0 < x \le 1, \\ x - 1, & \text{if } 1 < x \le 2, \end{cases}$ find $v_f(x)$ for x in $[0, 2]$.

4. Given $f(x) = \sin x$ on $[0, 2\pi]$, find $v_f(x)$ in $[0, 2\pi]$.

5. Prove: If g and h have bounded variation on $[a, b]$, then gh also has bounded variation on $[a, b]$.

6. Prove: If f has bounded variation on $[a, b]$, then f is (Riemann) integrable there. (*Hint:* See Theorem 7.8.)

7. Prove the converse of Inequality (3), that is, if $a \le x \le y \le b$ and f has bounded variation on $[a, b]$, then $V_a^y \le V_a^x + V_x^y$; hence $V_a^y f = V_a^x f + V_x^y$.

8. Prove: If f has bounded variation on $[a, b]$ and f is continuous at the number t in $[a, b]$, then v_f is also continuous at t.

10.3. Riemann-Stieltjes Sums and Integrals

Suppose that f and g are functions defined on the interval $[a, b]$. Let $\mathscr{P} = \{x_k\}_{k=0}^n$ be a partition of $[a, b]$ and let $\{\mu_k\}_{k=1}^n$ be a (finite) sequence such that each μ_k is in the kth subinterval determined by $\mathscr{P}$; that is, $x_{k-1} \le \mu_k \le x_k$. Then

$$\sum_{k=1}^n f(\mu_k)\{g(x_k) - g(x_{k-1})\}$$

is called a *Stieltjes sum* for f with respect to g.

DEFINITION 10.2. The function f is *Riemann-Stieltjes integrable with respect to g* on $[a, b]$ provided that there is a number J such that if $\varepsilon > 0$ then there is a positive number δ such that

$$\|\mathscr{P}\| < \delta \quad \text{implies} \quad \left| J - \sum_{k=1}^n f(\mu_k)\{g(x_k) - g(x_{k-1})\} \right| < \varepsilon$$

regardless of the choice of μ_k in $[x_{k-1}, x_k]$.

In this case, the number J is called the *Riemann-Stieltjes integral of f with respect to g*, and J is denoted by $\int_a^b f \, dg$. Also, f is called the *integrand* (function) and g is called the *integrator* (function).

Note: In case g is the identity function (that is, $g(x) = x$), the Stieltjes sums are precisely the Riemann sums for f, so the resulting limit would be the Riemann integral $\int_a^b f$. Thus we have introduced a concept that includes the Riemann integral as a special case.

EXAMPLE 10.6. Suppose f is a continuous function on $[a, b]$ and g is given by

$$g(x) = \begin{cases} 0, & \text{if } a \leq x \leq t, \\ p, & \text{if } t \leq x \leq b. \end{cases}$$

For any partition $\mathscr{P}$, the point of discontinuity t is in either one or two subintervals: either $x_{m-1} < t < x_m$, or else $t = x_m$, which gives $[x_{m-1}, t]$ and $[t, x_{m+1}]$ as subintervals corresponding to $\mathscr{P}$. All the terms of the Stieltjes sum that do not involve these subintervals are zero, so in the first case the sum reduces to $f(\mu_m)p$, and in the second case the sum reduces to

$$f(\mu_m) \cdot 0 + f(\mu_{m+1})p = f(\mu_{m+1})p. \tag{1}$$

As $\|\mathscr{P}\| \to 0$, both μ_m and μ_{m+1} approach t; therefore the continuity of f implies that the limit of the Stieltjes sums exists and equals $f(t)p$. Hence

$$\int_a^b f \, dg = f(t)p.$$

It is not difficult to see that the preceding example can be extended to the case where g is a step function with a finite number of jump points such as t in the above.

The next theorem gives the connection between the Riemann-Stieltjes integral and the Riemann integral suggested by the use of the differential notation dg:

THEOREM 10.3. If f and g are functions such that f is Riemann integrable and g' is continuous on $[a, b]$, then f is Riemann-Stieltjes integrable with respect to g, and

$$\int_a^b f \, dg = \int_a^b f g',$$

where the right-hand member denotes the Riemann integral of the product.

Proof. For any partition $\mathscr{P}$, g is differentiable on each subinterval, so by the Law of the Mean there is a number c_k in $[x_{k-1}, x_k]$ that satisfies

$$g'(c_k)(x_k - x_{k-1}) = g(x_k) - g(x_{k-1}).$$

This yields

$$\sum_{k=1}^{n} f(\mu_k)\{g(x_k) - g(x_{k-1})\} = \sum_{k=1}^{n} f(\mu_k)g'(c_k)(x_k - x_{k-1})$$

$$= \sum_{k=1}^{n} f(\mu_k)g'(\mu_k)(x_k - x_{k-1}) \qquad (2)$$

$$+ \sum_{k=1}^{n} f(\mu_k)\{g'(c_k) - g'(\mu_k)\}(x_k - x_{k-1}).$$

The first sum of the right-hand member of (2) is a Riemann sum for the product function fg', which must be Riemann integrable since both f and g' are integrable. Therefore as $\|\mathscr{P}\|$ tends to zero this sum approaches the Riemann integral $\int_a^b fg'$. Thus we can complete our proof by showing that the second sum in (2) approaches the limit zero as $\|\mathscr{P}\|$ tends to zero. To achieve this goal we refer to the *uniform* continuity of g' (Theorem 5.5): If $\varepsilon > 0$, choose $\delta > 0$ so that $|\mu - c| < \delta$ implies

$$|g'(c) - g'(\mu)| < \frac{\varepsilon}{(b - a)\text{lub}\{|f(x)|\}}.$$

We may assume $\text{lub}\{|f(x)|\} \neq 0$, for if f were identically zero the assertion would be trivial. Now $\|\mathscr{P}\| < \delta$ implies that for each k, $|c_k - \mu_k| < \delta$, so the second sum in (2) is less than or equal to

$$\sum_{k=1}^{n} |f(\mu_k)| \, |g'(c_k) - g'(\mu_k)|(x_k - x_{k-1})$$

$$< \max\{|f(\mu_k)|\} \frac{\varepsilon}{(b - a)\text{lub}\{|f(x)|\}} \sum_{k=1}^{n} (x_k - x_{k-1})$$

$$\leq \frac{\varepsilon}{(b - a)} \sum_{k=1}^{n} (x_k - x_{k-1})$$

$$= \varepsilon.$$

Hence as $\|\mathscr{P}\|$ tends to zero, the Stieltjes sums approach the limit $\int_a^b fg'$.

In the next theorem the underlying interval $[a, b]$ is not explicitly indicated, but it remains the same throughout. The proofs of the several parts are straightforward arguments based on the Stieltjes sums.

THEOREM 10.4. If each of f_1 and f_2 is Riemann-Stieltjes integrable with respect to each of g_1 and g_2, and p is a real number, then each of the integrals in the left-hand members of the following equations exists and its value is given by

(i) $\int (f_1 + f_2) dg_1 = \int f_1 \, dg_1 + \int f_2 \, dg_1,$

(ii) $\int f_1 d(g_1 + g_2) = \int f_1 \, dg_1 + \int f_1 \, dg_2,$

(iii) $\int (p f_1) dg_1 = p \int f_1 \, dg_1,$

(iv) $\int f_1 \, d(p g_1) = p \int f_1 \, dg_1.$

Proof. This is left as Exercises 10.3.7–10.3.10.

In working with the Riemann integral one becomes accustomed to the fact that step functions are very simple to integrate. Although step functions still provide simple examples in Riemann-Stieltjes integration, more care must be taken, because it happens that some very simple functions are not integrable.

EXAMPLE 10.7. If

$$f(x) = g(x) = \begin{cases} 0, & \text{if } 0 \le x \le 1, \\ 1, & \text{if } 1 < x \le 2, \end{cases}$$

then f is *not* Riemann-Stieltjes integrable with respect to g on $[0, 2]$; for if $\mathcal{P}$ is a partition of $[0, 2]$ containing 1 and having an arbitrarily small norm, suppose $x_{m-1} = 1$. Then $g(x_m) - g(x_{m-1}) = 1$, and we can choose either $\mu_m = 1$ or $\mu_m > 1$. If $\mu_m = 1$, we get

$$\sum_{k=1}^{n} f(\mu_n)\{g(x_k) - g(x_{k-1})\} = f(\mu_m)\{g(x_m) - g(x_{m-1})\} = 0,$$

and if $\mu_m > 1$, we get

$$\sum_{k=1}^{n} f(\mu_k)\{g(x_k) - g(x_{k-1})\} = 1 \cdot 1 = 1.$$

Since $\|\mathcal{P}\|$ can be as small as we wish, this shows that the Stieltjes sums cannot approach a limit as $\|\mathcal{P}\|$ tends to zero.

This example illustrates a general result: If the integrand and the integrator have a common point of discontinuity, then the integral does not exist. That is the next theorem.

THEOREM 10.5. If f and g are both discontinuous at the point c in their domain $[a, b]$, then f cannot be Riemann-Stieltjes integrable with respect to g on $[a, b]$.

Proof. We consider two cases; first suppose $\lim_c g(x)$ does not exist. Then there is an $\varepsilon_g > 0$ such that for any $\delta > 0$ we can choose numbers x_{m-1} and x_m in $[a, b]$ that satisfy

$$x_{m-1} < c < x_m, \quad x_m - x_{m-1} < \delta, \quad \text{and} \quad |g(x_m) - g(x_{m-1})| \geq \varepsilon_g.$$

Now let $\mathscr{P}$ be a partition of $[a, b]$ with $\|\mathscr{P}\| < \delta$ whose mth subinterval is $[x_{m-1}, x_m]$. The discontinuity of f at c implies the existence of an $\varepsilon_f > 0$ and μ_m', μ_m'' in $[x_{m-1}, x_m]$ such that

$$|f(\mu_m') - f(\mu_m'')| \geq \varepsilon_f.$$

If we choose $\mu_k' = \mu_k''$ for $k \neq m$, then the two Stieltjes sums differ by at least $\varepsilon_f \varepsilon_g$. Since $\|\mathscr{P}\|$ is arbitrarily small, this shows that the Stieltjes sums cannot approach a limit as $\|\mathscr{P}\|$ tends to zero.

Now consider the case in which $\lim_c g(x)$ exists but does not equal $g(c)$, say,

$$|g(c) - \lim_c g(x)| = \varepsilon_g > 0.$$

For a given $\delta > 0$, choose a partition $\mathscr{P}$ of $[a, b]$ such that $\|\mathscr{P}\| < \delta$, $x_m = c$, and either

$$|g(x_{m+1}) - g(c)| \geq \frac{\varepsilon_g}{2} \quad \text{or} \quad |g(c) - g(x_{m-1})| \geq \frac{\varepsilon_g}{2}.$$

The discontinuity of f implies the existence of an $\varepsilon_f > 0$ such that either $[x_{m-1}, x_m]$ or $[x_m, x_{m+1}]$ contains points μ' and μ'' such that $|f(\mu') - f(\mu'')| \geq \varepsilon_f$. As above, this yields two Stieltjes sums for $\mathscr{P}$ that differ by at least $\varepsilon_f \varepsilon_g/2$, so the sums do not approach a limit as $\|\mathscr{P}\|$ tends to zero.

The foregoing work is based on the definition of a partition that requires the tacit assumption that $a \leq b$. In order to consider an integral such as $\int_b^a f \, dg$, where $a < b$, we adopt the familiar convention used for the Riemann integral, namely, we *define*

$$\int_b^a f \, dg = -\int_a^b f \, dg. \tag{3}$$

Next we prove the Riemann-Stieltjes analogue of Theorem 7.3:

THEOREM 10.6. If f is Riemann-Stieltjes integrable with respect to g on the intervals $[a, b]$, $[b, c]$, and $[a, c]$, then

$$\int_a^c f\, dg = \int_a^b f\, dg + \int_b^c f\, dg. \tag{4}$$

Proof. With the defining formula (3), it is sufficient to prove that (4) holds under the assumption $a < b < c$ (see Exercise 10.3.12). Given $\varepsilon > 0$, choose $\delta > 0$ so that if $\mathcal{P}$ is a partition of either $[a, b]$, $[b, c]$, or $[a, c]$ satisfying $\|\mathcal{P}\| < \delta$, then any choice of μ_k's yields a Stieltjes sum that is within $\varepsilon/3$ of the integral:

$$\left| \int_a^b - \Sigma_a^b \right| < \frac{\varepsilon}{3}, \quad \left| \int_b^c - \Sigma_b^c \right| < \frac{\varepsilon}{3}, \quad \text{and} \quad \left| \int_a^c - \Sigma_a^c \right| < \frac{\varepsilon}{3} \tag{5}$$

whenever $\|\mathcal{P}\| < \delta$. (The meaning of the abbreviated notation in (5) is obvious.) Since $a < b < c$, the sum $\Sigma_a^b + \Sigma_b^c$ is a Stieltjes sum Σ_a^c for the interval $[a, c]$. Thus we can write

$$\left| \int_a^c - \left(\int_a^b + \int_b^c \right) \right| = \left| \left(\int_a^c - \Sigma_a^c \right) - \left(\int_a^b - \Sigma_a^b \right) - \left(\int_b^c - \Sigma_b^c \right) \right|$$

$$\leq \left| \int_a^c - \Sigma_a^c \right| + \left| \int_a^b - \Sigma_a^b \right| + \left| \int_b^c - \Sigma_b^c \right|$$

$$< \frac{\varepsilon}{3} + \frac{\varepsilon}{3} + \frac{\varepsilon}{3} \tag{6}$$

$$= \varepsilon.$$

We have shown that $\left| \int_a^c - \left(\int_a^b + \int_b^c \right) \right|$ is less than any positive number, so it must equal zero (the absolute value ensures that it is ≥ 0). Hence

$$\int_a^c = \int_a^b + \int_b^c.$$

Note that in Theorem 10.6 we assume that *all three* integrals exist and then conclude that (4) holds. In the Riemann integral analogue (Theorem 7.3) it was sufficient to assume only the existence of $\int_a^b$ and $\int_b^c$, which implies the existence of $\int_a^c$. That implication is not valid for the Riemann-Stieltjes integral, as we see in the following example:

EXAMPLE 10.8. Let

$$f(x) = \begin{cases} 0, & \text{if } 0 \leq x \leq 1, \\ 1, & \text{if } 1 < x \leq 2; \end{cases} \quad g(x) = \begin{cases} 0, & \text{if } 0 \leq x < 1, \\ 1, & \text{if } 1 \leq x \leq 2. \end{cases}$$

Then $\int_0^1 f\,dg = 0$, because f is identically zero on $[0, 1]$, and $\int_1^2 f\,dg = 0$, because g is constant on $[1, 2]$. But f and g are both discontinuous at 1, so by Theorem 10.5, $\int_0^2 f\,dg$ does not exist.

Exercises 10.3

In Exercises 1–6, evaluate the Riemann-Stieltjes integral.

1. $\int_0^4 x^2 d([x])$, where $[x]$ denotes the greatest integer function

2. $\int_0^\pi x\,d(\cos x)$

3. $\int_0^1 x^3 d(x^2)$

4. $\int_{-1}^2 \sqrt{x + 2}\,d([x])$

5. $\int_a^c f\,dg$, where f is continuous and $g(x) = \begin{cases} 0, & \text{if } a \leq x \leq b \\ 1, & \text{if } b < x \leq c \end{cases}$

6. $\int_a^c f\,dg$, where f is continuous and $g(x) = \begin{cases} 1, & \text{if } x = b \in (a, c) \\ 0, & \text{otherwise} \end{cases}$

7. Prove Theorem 10.4(i).

8. Prove Theorem 10.4(ii).

9. Prove Theorem 10.4(iii).

10. Prove Theorem 10.4(iv).

11. Suppose that f is continuous on $[a, b]$ and g is a step function such that $g(x) = \sigma_k$, if $a_{k-1} < x < a_k$, where $a = a_0 < a_1 < \cdots < a_m = b$. Prove that

    $$\int_c^b f\,dy = \sum_{k=1}^m f(a_k)(\sigma_k - \sigma_{k-1}) + f(a)[\sigma_1 - g(a)]$$
    $$+ f(b)[g(b) - \sigma_m].$$

    (Compare Exercises 10.3.5 and 10.3.6.)

12. Prove Theorem 10.6 for the case $a < c < b$. (Use Equation (3) and the case $a < b < c$ proved in Theorem 10.6.)

13. For the functions in Example 10.8, show that $\int_0^2 f\,dg$ does not exist by arguing directly from the definition as in Example 10.7.

10.4. Integration by Parts

The content of this short section is the Riemann-Stieltjes analogue of the Integration by Parts Theorem (Exercise 7.6.8).

THEOREM 10.7: INTEGRATION BY PARTS. If f is (Riemann-Stieltjes) integrable with respect to g on $[a, b]$, then g is integrable with respect to f on $[a, b]$, and

$$\int_a^b f \, dg + \int_a^b g \, df = f(b)g(b) - f(a)g(a). \tag{1}$$

Proof. Any Stieltjes sum for g with respect to f can be expanded and rewritten in the following way:

$$\sum_{k=1}^n g(\mu_k)\{f(x_k) - f(x_{k-1})\}$$

$$= g(\mu_1)\{f(x_1) - f(a)\} + g(\mu_2)\{f(x_2) - f(x_1)\}$$
$$+ \cdots + g(\mu_n)\{f(b) - f(x_{n-1})\}$$
$$= -f(a)g(\mu_1) - f(x_1)\{g(\mu_2) - g(\mu_1)\}$$
$$- \cdots - f(x_{n-1})\{g(\mu_n) - g(\mu_{n-1})\} + f(b)g(\mu_n)$$
$$= -f(a)g(a)$$
$$- \sum_{k=0}^n f(x_k)\{g(\mu_{k+1}) - g(\mu_k)\} + f(b)g(b),$$

where we have introduced the numbers $\mu_0 = a$ and $\mu_{n+1} = b_0$ in the last line. The sum in the last line is a Stieltjes sum for f with respect to g on the partition $\{\mu_k\}_{k=0}^{n+1}$ with x_k belonging to the subinterval $[\mu_k, \mu_{k+1}]$. This summation identity ensures that if the Stieltjes sums for $\int_a^b f \, dg$ converge as $\|\mathcal{P}\|$ tends to zero, then the Stieltjes sums for $\int_a^b g \, df$ also converge, and their limit values are related by Equation (1).

EXAMPLE 10.9. Evaluate $\int_{-1}^2 x \, d(|x|)$. Since $|x|$ is an awkward integrator function over an interval containing zero, we use integration by parts to write

$$\int_{-1}^2 x \, d(|x|) = [x \cdot |x|]_{-1}^2 - \int_{-1}^2 |x| d(x). \tag{2}$$

By Theorem 10.3, the right-hand integral of (2) is equal to the Riemann integral $\int_{-1}^2 |x|$, which has a value of 5/2. Using this value in (2), we get

$$\int_{-1}^2 x \, d(|x|) = 4 - (-1) - \frac{5}{2} = \frac{5}{2}.$$

10.5. Integrability of Continuous Functions

The final result of this chapter is the major existence theorem for Riemann-Stieltjes integrals. Such a theorem, which guarantees the existence of the integral for a large class of frequently encountered functions, is a part of the theory of every type of integral. For example, recall that for the Riemann integral we proved (Theorem 7.7) that every continuous function is Riemann integrable. For the Riemann-Stieltjes integral, we must make assumptions about both the integrand and integrator functions.

THEOREM 10.8. If one of the functions f and g is continuous and the other has bounded variation on $[a, b]$, then each is Riemann-Stieltjes integrable with respect to the other on $[a, b]$.

Proof. By Theorem 10.7, we may, for the sake of definiteness, assume that f is continuous and g has bounded variation. By Theorem 10.2, we know that $g = g_1 - g_2$, where g_1 and g_2 are nondecreasing functions. If we prove that $\int_a^b f \, dg_1$ exists for *any* nondecreasing g_1, then $\int_a^b f \, dg_2$ exists also. This implies the existence of $\int_a^b f \, d(g_1 - g_2)$, which is $\int_a^b f \, dg$. For the sake of simplifying notation, we write g in place of g_1.

For any partition $\mathscr{P} = \{x_k\}_0^n$, let I_k denote the kth subinterval $[x_{k-1}, x_k]$, and for $k = 1, \ldots, n$, define m_k and M_k by

$$m_k = \text{glb}\{f(x): x \in I_k\} \quad \text{and} \quad M_k = \text{lub}\{f(x): x \in I_k\}.$$

We also define the upper and lower sums $S_\mathscr{P}$ and $s_\mathscr{P}$ by

$$S_\mathscr{P} = \sum_{k=1}^n M_k\{g(x_k) - g(x_{k-1})\}$$

and

$$s_\mathscr{P} = \sum_{k=1}^n m_k\{g(x_k) - g(x_{k-1})\}.$$

If μ_k is in I_k, then $m_k \leq f(\mu_k) \leq M_k$, so it is clear that

$$s_\mathscr{P} \leq \sum_{k=1}^n f(\mu_k)\{g(x_k) - g(x_{k-1})\} \leq S_\mathscr{P}.$$

If $\mathscr{P}$ and $\mathscr{P}^*$ are two partitions of $[a, b]$, then $\mathscr{P} \cup \mathscr{P}^*$ is a refinement of both. It is easy to verify that refining a partition increases its lower sums and decreases its upper sums (Exercise 10.5.4), so it follows that $s_\mathcal{P} \leq s_{\mathcal{P} \cup \mathcal{P}*} \leq S_{\mathcal{P} \cup \mathcal{P}*} \leq S_{\mathcal{P}*}$; therefore *any* lower sum is less than or equal to *any* upper sum. It follows that the least upper bound of the set of all lower sums cannot exceed the greatest lower bound of the set of all upper sums:

$$\underline{J} \equiv \mathrm{lub}_\mathcal{P}\{s_\mathcal{P}\} \leq \mathrm{glb}_\mathcal{P}\{S_\mathcal{P}\} \equiv \overline{J}.$$

We can prove that $\int_a^b f\, dg = J$ by showing that for any positive ε there is a positive δ such that $\|\mathcal{P}\| < \delta$ implies $S_\mathcal{P} - s_\mathcal{P} < \varepsilon$. Since any Stieltjes sum using $\mathcal{P}$ is in $[s_\mathcal{P}, S_\mathcal{P}]$, and J is also in $[s_\mathcal{P}, S_\mathcal{P}]$, it follows that any such Stieltjes sum must also be within ε of J. Thus we establish that the Stieltjes sums converge to J as $\|\mathcal{P}\|$ tends to zero. To choose δ, we again rely on the uniform continuity of f: choose $\delta > 0$ so that for any x and y in $[a, b]$,

$$|x - y| < \delta \quad \text{implies} \quad |f(x) - f(y)| < \frac{\varepsilon/2}{g(b) - g(a)}.$$

(Note: We assume that $g(b) \neq g(a)$, for otherwise the monotonic function g would be constant and the conclusion trivial.) If $\|\mathcal{P}\| < \delta$, then for $k = 1, \ldots, n$,

$$M_k - m_k < \frac{\varepsilon}{g(b) - g(a)}.$$

Therefore $\|\mathcal{P}\| < \delta$ implies

$$S_\mathcal{P} - s_\mathcal{P} = \sum_{k=1}^n \{M_k - m_k\}\{g(x_k) - g(x_{k-1})\}$$

$$< \frac{\varepsilon}{g(b) - g(a)} \sum_{k=1}^n \{g(x_k) - g(x_{k-1})\}$$

$$= \varepsilon.$$

Exercises 10.5

1. Evaluate $\int_0^2 x\, d(x - [x])$.

2. Evaluate $\int_0^{\pi/2} x\, d(\sin x)$.

3. Evaluate $\int_{-2}^2 x^2\, d(|x|)$.

4. Using the notation as in the proof of Theorem 10.8, prove that if $\mathcal{P}'$ is a refinement of $\mathcal{P}$, then

$$s_{\mathcal{P}} \leq s_{\mathcal{P}'} \quad \text{and} \quad S_{\mathcal{P}'} \leq S_{\mathcal{P}}.$$

5. Suppose $\int_a^b f\, dg$ exists, f is continuous on $[a, b]$, and we replace g by a new integrator function h, where $h(x) = g(x)$ if $x \neq c$ for some c in (a, b). Prove that

$$\int_a^b f\, dh = \int_a^b f\, dg.$$

6. Let $\psi(x) = \begin{cases} 1, & \text{if } x \text{ is a rational number,} \\ 0, & \text{if } x \text{ is an irrational number;} \end{cases}$

for what integrator functions g does $\int_0^1 \psi\, dg$ exist?

7. Compare the class of Riemann integrable functions with the class of functions that have bounded variation; that is, prove or disprove each of the following assertions: If f is Riemann integrable on $[a, b]$, then f has bounded variation on $[a, b]$; if f has bounded variation on $[a, b]$, then f is Riemann integrable on $[a, b]$.

11

FUNCTION SEQUENCES

11.1. Pointwise Convergence

In the previous chapters the term *sequence* was used exclusively to mean "number sequence." We did encounter the concept of a sequence of intervals in the Nested Intervals Theorem, but that was the only situation in which we considered a sequence of objects that were not elements of $\mathbb{R}$. We now introduce the concept of a sequence of functions. Suppose that D is a nonempty subset of $\mathbb{R}$ and for each n in $\mathbb{N}$, let f_n be a function from D into $\mathbb{R}$; that is, the domain of each f_n is D, and its range is $\mathbb{R}$. Then we say that $\{f_n\}$ is a *function sequence*.

The stipulation that every function f_n has the same domain is somewhat stronger than we need. But we must have some points where all the functions in $\{f_n\}$ are defined. The set of all such points is the intersection of the domains: $\cap_{n=1}^{\infty} \mathrm{dom}\, f_n$. We could call this D and simply ignore those parts of the domains that are not in D. But it would still be necessary to know that D is not empty, so for the time being we simply stipulate that D is nonempty and each f_n is defined on D.

DEFINITION 11.1. The function sequence $\{f_n\}$ is *pointwise convergent* on D provided that for each x in D the number sequence $\{f_n(x)\}_{n=1}^{\infty}$ is convergent.

In case $\{f_n\}$ is pointwise convergent on D, the set of limit values determine a function F as follows:

$$F(x) = \lim_n f_n(x), \quad \text{for each } x \text{ in } D.$$

Thus the domain of F is also D, and F is called the *limit function* of $\{f_n\}$. This situation is also described by the phrase "$\{f_n\}$ converges pointwise to F on D," which is denoted symbolically by "$f_n \to F$ on D."

EXAMPLE 11.1. Suppose

$$f_n(x) = 3x + \frac{x^2}{n} \quad \text{on } \mathbb{R};$$

then for each x in $\mathbb{R}$,

$$\lim_n \left(3x + \frac{x^2}{n} \right) = 3x.$$

Therefore $F(x) = 3x$ on $\mathbb{R}$.

EXAMPLE 11.2. Suppose

$$f_n(x) = x^n, \quad \text{if} \quad 0 \le x \le 1.$$

Then

$$\lim_n x^n = 0, \quad \text{if} \quad 0 \le x < 1,$$

and

$$f_n(1) = 1 \quad \text{for every } n;$$

hence $\{x^n\} \to F$, where

$$F(x) = \begin{cases} 0, & \text{if } 0 \le x < 1, \\ 1, & \text{if } x = 1. \end{cases}$$

The definition of pointwise convergence can be stated precisely in basic terms as follows:

$f_n \to F$ on D provided that for each x in D and each positive number ε, there is a number N_x such that $n > N_x$ (1) implies $|f_n(x) - F(x)| < \varepsilon$.

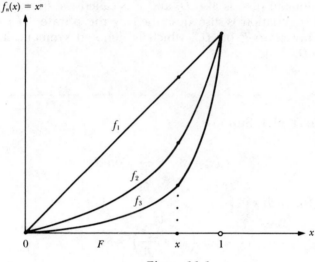

Figure 11.1

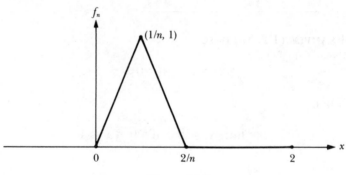

Figure 11.2

The concept of a convergent function sequence can be illustrated pictorially by drawing the graphs of the functions on a common set of axes. This is done in Figure 11.1 for the sequence $\{x^n\}$ of Example 11.2. For each x in $[0, 1)$, the number sequence $\{f_n(x)\}_{n=1}^{\infty}$ gives the vertical coordinates of the decreasing sequence of points that approaches the x-axis, which is the graph of F on $[0, 1)$. For each n, $f_n(1) = 1$, so all the graphs contain the point $(1, 1)$, and $F(1) = 1$. Of course, this example was chosen because the functions behave in a very orderly manner and are therefore easy to picture. There are cases for which it is easier to describe a function sequence by drawing the graph of f_n rather than giving its formula. Figure 11.2 displays such an example (see Exercise 11.2.6).

EXAMPLE 11.3. If f_n is the function whose graph is shown in Figure 11.2, then $f_n \to \Phi$ on $[0, 2]$, where $\Phi(x)$ is identically zero. For, if $x > 0$, then $n > 2/x$ implies that $x > 2/n$, so $f_n(x) = 0$. And since $f_n(0) = 0$ for every n, we see that $\lim_n f_n(x) = 0$ for every x in $[0, 2]$.

11.2. Uniform Convergence

In Figures 11.1 and 11.2, it is clear that for every n there are points x in the domain where $|f_n(x) - F(x)| > 1/2$. Of course, if this x is held fixed while $n \to \infty$, then this difference becomes arbitrarily small for sufficiently large n. But there will always be other points in the domain where $|f_n(y) - F(y)| > 1/2$. This phenomenon may be described by saying that $\{f_n(x)\}$ converges at different rates for different values of x. It would perhaps be convenient if $\{f_n(x)\}$ converged at a uniform rate throughout the domain.

DEFINITION 11.2. The function sequence $\{f_n\}$ is *uniformly convergent* on D to the function F provided that if $\varepsilon > 0$, then there is a number N such that for every x in D

$$n > N \quad \text{implies} \quad |f_n(x) - F(x)| < \varepsilon. \tag{2}$$

This statement looks very much like (1), the epsilon terminology definition of pointwise convergence. The only substantive difference is the location of the phrases "for each x in D" and "for every x in D." (The word "each" is replaced by "every" merely for emphasis; it is only the location in the sentence that changes its effect.) In the statement of pointwise convergence, the phrase is located *before* the guarantee that such an N exists. Thus the number N_x depends on x as well as ε, which is the reason that N_x is written with the subscript. But in the definition of uniform convergence, N is asserted to exist before any mention is made of the number x. Therefore N must be large enough to guarantee the defining implication (2) irrespective of the point in D at which the functions are evaluated.

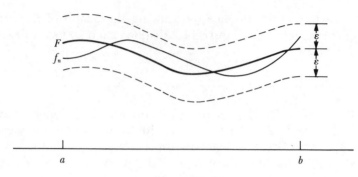

Figure 11.3

The difference between the two concepts of convergence of function sequences is illustrated by the sequences in Figures 11.1 and 11.2. Although both sequences are pointwise convergent on [0, 1], neither is uniformly convergent there. For as we have already observed, choosing n large is not enough to ensure that $|f_n(x) - F(x)| < \varepsilon$ for *every* x in [0, 1].

In Figure 11.3 there is a pictorial illustration of the definition of uniform convergence. For purposes of illustration, the functions $\{f_n\}$ and F are shown as continuous functions, and D is an interval $[a, b]$. For a given positive number ε, a band is drawn around the graph of F; the band contains all points that lie within ε units of the graph of F. The definition of uniform convergence says that when $n > N$, the graph of f_n lies entirely within that band. Thus, at most a finite number of the functions (N, to be precise) can have some portion of their graphs outside the band. If this suggests that uniform convergence is a very strong condition that is hard to achieve, then this is not overstating the case.

If $\{f_n\}$ converges uniformly on D to F, we abbreviate this by "$f_n \rightrightarrows F$ on D." Note the double arrow as contrasted with the single arrow for denoting pointwise convergence. Although pointwise convergence does not imply uniform convergence (as in Examples 11.2 and 11.3), the converse is easily verified.

PROPOSITION 11.1. If the function sequence $\{f_n\}$ converges uniformly on D to F, then $\{f_n\}$ converges pointwise on D to F.

Proof. If $\varepsilon > 0$, then the number N that is guaranteed by uniform convergence can be used as N_x to satisfy the definition of pointwise convergence.

EXAMPLE 11.4. If $f_n(x) = x^n$, then $\{f_n\}$ converges uniformly on $[0, 1/2]$ to Φ, where $\Phi(x) = 0$ for all x. This can be verified as follows: If $\varepsilon > 0$, choose N so that $(1/2)^N < \varepsilon$. Then for any x in $[0, 1/2]$, $n > N$ implies

$$|f_n(x) - F(x)| = |x^n| \le (1/2)^n < (1/2)^N < \varepsilon.$$

EXAMPLE 11.5. If $f_n(x) = x + 3/n$, then $\{f_n\} \rightrightarrows F$ on $\mathbb{R}$, where $F(x) = x$. For,

$$|f_n(x) - F(x)| = |(x + 3/n) - x| = 3/n,$$

for every x in $\mathbb{R}$ and every n in $\mathbb{N}$.

There is another way of characterizing uniform convergence that is often convenient for checking specific examples. Referring to Figure 11.3, we recall that the defining inequality $|f_n(x) - F(x)| < \varepsilon$ means that the entire graph of f_n lies within the band of width 2ε about the graph of F. This means that on D the maximum difference between $f_n(x)$ and $F(x)$ cannot exceed ε, and we write this as $\mathrm{lub}_{x \in D}|f_n(x) - F(x)| < \varepsilon$. (We use "lub" instead of "max" since we cannot assume that $f_n - F$ takes on its maximum value on D.) These observations are stated formally in the following lemma:

But test

LEMMA 11.1. The function sequence $\{f_n\}$ converges uniformly on D to F if and only if

$$\lim_n \{\mathrm{lub}_{x \in D}|f_n(x) - F(x)|\} = 0.$$

In many cases the functions f_n and F are differentiable, and then $\mathrm{lub}_{x \in D}|f_n(x) - F(x)|$ can be calculated explicitly using the methods of elementary calculus for finding maximum values. The next example illustrates this technique:

EXAMPLE 11.6. If $f_n(x) = nxe^{-nx}$ and $\Phi(x) \equiv 0$, then $f_n \rightrightarrows \Phi$ on $[1, \infty)$. Also, $f_n \to \Phi$ on $(0, \infty)$, but the convergence is *not* uniform on $(0, \infty)$. To verify these claims, we first calculate

$$f'_n(x) = ne^{-nx} - n^2xe^{-nx} = ne^{-nx}(1 - nx).$$

It is now plain that f_n has a maximum value given by

$$f_n(1/n) = n(1/n)e^{-n(1/n)} = 1/e,$$

and f_n is decreasing on $[1/n, \infty)$ because f'_n is negative there. Therefore

$$\text{lub}_{x \geq 1}|f_n(x) - F(x)| = \max_{x \geq 1}|f_n(x)| = f_n(1) = ne^{-n}.$$

In studying L'Hôpital's Rule, we showed that $\lim_{x \to \infty} xe^{-x} = 0$, so it follows that $\lim_n ne^{-n} = 0$. Hence, $f_n \rightrightarrows \Phi$ on D. The same L'Hôpital's Rule example shows that $f_n \to \Phi$ on $(0, \infty)$, but the convergence is not uniform on $(0, \infty)$, because each f_n takes on the value $1/e$ for some x in $(0, \infty)$.

Since the concept of uniform convergence is somewhat subtle, it is helpful to have a precise statement of its negation:

DEFINITION 11.3. The function sequence $\{f_n\}$ fails to converge uniformly on D to the function F provided that there exists a positive number ε such that for infinitely many n in $\mathbb{N}$ there is a number x_n in D satisfying $|f_n(x_n) - F(x_n)| \geq \varepsilon$.

In Exercises 11.2, this definition, or Lemma 11.1 can be used to show nonuniformity of convergence. For proving that uniform convergence does hold, Lemma 11.1 is frequently the best tool.

Exercises 11.2

In Exercises 1–5, show that the given function sequence converges uniformly on D but not uniformly on D'.

1. $f_n(x) = \dfrac{x}{x + n}$; $D = [0, 1]$; $D' = [0, \infty)$.

2. $f_n(x) = \sin^n x$; $D = [0, \pi/4]$; $D' = [0, \pi/2)$.

3. $f_n(x) = \dfrac{x^n}{1 + x^n}$; $D = [0, 1/2]$; $D' = [0, 1)$.

4. $f_n(x) = \dfrac{nx}{1 + n^2x^2}$; $D = [1, \infty)$; $D' = (0, \infty)$.

5. $f_n(x) = n^2x^2e^{-nx}$; $D = [1, \infty)$; $D' = (0, \infty)$.

6. Find an explicit formula for the function f_n shown in Figure 11.2.

In Exercises 1–5, the interval D of uniform convergence is not the largest possible. For example, in Exercise 4, D can be replaced by $D'' = [\delta, \infty)$, for any positive δ. In Exercises 7–11 you are asked to prove this and make similar extensions of D for Exercises 1–5.

Note that it is possible for $\{f_n\}$ to converge uniformly on $[\delta, \infty)$ for *every* $\delta > 0$ and still not converge uniformly on $(0, \infty)$.

7. Prove: If $b > 0$ and $f_n(x) = \dfrac{x}{x + n}$, then $\{f_n\}$ converges uniformly on $[0, b]$.

8. Prove: If $0 < \delta < \pi/2$ and $f_n(x) = \sin^n x$, then $\{f_n\}$ converges uniformly on $[0,(\pi/2 - \delta)]$.

9. Prove: If $0 < \delta < 1$ and $f_n(x) = \dfrac{x^n}{1 + x^n}$, then $\{f_n\}$ converges uniformly on $[0, 1 - \delta]$.

10. Prove: If $\delta > 0$ and $f_n(x) = \dfrac{nx}{1 + n^2x^2}$, then $\{f_n\}$ converges uniformly on $[\delta, \infty)$.

11. Prove: If $\delta > 0$ and $f_n(x) = n^2x^2e^{-nx}$, then $\{f_n\}$ converges uniformly on $[\delta, \infty)$.

11.3. Sequences of Continuous Functions

In Sections 11.3, 11.4, and 11.5, we consider convergent function sequences in which each f_n exhibits a certain property and ask whether the limit function F must inherit that property. For example, if each f_n is continuous on D, does it follow that F is also continuous? The answer is no, and we can cite Example 11.2 as a counterexample. Each f_n is continuous on $[0, 1]$, but the limit function is discontinuous at 1. On the other hand, if $D = [0, 1/2]$, as in Example 11.4, then the limit function is identically zero on D, and so it is certainly continuous there. As we saw in Example 11.4, the sequence $\{x^n\}$ converges uniformly on $[0, 1/2]$, whereas the convergence is not uniform on $[0, 1]$. The next theorem shows that it is the uniform convergence that makes all the difference.

THEOREM 11.1. If the function sequence $\{f_n\}$ converges uniformly on D to the function F and each f_n is continuous on D, then F is continuous on D.

Proof. Let a be an arbitrary point in D, and suppose $\varepsilon > 0$. We want to establish the continuity of F at a by showing that $|F(x) - F(a)| < \varepsilon$ when x is sufficiently close to a. First consider the inequality

$$|F(x) - F(a)| = |F(x) - f_N(x) + f_N(x) - f_N(a) + f_N(a) - F(a)|$$
$$\leq |F(x) - f_N(x)| + |f_N(x) - f_N(a)| + |f_N(a) - F(a)|, \tag{1}$$

which is obtained by adding and subtracting terms and using the Triangle Inequality. Since $f_n \rightrightarrows F$ on D, we can choose N so that for any x in D (including $x = a$), $|f_N(x) - F(x)| < \varepsilon/3$. Since f_N is continuous at a, there is a positive number δ such that when x is in D,

$$|x - a| < \delta \quad \text{implies} \quad |f_N(x) - f_N(a)| < \varepsilon/3. \tag{2}$$

Combining (1) and (2), we see that $|x - a| < \delta$ implies

$$|F(x) - F(a)| < \frac{\varepsilon}{3} + \frac{\varepsilon}{3} + \frac{\varepsilon}{3} = \varepsilon,$$

and we have proved that F is continuous at a.

Theorem 11.1 provides a handy tool for showing that the convergence of a sequence of continuous functions is not uniform: If the limit function is not continuous on D, then the convergence can not be uniform. Theorem 11.1 can also be viewed as an example of "interchanging the order of limit processes," because continuity of F at a means that

$$F(a) = \lim_a F(x) = \lim_{x \to a} \{\lim_n f_n(x)\}, \tag{3}$$

whereas the definitions of $f_n \to F$ and continuity of f_n mean that

$$F(a) = \lim_n f_n(a) = \lim_n \{\lim_{x \to a} f_n(x)\}. \tag{4}$$

When the right-hand members of (3) and (4) are compared, we see that they differ only by the order in which the limits are taken. Theorem 11.1 ensures that with uniform convergence the order of these limit processes is immaterial.

In Theorem 11.1 we saw that the assumption of uniform convergence allowed us to conclude that continuity of the limit function is inherited from the functions in the converging sequence. We now ask whether a converse implication can be proved; that is, if we are *given* that the limit function is continuous on D, can we *conclude* that the convergence must be uniform? Such a converse is not true in general, as we can see by Example 11.3. But the implication can be proved in this direction for function sequences that behave in a monotonic fashion. The resulting theorem is credited to the Italian mathematician Dini.

DEFINITION 11.4. The function sequence $\{f_n\}$ is a *nondecreasing function sequence* on D provided that for each x in D, $\{f_n(x)\}$ is a nondecreasing number sequence. Similarly, $\{f_n\}$ is a *nonincreasing function sequence* on D if, for each x in D, $\{f_n(x)\}$ is a nonincreasing number sequence. If $\{f_n\}$ is either nondecreasing or nonincreasing, it is said to be a *monotonic function sequence*.

A nonincreasing function sequence is shown in Figure 11.1 (see Example 11.2), and it should be noted that for each n the *entire* graph of f_{n+1} lies below or on the graph of f_n. If $\{f_n(x')\}$ were to increase for one value x' while $\{f_n(x'')\}$ were to decrease for another value x'', $\{f_n\}$ would *not* be a monotonic function sequence.

[margin: Conditions (on [a, b])]

THEOREM 11.2: DINI'S THEOREM. Suppose that $\{f_n\}$ is a monotonic function sequence that converges pointwise on $[a, b]$ to the function F. If F and every f_n are continuous on $[a, b]$, then $\{f_n\}$ converges uniformly on $[a, b]$ to F.

[margin: (1) $f_n \to F$ on D; (2) f_m is monotone (strictly increasing or "decreasing" or nondecreasing or nonincreasing) (or after some finite # of terms); (3) F has to be continuous; (4) f_n is cont $\forall n$ $\Rightarrow$ $f_n \rightrightarrows F$]

[margin: $f_n(x) = n x e^{-nx}$]

Proof. Case (i): Assume that $\{f_n\}$ is nonincreasing and $F = \Phi$ (identically zero). Suppose the convergence is not uniform on $[a, b]$. Then there exists a positive number ε with the property that for every N there is an integer $n(N) \geq N$ and a number $x_{n(N)}$ in $[a, b]$ such that $f_{n(N)}(x_{n(N)}) \geq \varepsilon$. By the Bolzano-Weierstrass Theorem, the bounded sequence $\{x_{n(N)}\}$ has a convergent subsequence, and since $a \leq x_{n(N)} \leq b$, its limit must also be in the closed interval $[a, b]$, say, $\lim_k x_{n(k)} = c$, where $a \leq c \leq b$. Consider an arbitrary N; if $n(k) \geq N$, then the monotonicity of $\{f_n\}$ ensures that

$$f_N(x_{n(k)}) \geq f_{n(k)}(x_{n(k)}) \geq \varepsilon. \qquad (5)$$

[margin: these insure $\Rightarrow$ $f_n \rightrightarrows F$]

Now let k tend to ∞ so that $x_{n(k)}$ approaches c. The continuity of f_N and the SCC imply that

$$\lim_k f_N(x_{n(k)}) = f_N(c). \qquad (6)$$

But according to (5) the function values $f_N(x_{n(k)})$ are not less than ε whenever $n(k) \geq N$. Hence (6) implies that $f_N(c) \geq \varepsilon$. This holds for an *arbitrary* N, so we must conclude that $\lim_N f_N(c) \geq \varepsilon$, that is, $F \neq \Phi$. This contradiction completes the proof of Case (i).

Case (ii): Assume that $\{f_n\}$ converges to an arbitrary function F; that is, $\{|f_n - F|\}$ is a nonincreasing function sequence converging (pointwise) to Φ. Therefore by Case (i), $|f_n - F| \rightrightarrows \Phi$, and it follows from Lemma 11.1 that $f_n \rightrightarrows F$. Thus the theorem is proved.

Note that in Dini's Theorem we assume the domain to be a closed interval $[a, b]$, and this plays a crucial role in the proof. The boundedness of $[a, b]$ is needed in order to apply the Bolzano-Weierstrass Theorem, and we also need to know that $\{x_{n(k)}\}$ being in D guarantees that $\lim x_{n(k)}$ is in D. For the latter part, we can assume less than $D = [a, b]$; for example, D could be the union of finitely many closed intervals. But this seems unduly awkward, and even this is not the most general type of domain on which Dini's Theorem holds. Since our purpose is not to study the structure of subsets of $\mathbb{R}$, we can let well enough alone and be content with the case in which D is a closed interval.

11.4. Sequences of Integrable Functions

The question of interchanging order of limits can be raised for other types of limit processes. The next one we consider is the integral. We assume that each f_n is Riemann integrable and $f_n \to F$ on $[a, b]$. Then we ask whether it follows that F is integrable and, if so, are the two values the same:

$$\int_a^b F = \int_a^b \lim_n f_n \overset{?}{=} \lim_n \left\{ \int_a^b f_n \right\}.$$

In general, the answer is no, and we can give a simple counterexample by modifying the sequence shown in Figure 11.2 (Example 11.3):

EXAMPLE 11.7. In Figure 11.4, the graph of f_n shows a maximum height of $f_n(1/n) = n$, but we still have $f_n \to 0$ on $[0, 2]$ for the same reasons as in Example 11.3. Also, each f_n is continuous, and by calculating the area of the triangle under the graph, we see that for each n, $\int_0^2 f_n = 1$. But $F(x) = 0$ for every x, so $\int_0^2 F \neq \lim_n \left\{ \int_0^2 f_n \right\}$.

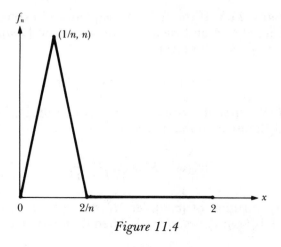

Figure 11.4

It could be observed that, in the preceding example, at least the limit function F was integrable, even though the value of the integral did not equal the limit of the integral values. This leaves open the conjecture that the limit of integrable functions is always an integrable function. But we disprove that conjecture with the following counterexample:

EXAMPLE 11.8. Let $\{r_k\}$ be a sequence such that each rational number in $[0, 1]$ appears exactly once as a term r_k. (See Appendix B to verify the existence of such a sequence.) Now define

$$f_n(x) = \begin{cases} 1, & \text{if } x = r_k \text{ for some } k \le n, \\ 0, & \text{otherwise.} \end{cases}$$

Then each rational number in $[0, 1]$ is eventually in the set for which $f_n(x) = 1$, and it remains so for all higher n. Hence $f_n \to \psi$, where $\psi(x) = 1$ if $x \in \mathbb{Q}$ and $\psi(x) = 0$ if x is an irrational number. The function ψ is a familiar example (Example 7.2) of a nonintegrable function.

As in Theorem 11.1, we now prove that the additional hypothesis of uniform convergence guarantees that integrability is inherited by the limit function.

Theorem 11.3. If the function sequence $\{f_n\}$ converges uniformly on $[a, b]$ to F and each f_n is integrable on $[a, b]$, then F is integrable on $[a, b]$. Moreover,

$$\int_a^b F = \int_a^b \lim_n f_n = \lim_n \left\{ \int_a^b f_n \right\}. \tag{1}$$

Proof. We first show that F is integrable on $[a, b]$. Suppose $\varepsilon > 0$, and choose N so that if x is in $[a, b]$, then

$$|f_N(x) - F(x)| < \frac{\varepsilon/3}{b - a}. \tag{2}$$

If $\mathscr{P}$ is any partition of $[a, b]$, let $M_k^N = \text{lub}\{f_N(x): x_{k-1} \leq x \leq x_k\}$ and $M_k = \text{lub}\{F(x): x_{k-1} \leq x \leq x_k\}$. Then (2) guarantees that

$$|M_k - M_k^N| \leq \frac{\varepsilon/3}{b - a},$$

so

$$\begin{aligned}
|U(f_N, \mathscr{P}) - U(F, \mathscr{P})| &= \left| \sum_{k=1}^n (M_k - M_k^N)(x_k - x_{k-1}) \right| \\
&\leq \sum_{k=1}^n |M_k - M_k|(x_k - x_{k-1}) \\
&\leq \frac{\varepsilon/3}{b - a} \sum_{k=1}^n (x_k - x_{k-1}) \\
&= \frac{\varepsilon}{3}.
\end{aligned} \tag{3}$$

Similarly, we have

$$|L(f_N, \mathscr{P}) - L(F, \mathscr{P})| \leq \frac{\varepsilon}{3}. \tag{4}$$

Since f_N is integrable on $[a, b]$, there is a partition $\mathscr{P}'$ such that

$$|U(f_N, \mathscr{P}') - L(f_N, \mathscr{P}')| < \frac{\varepsilon}{3}. \tag{5}$$

Combining (3), (4), and (5), we see that

$$\begin{aligned}
|U(F, \mathscr{P}') - L(F, \mathscr{P}')| &\leq |U(F, \mathscr{P}') - U(f_N, \mathscr{P}')| \\
&\quad + |U(f_N, \mathscr{P}') - L(f_N, \mathscr{P}')|
\end{aligned}$$

$$+ |L(f_N, \mathcal{P}') - L(F, \mathcal{P}')|$$

$$< \frac{\varepsilon}{3} + \frac{\varepsilon}{3} + \frac{\varepsilon}{3}$$

$$= \varepsilon.$$

Hence, by the Darboux Integrability Criterion (Lemma 7.3), F is integrable on $[a, b]$.

To prove that $\int_a^b F = \lim_n \{\int_a^b f_n\}$, we again let ε be an arbitrary positive number and choose N' so that $n > N'$ implies

$$\text{lub}_{x \in [a, b]} |f_n(x) - F(x)| < \frac{\varepsilon}{b - a}.$$

Then $n > N'$ implies

$$\left| \int_a^b f_n - \int_a^b F \right| = \left| \int_a^b (f_n - F) \right|$$

$$\leq \int_a^b |f_n - F|$$

$$< \int_a^b \frac{\varepsilon}{b - a}$$

$$\leq \frac{\varepsilon}{b - a} (b - a)$$

$$= \varepsilon.$$

Hence $\lim_n \left\{ \int_a^b f_n \right\} = \int_a^b F$.

11.5. Sequences of Differentiable Functions

The next property that we wish to investigate with respect to function sequences is differentiability. In Example 11.2 we saw a sequence $\{x^n\}$ that converges pointwise on $[0, 1]$ to a limit function that is not differentiable at $x = 1$. Of course, the convergence is not uniform on $[0, 1]$, and if we restrict the domain to $[0, 1/2]$, then the limit function is identically zero, which is certainly a differentiable function. The convergence of $\{x^n\}$ is uniform on $[0, 1/2]$, and by now we are conditioned to expect that uniform convergence guarantees that the limit function inherits the desired property from the sequence of functions. But that is not the case with differentiability, for in the next example we see a sequence of differentiable functions that converges uniformly on $\mathbb{R}$ to a non-differentiable limit function.

study

EXAMPLE 11.9.

> If $f_n(x) = \sqrt{x^2 + (1/n^2)}$, then $f_n \rightrightarrows |x|$ on $\mathbb{R}$.

We know that $|x|$ is not differentiable at zero, and before verifying that $f_n \rightrightarrows |x|$ on $\mathbb{R}$, we note that the graph of f_n is the upper wing of the hyperbola given by the equation

$$\frac{y^2}{1/n^2} - \frac{x^2}{1/n^2} = 1.$$

(See Figure 11.5.) To establish the claim of uniform convergence, we do some algebraic manipulations:

differentiable func. *converges to* *non-differentiable function*

$$\begin{aligned}
|f_n(x) - |x|\,| &= \sqrt{x^2 + (1/n^2)} - \sqrt{x^2} \\
&= \frac{(\sqrt{x^2 + (1/n^2)} - \sqrt{x^2})(\sqrt{x^2 + (1/n^2)} + \sqrt{x^2})}{\sqrt{x^2 + (1/n^2)} + \sqrt{x^2}} \\
&= \frac{x^2 + (1/n^2) - x^2}{\sqrt{x^2 + (1/n^2)} + \sqrt{x^2}} \\
&\leq \frac{1/n^2}{\sqrt{1/n^2}} \\
&= 1/n.
\end{aligned}$$

not differentiable at $x = 0$

It is now plain that $\lim_n \{\mathrm{lub}_{x \in \mathbb{R}} |f_n(x) - |x|\} = 0$, so by Lemma 11.1, $f_n \rightrightarrows |x|$.

$\lim_{n \to 0} \sqrt{n^2 - \frac{1}{n^2}} = \frac{1}{n}$

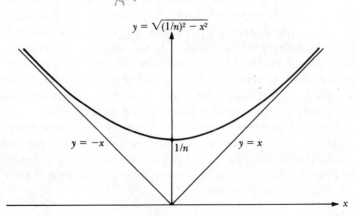

$y = \sqrt{(1/n)^2 - x^2}$

$y = -x$ $1/n$ $y = x$

Figure 11.5

Although the preceding example may have shattered our faith in our intuition, it is not the end of our troubles. We would also like to know that if a sequence of differentiable functions does converge to a differentiable function, then the limit of their derivatives is equal to the derivative of the limit function; that is,

$$\lim_n f_n' \overset{?}{=} (\lim_n f_n)'. \tag{1}$$

In the next example we see that this is not guaranteed even when differentiable functions converge *uniformly* to a differentiable function.

EXAMPLE 11.10. If $f_n(x) = x/(1 + nx^2)$, then $f_n \rightrightarrows \Phi$ on $[0, \infty)$, where $\Phi(x) \equiv 0$. To verify this assertion we first find the derivative

$$f_n'(x) = \frac{(1 + nx^2) - x \cdot 2nx}{(1 + nx^2)^2} = \frac{1 - nx^2}{(1 + nx^2)^2}.$$

Therefore f_n has a maximum value at $x = 1/\sqrt{n}$, and

$$f_n(1/\sqrt{n}) = \frac{1}{2\sqrt{n}}.$$

Thus it is clear that $\text{lub}_{x \geq 0} |f_n(x)|$ tends to 0 as $n \to \infty$, whence $f_n \rightrightarrows \Phi$ on $[0, \infty)$. Of course, $\Phi' = \Phi$, so Φ is differentiable. But now consider $\lim_n f_n'$: If $x \neq 0$, then

$$\lim_n f_n'(x) = \lim_n \frac{1 - nx^2}{(1 + nx^2)^2} = \lim_n \frac{1/n^2 - x^2/n}{1 + 2x^2/n + x^2} = 0;$$

if $x = 0$, then $\lim_n f_n' = 1$. Hence

$$\lim_n f_n'(x) = \begin{cases} 0, & \text{if } x > 0 \\ 1, & \text{if } x = 0 \end{cases} \neq \Phi(x).$$

By now we may wonder whether differentiability of the limit function is related in any way to uniform convergence, so it is time to state the theorem that gives such a relationship. The key is to assume uniform convergence of the derivatives $\{f_n'\}$ rather than $\{f_n\}$. The theorem that we prove here is not the most general, but to weaken the hypothesis we would need a much more sophisticated proof.

THEOREM 11.4. Suppose that $\{f_n\}$ is a function sequence such that each f_n has a continuous derivative on $[a, b]$, $\{f_n(c)\}$ converges for some c in $[a, b]$, and $\{f_n'\}$ converges uniformly on $[a, b]$ to some function g; then $\{f_n\}$ converges uniformly on $[a, b]$ to some function F, where F is differentiable and $F' = g$.

Proof. Since f_n' is continuous and therefore integrable, the Fundamental Theorem of Calculus ensures that

$$f_n(x) - f_n(c) = \int_c^x f_n'. \tag{2}$$

By Theorem 11.3, $f_n' \rightrightarrows g$ implies that

$$\lim_n\{f_n(x) - f_n(c)\} = \lim_n\left\{\int_c^x f_n'\right\} = \int_c^x \lim_n f_n' = \int_c^x g.$$

By hypothesis, $\{f_n(c)\}$ is convergent, say, $\lim_n f_n(c) = F(c)$; so it follows that $\{f_n(x)\}$ is also convergent. Let us define

$$F(x) = \lim_n f_n(x) \quad \text{for each } x \text{ in } [a, b].$$

Then (2) implies that $F(x) - F(c) = \int_c^x g$. By Theorem 11.1, $f_n' \rightrightarrows g$ implies that g is continuous, so by Theorem 7.13, F is a primitive of g; that is, $F' = g$ on $[a, b]$. It remains to be shown that $f_n \rightrightarrows F$ on $[a, b]$. To prove this we write

$$|f_n(x) - F(x)| = \left|\int_c^x f_n' + f_n(c) - \left\{\int_c^x F' + F(c)\right\}\right|$$

$$= \left|\int_c^x (f_n' - F') + f_n(c) - F(c)\right|$$

$$\leq \left|\int_c^x |f_n' - g|\right| + |f_n(c) - F(c)|$$

$$\leq \{\text{lub}_{c \leq t \leq x}|f_n'(t) - g(t)|\}|x - c| + |f_n(c) - F(c)|.$$

Since $f_n' \rightrightarrows g$ on $[a, b]$, if $\varepsilon > 0$ we can choose N so that $n > N$ implies

$$\text{lub}_{a \leq t \leq b}|f_n(t) - g(t)| < \frac{\varepsilon/2}{b - a},$$

and

$$|f_n(c) - F(c)| < \frac{\varepsilon}{2}.$$

Thus $n > N$ implies $\text{lub}_{x \in [a,b]}|f_n(x) - F(x)| < \varepsilon$, so by Lemma 11.1, $f_n \rightrightarrows F$.

Exercises 11.5

In Exercises 1–3, use Theorem 11.1 to prove that the given function sequence does not converge uniformly on the domain D.

1. $f_n(x) = 1 - x^n$; $D = [0, 1]$.

2. $f_n(x) = e^{-nx}$; $D = [0, \infty)$.

3. $f_n(x) = \tan^n x$; $D = [0, \pi/4]$.

In Exercises 4–6, use Theorem 11.3 to prove that $\{f_n\}$ does not converge uniformly on D. (It may help to sketch the graph of f_n.)

4. $f_n(x) = \begin{cases} 0, & \text{if } 0 \le x < 1/n, \\ 1/x, & \text{if } 1/n \le x \le 1; \end{cases}$ $D = [0, 1]$.

5. $f_n(x) = \begin{cases} n, & \text{if } 0 \le x < 1/n, \\ 0, & \text{if } 1/n \le x \le 1; \end{cases}$ $D = [0, 1]$.

6. $f_n(x) = \begin{cases} n(1 - nx), & \text{if } 0 \le x < 1/n, \\ 0, & \text{if } 1/n \le x \le 1; \end{cases}$ $D = [0, 1]$.

7. Prove: If $f_n \rightrightarrows F$ on D and $f_n \rightrightarrows F$ on D', then $f_n \rightrightarrows F$ on $D \cup D'$.

8. Prove: If $f_n \to F$ and each f_n is nondecreasing, then F is nondecreasing.

9. Suppose that $f_n \to F$ on D, $\{f_n(x)\}$ is nondecreasing for each x in D, and $\{f_n\}$ has a subsequence that converges uniformly on D. Prove that $f_n \rightrightarrows F$ on D.

10. Suppose that $f_n \rightrightarrows F$ on D, and each f_n is bounded on D, say, $\mathrm{lub}_{x \in D}|f_n(x)| \le B_n$. Prove that there is a uniform bound B such that for every n in $\mathbb{N}$ and every x in D, $|f_n(x)| \le B$.

11. Prove the Cauchy Criterion for Uniform Convergence: The function sequence $\{f_n\}$ converges uniformly on D if and only if for every positive number ε there is a number N such that for every x in D

$$m > n > N \quad \text{implies} \quad |f_m(x) - f_n(x)| < \varepsilon.$$

11.6. The Weierstrass Approximation Theorem

In this section we present one of the most useful theorems in applied analysis, the Weierstrass Approximation Theorem. In a sense, it concerns the uniform convergence of sequences of polynomials (see Corollary 11.5). Since polynomials are very simple to work with, it is often desirable to replace a more general function that is under consideration by a polynomial function that is a close approximation to it. Thus it becomes important to know when such a polynomial can be found that is arbitrarily close to a given function. The property that ensures that a function can be so approximated is *continuity*.

THEOREM 11.5: WEIERSTRASS APPROXIMATION THEOREM. If the function f is continuous on $[a, b]$ and $\varepsilon > 0$, then there is a polynomial P such that for every x in $[a, b]$,

$$|P(x) - f(x)| < \varepsilon.$$

Proof. We first show that it is sufficient to prove the assertion for the special case in which $[a, b]$ is the unit interval $[0, 1]$. Suppose the statement of the theorem is true on $[0, 1]$, and let f be continuous on $[a, b]$. Then the composition function g, given by $g(x) = f(a + \{b - a\}x)$, is continuous on $[0, 1]$. Therefore there exists a polynomial P such that

$$|g(u) - P(u)| < \varepsilon \quad \text{for every } u \text{ in } [0, 1]. \tag{1}$$

For any x in $[a, b]$, we make the substitutions

$$u = \frac{x - a}{b - a} \quad \text{and} \quad g(u) = f\left(a + \{b - a\} \frac{x - a}{b - a}\right) = f(x),$$

which yields

$$\left| f(x) - P\left(\frac{x - a}{b - a}\right) \right| < \varepsilon. \tag{2}$$

Since $P((x - a)/(b - a))$ is a polynomial in x, the assertion holds for the interval $[a, b]$.

To prove the assertion in case $[a, b] = [0, 1]$, we use an argument developed by Bernstein. This employs a particular class of polynomials known as *Bernstein polynomials*. The nth Bernstein polynomial for the function f is given by

$$B_n(x) = \sum_{k=0}^{n} \binom{n}{k} x^k (1 - x)^{n-k} f(k/n), \tag{3}$$

where $\binom{n}{k} = n!/(k!(n - k)!)$. Note that if the factor $f(k/n)$ were deleted, the sum in (3) would become the binomial expansion of

$\{x + (1 - x)\}^n = 1$. In order to use this observation, we consider the function ϕ given by

$$\phi(x) - \sum_{k=0}^{n}\binom{n}{k}x^k y^{n-k} = (x + y)^n, \tag{4}$$

where y is an arbitrary (but fixed) number in $[0, 1]$. Then

$$\phi'(x) = \sum_{k=0}^{n}\binom{n}{k}kx^{k-1}y^{n-k} = n(x + y)^{n-1},$$

so if we multiply through by x/n, we get

$$\sum_{k=0}^{n}\frac{k}{n}\binom{n}{k}x^k y^{n-k} = x(x + y)^{n-1}. \tag{5}$$

Differentiating both sides of (5), we get

$$\sum_{k=0}^{n}\frac{k^2}{n}\binom{n}{k}x^{k-1}y^{n-k} = (x + y)^{n-1} + x(n - 1)(x + y)^{n-2},$$

so after again multiplying through by x/n, we have

$$\sum_{k=0}^{n}\frac{k^2}{n^2}\binom{n}{k}x^k y^{n-k} = \frac{x}{n}(x + y)^{n-1} + \left(1 - \frac{1}{n}\right)x^2(x + y)^{n-2}. \tag{6}$$

Equations (4), (5), and (6) are valid for any y in $[0, 1]$, so in particular, we can replace y by $1 - x$ to get

$$\sum_{k=0}^{n}\binom{n}{k}x^k(1 - x)^{n-k} = 1, \tag{7a}$$

$$\sum_{k=0}^{n}\frac{k}{n}\binom{n}{k}x^k(1 - x)^{n-k} = x, \quad \text{and} \tag{7b}$$

$$\sum_{k=0}^{n}\frac{k^2}{n^2}\binom{n}{k}x^k(1 - x)^{n-k} = \frac{x}{n} + \left(1 - \frac{1}{n}\right)x^2. \tag{7c}$$

If we expand the binomial $[(k/n - x)]^2$ and use (7a), (7b), and (7c), we get

$$\sum_{k=0}^{n}\left(\frac{k}{n} - x\right)^2\binom{n}{k}x^k(1 - x)^{n-k}$$

$$= \sum_{k=0}^{n}\left(\frac{k^2}{n^2} - \frac{2kx}{n} + x^2\right)\binom{n}{k}x^k(1 - x)^{n-k}$$

$$= \frac{x}{n} + \left(1 - \frac{1}{n}\right)x^2 - 2x \cdot x + 1 \cdot x^2 \tag{8}$$

$1 - 2t + t^2) f(0) + (2t - 2t^2) f(\frac{1}{2}) + t^2 f(1)$

$[f(0) - 2f(\frac{1}{2}) + f(1)] t^2 + [-2f(0) + 2f(\frac{1}{2})] t + f(0)$

$$= \frac{x}{n} - \frac{x^2}{n}$$

$$= \frac{x(1-x)}{n}.$$

Now suppose that f is continuous on $[0, 1]$. Then f is uniformly continuous there, so if $\varepsilon > 0$, there is a positive number δ such that if x and y are in $[0, 1]$, then

$$|x - y| < \delta \quad \text{implies} \quad |f(x) - f(y)| < \varepsilon/2.$$

Define $M = 1 + \max_{x \in [0, 1]}|f(x)|$, and choose N so that

$$\frac{1}{N^{1/4}} < \delta \quad \text{and} \quad \frac{1}{\sqrt{N}} < \frac{\varepsilon}{4M}.$$

If we multiply (7a) by $f(x)$ and subtract (3) from the result, we get

$$f(x) - B_n(x) = \sum_{k=0}^{n}\{f(x) - f(k/n)\}\binom{n}{k}x^k(1 - x)^{n-k}. \tag{9}$$

The right-hand member of (9) is now separated into two sums Σ' and Σ'', where

$$\Sigma' \text{ is the sum over all } k \text{ such that } |k/n - x| < 1/n^{1/4},$$

and

$$\Sigma'' \text{ is the sum of the remaining terms.}$$

First consider Σ': If $n \geq N$ and $|(k/n) - x| < 1/n^{1/4}$, then $|(k/n) - x| < \delta$, so $|f(x) - f(k/n)| < \varepsilon/2$. Therefore

$$\begin{aligned}
|\Sigma'| &\leq \Sigma'|f(x) - f(k/n)|\binom{n}{k}x^k(1 - x)^{n-k} \\
&< (\varepsilon/2)\sum_{k=0}^{n}\binom{n}{k}x^k(1 - x)^{n-k} \\
&= \frac{\varepsilon}{2}.
\end{aligned} \tag{10}$$

Next we consider Σ'': If $|(k/n) - x| \geq 1/n^{1/4}$, then $1/n^{1/2} \leq |(k/n) - x|^2$, or

$$n^{3/2} \leq n^2|(k/n) - x|^2 = \{n[(k/n) - x]\}^2 = \{k - nx\}^2.$$

Therefore

$$
\begin{aligned}
|\Sigma''| &= |\Sigma'' \{f(x) - f(k/n)\}\binom{n}{k}x^k(1-x)^{n-k}| \\
&\leq \Sigma'' \{|f(x)| + |f(k/n)|\}\binom{n}{k}x^k(1-x)^{n-k} \\
&\leq 2M\Sigma'' \binom{n}{k}x^k(1-x)^{n-k} \\
&\leq 2M\Sigma'' \frac{(k-nx)^2}{n^{3/2}}\binom{n}{k}x^k(1-x)^{n-k} \\
&\leq \frac{2M}{n^{3/2}}\Sigma_{k=0}^{n}(k-nx)^2\binom{n}{k}x^k(1-x)^{n-k}.
\end{aligned}
\tag{11}
$$

Substituting (8) into (11), we have

$$
\begin{aligned}
|\Sigma''| &\leq \frac{2M}{n^{3/2}}\Sigma_{k=0}^{n}n^2\left(\frac{k}{n}-x\right)^2\binom{n}{k}x^k(1-x)^{n-k} \\
&< \frac{2M}{n^{3/2}}\frac{n^2x(1-x)}{n} \\
&= \frac{2M}{\sqrt{n}}.
\end{aligned}
$$

If $n \geq N > (4M/\varepsilon)^2$, then we have $|\Sigma''| < \varepsilon/2$. Hence, if $n \geq N$ and x is any number in $[0, 1]$, then

$$
|f(x) - B_n(x)| < \varepsilon. \tag{12}
$$

(Note that the choice of N did not depend on x, so the approximating inequality holds uniformly throughout $[0, 1]$.)

The Weierstrass Approximation Theorem can be stated in terms of uniformly convergent function sequences, because the proof that we have given shows that $B_n \rightrightarrows f$ on $[0, 1]$. We give the formal statement of this observation in the following corollary:

COROLLARY 11.5. If the function f is continuous on $[a, b]$, then there exists a sequence of polynomials $\{P_n\}$ that converges uniformly on $[a, b]$ to f.

Exercises 11.6

1. Find the third Bernstein polynomial B_3 for $f(x) = \sin(\pi x/2)$.

2. Find the fourth Bernstein polynomial B_4 for $f(x) = \sqrt{x}$.

3.) Find a polynomial $P(x)$ such that

$$\text{lub}_{x \in [-1, 1]} \left| P(x) - |x| \right| < 1/5.$$

4.) By considering the function $f(x) = 1/x$ on $(0, 1)$, show that the Weierstrass Approximation Theorem would not hold if $[a, b]$ were replaced by (a, b).

11.7. Function Series

In the final section of this chapter we shift our emphasis from the terms of a function sequence to the corresponding sequence of partial sums. It is helpful to restate the results of Sections 11.1–11.5, this time as series statements. For example, if $\{f_k\}$ is a function sequence and for each x in some domain D the number sequence $\{\sum_{k=0}^{n} f_k(x)\}_{n=0}^{\infty}$ is convergent, then the function series $\sum f_k$ is said to be (pointwise) convergent on D to the function $\sum_{k=0}^{\infty} f_k$. Similarly, if the sequence $\{\sum_{k=0}^{n} f_k(x)\}_{n=0}^{\infty}$ is uniformly convergent on D, then we say that the series $\sum f_k$ is uniformly convergent on D. Since $\sum_{k=0}^{\infty} f_k$ is the limit of the sequence $\{\sum_{k=0}^{n} f_k\}_{n=0}^{\infty}$, it is clear that

$$\sum_{k=0}^{\infty} f_k - \sum_{k=0}^{n} f_k = \sum_{k>n} f_k.$$

Therefore we can restate Lemma 11.1 in the following way:

LEMMA 11.2. The function series $\sum f_k$ is uniformly convergent on D if and only if

$$\lim_n \{\sup_{x \in D} |\sum_{k>n} f_k(x)|\} = 0.$$

Theorems 11.1, 11.3, and 11.4 on hereditary properties each has an analogue for function series. These are stated without proof, because if each f_k is continuous, integrable, or differentiable, respectively, then so is each partial sum $\sum_{k=0}^{n} f_k$. Thus the three hereditary theorems can be applied to the sequences of partial sums, and the conclusions follow immediately.

THEOREM 11.6. If each f_k is a continuous function on the domain D and $\sum f_k$ is uniformly convergent on D, then $\sum_{k=0}^{\infty} f_k$ is continuous on D.

THEOREM 11.7. If each f_k is integrable on $[a, b]$ and $\sum f_k$ is uniformly convergent on $[a, b]$, then $\sum_{k=0}^{\infty} f_k$ is integrable on $[a, b]$ and

$$\int_a^b \left(\sum_{k=0}^{\infty} f_k \right) = \sum_{k=0}^{\infty} \int_a^b f_k.$$

THEOREM 11.8. Suppose that each f_k has a continuous derivative on $[a, b]$, $\Sigma f_k(c)$ is convergent for some c in $[a, b]$, and $\Sigma f_k'$ is uniformly convergent on $[a, b]$. Then Σf_k is uniformly convergent on $[a, b]$ to the differentiable function $\Sigma_{k=0}^{\infty} f_k$, and

$$\left(\Sigma_{k=0}^{\infty} f_k\right)' = \Sigma_{k=0}^{\infty} f_k'.$$

The next result is customarily stated only in the series form, and it provides an extremely useful means of proving that a function series is uniformly convergent.

THEOREM 11.9: WEIERSTRASS M-TEST. Suppose that Σf_k is a function series and ΣM_k is a convergent positive number series such that for each k, $\text{lub}_{x \in D} |f_k(x)| \leq M_k$; then Σf_k is uniformly convergent on D.

Proof. Suppose $\varepsilon > 0$. Since ΣM_k is convergent, there exists a number N such that $n > N$ implies

$$\left| \Sigma_{k=0}^{\infty} M_k - \Sigma_{k=0}^{n} M_k \right| = \left| \Sigma_{k>n} M_k \right| < \varepsilon.$$

But

$$\text{lub}_{x \in D} \left| \Sigma_{k>n} f_k(x) \right| \leq \text{lub}_{x \in D} \Sigma_{k>n} |f_k(x)|$$
$$\leq \Sigma_{k>n} \text{lub}_{x \in D} |f_k(x)|$$
$$\leq \Sigma_{k>n} M_k,$$

so $n > N$ implies $\text{lub}_{x \in D} \left| \Sigma_{k>n} f_k(x) \right| < \varepsilon$. Hence

$$\lim_n \{ \text{lub}_{x \in D} | \Sigma_{k>n} f_k(x) | \} = 0,$$

so Σf_k is uniformly convergent on D.

The preceding argument proves a stronger statement than that given in Theorem 11.9. Upon closer examination we see that we have proved that the series $\Sigma |f_k|$ is uniformly convergent on D. This implies not only that Σf_k is uniformly convergent—which is the conclusion of Theorem 11.9—but also that $\Sigma |f_k|$ is pointwise convergent on D. This raises a natural question: Is it possible for a function series Σf_k to converge uniformly on some D and also absolutely at each point of D while the absolute series $\Sigma |f_k|$ is *not* uniformly convergent? The answer is yes, and the next example illustrates this fact. It also shows that a function series can be uniformly convergent even though the M-Test cannot be applied to it.

EXAMPLE 11.11. If $f_{2k}(x) = -f_{2k+1}(x) = x^k - x^{k+1}$ for $k \geq 0$, then Σf_k converges uniformly on $[0, 1]$ to the identically zero function Φ. Also, $\Sigma |f_k(x)|$ converges for each x in $[0, 1]$, but $\Sigma |f_k|$ does not converge uniformly on $[0, 1]$. First, let us display this series:

$$\Sigma f_k(x) = (1 - x) - (1 - x) + (x - x^2) - (x - x^2)$$
$$+ \cdots + (x^k - x^{k+1}) - (x^k - x^{k+1}) + \cdots .$$

To establish the uniform convergence of Σf_k, we note that for every n,

$$\Sigma_{k=0}^{2n+1} f_k(x) = 0$$

and

$$\Sigma_{k=0}^{2n} f_k(x) = f_{2n}(x) = x^n - x^{n+1} = x^n(1 - x).$$

We can find $\max_{x \in [0,1]} |\Sigma_{k=0}^{2n} f_k(x)|$ using elementary techniques:

$$f'_{2n}(x) = 2\{nx^{n-1} - [n + 1]x^n\} = 2x^{n-1}(n - [n + 1]x).$$

Therefore

$$\max_{x \in [0,1]} |f_{2n}(x)| = f_{2n}\left(\frac{n}{n + 1}\right)$$

$$= 2\left(\frac{n}{n + 1}\right)^n \left(1 - \frac{n}{n + 1}\right)$$

$$= 2\left(\frac{n}{n + 1}\right)^n \left(\frac{1}{n + 1}\right)$$

$$\leq 2\left(\frac{1}{n + 1}\right).$$

Since $\lim_n 1/(n + 1) = 0$, it follows that $\lim_n \{\max_{x \in [0,1]} |f_{2n}(x)|\} = 0$; so by Lemma 11.2, Σf_k is uniformly convergent on $[0, 1]$. The examination of $\Sigma |f_k|$ is easier. The partial sums collapse, yielding

$$\Sigma_{k=0}^{2n-1} |f_k(x)| = (1 - x) + (1 - x) + (x - x^2) + (x - x^2)$$
$$+ \cdots + (x^{n-1} - x^n)$$

$$= 2(1 - x) + 2(x - x^2) + \cdots + 2(x^{n-1} - x^n)$$

$$= 2(1 - x^n).$$

Therefore

$$\Sigma_{k=0}^{\infty}|f_k(x)| = \begin{cases} 2, & \text{if } x \in [0, 1), \\ 0, & \text{if } x = 1. \end{cases}$$

(We do not have to consider the other partial sums $\Sigma_{k=0}^{2n}|f_k|$, because $\lim_n\{\Sigma_{k=0}^{2n}|f_k(x)| - \Sigma_{k=0}^{2n-1}|f_k(x)|\} = \lim_n|f_{2n}(x)| = 0$.) Since each f_k is continuous on $[0, 1]$ and the limit function is not continuous there, we conclude from Theorem 11.6 that $\Sigma|f_k|$ does not converge uniformly on $[0, 1]$.

We made the remark that the M-Test could not be applied to the series in Example 11.11, and this must be the case because the M-Test would imply that $\Sigma|f_k|$ is uniformly convergent. To see explicitly why the M-Test fails, we note that M_{2n} would have to be at least as large as

$$\max_{x \in [0, 1]}|f_{2n}(x)| = 2\left(\frac{n}{n + 1}\right)^n\left(\frac{1}{n + 1}\right). \tag{1}$$

But the series

$$\Sigma\left(\frac{n}{n + 1}\right)^n\left(\frac{1}{n + 1}\right)$$

is divergent because it dominates $\Sigma 1/(n + 1)$. To verify this dominance, it must be shown that $[n/(n + 1)]^n$ is bounded away from zero. This can be shown by using L'Hôpital's Rule to evaluate the limit:

$$\lim_{x \to \infty}\left(\frac{x}{x + 1}\right)^x = \frac{1}{e}. \tag{2}$$

These details are requested in Exercise 11.7.1.

Exercises 11.7

1. Prove the assertions in the paragraph following Example 11.11; that is, verify Equations (1) and (2).

2. Prove: If Σf_k is convergent on D, then $\lim_k f_k = \Phi$ (identically zero) on D.

In Exercises 3–9, prove that Σf_k is uniformly convergent on the domain D.

3. $f_k(x) = \dfrac{\sin kx}{k^2}; \quad D = \mathbb{R}.$

4. $f_k(x) = x^k; \quad D = \left[-\dfrac{1}{2}, \dfrac{1}{2}\right].$

5. $f_k(x) = \left(\dfrac{\tan x}{2}\right)^k; \quad D = \left[0, \dfrac{\pi}{4}\right].$

6. $f_k(x) = \dfrac{\sin kx + \cos kx}{k(\log k)^2}; \quad D = \mathbb{R}.$

7. $f_k(x) = kx^k; \quad D = \left[-\dfrac{1}{2}, \dfrac{1}{2}\right].$

8. $f_k(x) = \left(\dfrac{3}{2}\sin x\right)^k; \quad D = \left[-\dfrac{\pi}{6}, \dfrac{\pi}{6}\right].$

9. $f_k(x) = \dfrac{k+1}{k} e^{-kx}; \quad D = [1, \infty).$

Handwritten annotations:

$\sum_{k=1}^{\infty} f_k(x) = \sum_{k=1}^{\infty} x^k$ over $\left[-\frac{1}{2}, \frac{1}{2}\right]$

Apply Weierstrass M-Test
1) Prove bounded
$|x^k| \le M_k$ $x + x^2 + x^3 + x^4$

look at some power series

theorem   (1) $|f_k(x)| \le M_k$
                    $\forall k \; \forall x \in [a,b]$
            (2) $\sum M_k < \infty$
            $\Rightarrow F_n \rightarrow F$ on $[a,b]$

#6 $f_k(x) = \frac{1}{k(\ln k)^2}|\sin kx + \cos kx|$
$\le \frac{1}{k(\ln k)^2}(|\sin kx| + |\cos kx|)$ bounded
$\le \frac{2}{k(\ln k)^2} = M_k$

$\sum \frac{2}{k(\ln k)^2}$ p-series $\sum \frac{1}{k(\ln k)^8}$ if $p > 1$

therefore you could integrate or differentiate to remove(?)

$|x^1| \le \frac{1}{2} = M_1$
$|x^2| \le \frac{1}{4} = M_2$
$|x^3| \le \frac{1}{8} = M_3$
$|x^4| \le \frac{1}{16} = M_4$
$|x^n| \le \left(\frac{1}{2}\right)^n = M_n$

true for all $x \left[-\frac{1}{2}, \frac{1}{2}\right]$

$\lim \sum_{n=1}^{\infty} x^n = \frac{x}{1-x} f(x)$ $x \in \left[-\frac{1}{2}, \frac{1}{2}\right]$

(so converges)

$\frac{1}{2} \sum_{k=1}^{\infty} \left(\frac{1}{2}\right)^{k-1} = \frac{1}{2}\left(\frac{1}{1-\frac{1}{2}}\right)$

$\sum_{k=1}^{\infty} x^k = x + x^2 + x^3 + x^4$
$= x(1 + x + x^2 + x^3)$
$x \sum_{k=1}^{\infty} x^{k-1} = x\left(\frac{1}{1-x}\right)$
$= \frac{x}{1-x} = F($

POWER SERIES

12.1. Convergence of Power Series

Perhaps the most elementary class of functions is that consisting of powers of the identity function, for example, $f(x) = x^n$. It is these functions that are combined arithmetically to form polynomials and rational functions. And it is these functions that have been used most successfully in forming function series that yield a rich theory and lend themselves to many applications. In the following definition and throughout this chapter, the series we study has a zero-th term as the initial term of the series. Thus $\{a_k\}$ is a brief form of $\{a_k\}_{k=0}^{\infty}$.

DEFINITION 12.1. If $\{a_k\}$ is a number sequence and a is a number, let f_k be the function given by $f_k(x) = a_k(x - a)^k$; then the function series Σf_k is called a *power series*. If x_1 is a number such that $\Sigma a_k(x_1 - a)^k$ is convergent, then we say that the power series $\Sigma a_k(x - a)^k$ is "convergent at x_1." The phrase "nonconvergent at x_1" is used similarly.

We begin by establishing some facts about the set of numbers for which a given power series is convergent.

LEMMA 12.1. If the power series $\Sigma a_k(x - a)^k$ is convergent at x_1 and if x_2 is a number satisfying $|x_2 - a| < |x_1 - a|$, then $\Sigma a_k(x - a)^k$ is convergent at x_2.

I finite interval

I infinite interval $= (-\infty, \infty) = \mathbb{R}$

$I = \{a\}$

221

Proof. If $x_1 = a$, there is nothing to prove, so assume that $x_1 \neq a$ and let x_2 be a number satisfying $|x_2 - a| < |x_1 - a|$. Define

$$r = \frac{|x_2 - a|}{|x_1 - a|}.$$

Since $\Sigma a_k(x_1 - a)^k$ is convergent, we know that $\lim_k a_k(x_1 - a)^k = 0$; so the sequence $\{a_k(x_1 - a)^k\}$ is bounded, say,

$$B = \mathrm{lub}_k \{|a_k(x_1 - a)^k|\}.$$

Then for every k, we have

$$|a_k(x_2 - a)^k| = |a_k| \, |x_1 - a|^k \left| \frac{x_2 - a}{x_1 - a} \right|^k \leq Br^k.$$

Hence $\Sigma a_k(x - a)^k$ is dominated by the convergent geometric series Σr^k, and therefore it too is convergent.

Lemma 12.1 is sometimes used in its contrapositive form: If $|x_2 - a| < |x_1 - a|$ and $\Sigma a_k(x - a)^k$ is nonconvergent at x_2, then the power series is nonconvergent at x_1 as well.

Note that in Lemma 12.1 we assumed strict inequality between $|x_2 - a|$ and $|x_1 - a|$. The conclusion does not necessarily hold for all x_2 such that $|x_2 - a| \leq |x_1 - a|$. The following example illustrates this possibility.

EXAMPLE 12.1. The power series $\Sigma x^k/k$ is convergent if and only if $-1 \leq x < 1$. This is shown by using the Comparison Test and the Alternating Series Theorem. But if we take $x_1 = -1$, then it is not true that $\Sigma x^k/k$ is convergent at every x_2 such that $|x_2| \leq |x_1| = |-1|$, because when $x_2 = 1$ the power series is divergent.

With the aid of Lemma 12.1 we can describe the set of all numbers x for which the power series $\Sigma a_k(x - a)^k$ is convergent. The following theorem tells us that this set is either the entire real line $\mathbb{R}$ or an interval centered at a. Therefore this set is called the *interval of convergence* of the power series. The number R in this theorem is called the *radius of convergence* of the power series.

THEOREM 12.1. Let I be the set $\{x_1 \in \mathbb{R}: \Sigma a_k(x_1 - a)^k$ is convergent$\}$; then I is either $\mathbb{R}$ or an interval of the form $[a - R, a + R]$, $[a - R, a + R)$, $(a - R, a + R]$, or $(a - R, a + R)$.

Proof. If $I \neq \mathbb{R}$, then Lemma 12.1 assures us that I is bounded, so let $R = \mathrm{lub}\{|x_1 - a|: x_1 \in I\}$. If x_2 is in $(a - R, a + R)$, then $|x_2 - a| < R$, and there is some x_1 in I such that $|x_2 - a| < |x_1 - a|$. Then Lemma 12.1 ensures that $\Sigma a_k (x_2 - a)^k$ is convergent, so x_2 is in I. This proves that $(a - R, a + R) \subseteq I$. Now suppose x_1 is in I. Then our choice of R implies that $|x_1 - a| \leq R$, so x_1 is in $[a - R, a + R]$. Hence, $(a - R, a + R) \subseteq I \subseteq [a - R, a + R]$, and it follows that I must be one of the four intervals centered at a.

It may be recalled from elementary calculus that one can usually determine the interval of convergence by relatively easy methods. The first thing to do is to find the radius of convergence, and the next theorem gives a very simple tool for this task. Although it does not give an answer for every power series, it is nevertheless very useful.

THEOREM 12.2. If $\{a_k\}$ is a number sequence such that

$$\lim_k \left| \frac{a_k}{a_{k+1}} \right| = R,$$

and a is any number, then R is the radius of convergence of the power series $\Sigma a_k (x - a)^k$.

Proof. If we apply the Ratio Test to the power series $\Sigma a_k (x - a)^k$, we get

$$\lim_k \left| \frac{a_{k+1}(x - a)^{k+1}}{a_k(x - a)^k} \right| = \lim_k \left| \frac{a_{k+1}}{a_k} \right| |x - a| = \frac{1}{R} |x - a|.$$

By the Ratio Test, the series is (absolutely) convergent if the limit value $|x - a|/R$ is less than 1, which is equivalent to $|x - a| < R$. Similarly, the Ratio Test implies nonconvergence if $|x - a| > R$. Hence the number R is the radius of convergence.

It should be noted that the radius of convergence R in Theorem 12.2 can be interpreted in the extended sense; that is, if $\lim_k |a_k/a_{k+1}| = \infty$, then the interval of convergence is $\mathbb{R}$. Also, if $\lim_k |a_k/a_{k+1}| = 0$, then the interval of convergence is the degenerate closed interval $[a, a]$, consisting of the single point a.

In finding the interval of convergence for a particular power series, one first applies Theorem 12.2 or one of its variants (see Example 12.3) to find R; then one checks the two "endpoint" series at $x = a + R$ and $x = a - R$ to determine which of the four possible intervals is the desired one. Since this is a standard exercise in elementary courses, we review it here only briefly with two examples.

EXAMPLE 12.2. The interval of convergence of $\Sigma(x - 5)^k/(k + 1)$ is $[4, 6)$. First,

$$R = \lim_k \frac{a_k}{a_{k+1}} = \lim_k \frac{k + 2}{k + 1} = 1.$$

When $x = 6$, the power series becomes $\Sigma 1/(k + 1)$, which is divergent, and when $x = 4$, the series becomes $\Sigma(-1)^{k+1}/(k + 1)$, which is convergent.

EXAMPLE 12.3. The interval of convergence of $\Sigma 2^k x^{2k}/(k + 1)^2$ is $[-1/\sqrt{2}, 1/\sqrt{2}]$. Because $a_{2k+1} = 0$, we cannot form the ratio to use Theorem 12.2 directly, but we can treat this as a series of powers of x^2. Thus the coefficient of the kth power of x^2 is $b_k = 2^k(k + 1)^{-2}$. Applying Theorem 12.2 to the coefficient sequence $\{b_k\}$, we get

$$\lim_k \frac{b_k}{b_{k+1}} = \lim_k \frac{2^k(k + 1)^{-2}}{2^{k+1}(k + 2)^{-2}} = \lim_k \frac{1}{2}\left(\frac{k + 2}{k + 1}\right)^2 = \frac{1}{2}.$$

Thus the series is convergent when $|x^2 - 0| < 1/2$, which is equivalent to $-1/\sqrt{2} < x < 1/\sqrt{2}$. At the endpoints we get

$$\Sigma \frac{2^k(\pm 1/\sqrt{2})^{2k}}{(k + 1)^2} = \Sigma \frac{2^k(1/2)^k}{(k + 1)^2} = \Sigma \frac{1}{(k + 1)^2},$$

which is convergent.

Exercises 12.1

Find the interval of convergence for each of the following power series.

1. $\Sigma \dfrac{(-1)^{k+1}}{k + 1}(x + 2)^k$

2. $\Sigma \dfrac{(-2)^{k+1}}{k + 1}(x - 1)^k$

3. $1 - \dfrac{1}{2}x + \dfrac{1 \cdot 3}{2 \cdot 4}x^2 - \dfrac{1 \cdot 3 \cdot 5}{2 \cdot 4 \cdot 6}x^3 + \cdots$

4. $\sum \dfrac{x^{2k}}{k!}$

5. $\sum \dfrac{(-1)^{k+1}}{2k+1} x^{2k+1}$

6. $\sum \dfrac{(-1)^{k+1}}{k^2} (x+1)^k$

7. $\dfrac{1}{2} + \dfrac{x}{3} + \dfrac{x^2}{2^2} + \dfrac{x^3}{3^2} + \dfrac{x^4}{2^3} + \dfrac{x^5}{3^3} + \dfrac{x^6}{2^4} + \dfrac{x^7}{3^4} + \cdots$ *no limit of ratio exists*

8. $\sum \dfrac{1}{k} (x+2)^{2k+1}$

9. $x + 1^2 x^2 + \sqrt{1}\, x^3 + 2^2 x^4 + \sqrt{2}\, x^5 + 3^2 x^6 + \sqrt{3}\, x^7 + \cdots$

10. $\sum \dfrac{k!}{k^k} x^k$ (The series starts with $k=1$.)

11. $\sum (\log k)^k x^k$ (The series starts with $k=1$. *Hint:* Compare this series to $\sum R^k x^k$.)

12. $\sum k^{\log k} x^k$ (The series starts with $k=1$. *Hint:* $k^{\log k} = e^{(\log k)^2}$ and $\lim_{\infty}\{[\log x]^2 - [\log(x+1)]^2\}$ can be found by using the Law of the Mean.)

12.2. Integration and Differentiation of Power Series

In the interval of convergence, the sum of a power series determines a function f given by

$$f(x) = \sum_{k=0}^{\infty} a_k(x-a)^k.$$

Since this function is the limit of a sequence of polynomials, it is reasonable to expect that f will inherit some of the nice properties of polynomials such as continuity, integrability, and differentiability. In fact, each of these properties holds in the interior of the interval of convergence; these are consequences of the uniform convergence that is exhibited there. That uniformity of convergence is the object of the next theorem.

THEOREM 12.3. If the power series $\sum a_k(x-a)^k$ has radius of convergence R, then it is uniformly convergent on every closed subinterval of $(a-R, a+R)$.

when $x=1$ $\dfrac{2k-1}{2k}$ converges over [-1,1] interval when $a=0$

if $a=5$ $|x-5|<1$ $-1<x-5<1$ $4<x<6$

interval of convergence will always be $[a-1, a+1]$

Proof. Suppose that $[c - d] \subset (a - R, a + R)$. Then $\Sigma a_k(x - a)^k$ is convergent at c and d. If $|c - a| \le |d - a|$, then we define $M_k = |a_k(d - a)^k|$; if $|c - a| \ge |d - a|$, then we define $M_k = |a_k(c - a)^k|$. In either case, ΣM_k is convergent and $\operatorname{lub}_{x \in [c, d]} |a_k(x - a)^k| \le M_k$. Therefore the M-Test applies to give the asserted uniform convergence.

Note that in Theorem 12.3 the interval on which it is concluded that the power series is uniformly convergent does *not* include either endpoint of the interval of convergence. This is not the best possible result (for example, Abel proved that if Σa_k is convergent, then $\Sigma a_k x^k$ is uniformly convergent on $[0, 1]$), but with Theorem 12.3 we can easily prove that the sum of a power series is an integrable function.

THEOREM 12.4. Suppose that $\Sigma a_k(x - a)^k$ has radius of convergence R and $f(x) = \Sigma_{k=0}^{\infty} a_k(x - a)^k$ when x is in $(a - R, a + R)$. If $[c, d] \subset (a - R, a + R)$, then f is integrable on $[c, d]$ and term-by-term integration is valid:

$$\int_c^d \sum_{k=0}^{\infty} a_k(x - a)^k dx = \sum_{k=0}^{\infty} a_k \int_c^d (x - a)^k dx. \tag{1}$$

Proof. This result is an immediate consequence of Theorem 11.7 and Theorem 12.3.

The technique of term-by-term integration can be used to evaluate some interesting sums. This is illustrated in the following example.

EXAMPLE 12.4. Evaluate $\Sigma_{k=1}^{\infty} 1/(k2^k)$. We start with the known geometric series: $\sum \frac{1}{k(2^k)} = \sum \frac{1}{k} \left(\frac{1}{2}\right)^k$

$$\sum_{k=0}^{\infty} x^k = \frac{1}{1 - x} \quad \text{if} \quad -1 < x < 1. \tag{2}$$

By Theorem 12.4 we can integrate term by term:

$$\sum_{k=0}^{\infty} \frac{x^{k+1}}{k + 1} = -\log|1 - x| \quad \text{if} \quad -1 < x < 1. \tag{3}$$

Since this is valid for any x in $(-1, 1)$, we can substitute $1/2$ for x in (3) and get

$$\sum_{k=0}^{\infty} \frac{(1/2)^{k+1}}{k + 1} = -\log\frac{1}{2}.$$

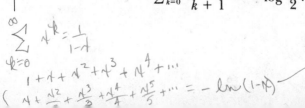

Replacing $k + 1$ with j, we conclude that

$$\sum_{j=1}^{\infty} \frac{1}{j2^j} = \log 2,$$

which is clearly equivalent to the desired sum.

Next we investigate the differentiability of the sum of a power series. In order to use Theorem 11.8, we need to know that the series of derivatives is uniformly convergent, and for that we must prove a preliminary result that compares a power series with its derived series.

LEMMA 12.2. For any sequence $\{a_k\}$ and any number a, the two power series $\sum a_k(x - a)^k$ and $\sum ka_k(x - a)^{k-1}$ have the same radius of convergence.

Proof. Let R and R' be the radii of convergence of $\sum a_k(x - a)^k$ and $\sum ka_k(x - a)^{k-1}$, respectively. If $|x_1 - a| < R'$, then $\sum ka_k(x_1 - a)^{k-1}$ is absolutely convergent; therefore

$$\sum a_k(x_1 - a)^k = (x_1 - a) \sum a_k(x_1 - a)^{k-1}$$

is also absolutely convergent, because the right-hand series is dominated by $\sum ka_k(x_1 - a)^{k-1}$. Since $\sum a_k(x_1 - a)^k$ is convergent, $|x_1 - a| \leq R$. Hence $R' \leq R$.

To show that $R \leq R'$, we assume that $\sum a_k(x_1 - a)^k$ is absolutely convergent and deduce that $\sum ka_k(x_2 - a)^{k-1}$ is absolutely convergent for *every* x_2 satisfying $|x_2 - a| < |x_1 - a|$. Consider the following:

$$\left| \frac{ka_k(x_2 - a)^{k-1}}{a_k(x_1 - a)^k} \right| = k \left| \frac{x_2 - a}{x_1 - a} \right|^{k-1} \frac{1}{|x_1 - a|} = \frac{1}{|x_2 - a|} kr^{k-1},$$

where

$$r = \frac{|x_2 - a|}{|x_1 - a|} < 1.$$

Since $\lim_k kr^{k-1} = 0$ (why?), we see that $\sum |ka_k(x_2 - a)^{k-1}|$ is dominated by $\sum |a_k(x_1 - a)^k|$. Hence the derived series is absolutely convergent whenever the latter series is absolutely convergent; that is, $R \leq R'$. This completes the proof that $R = R'$.

Note that we have not proved that a power series and its derived series have exactly the same interval of convergence. We can conclude only that their radii of convergence are the same; their behavior at the endpoints may differ.

EXAMPLE 12.5. The series $\Sigma x^k/k^2$ and $\Sigma x^{k-1}/k$ both have radius of convergence $R = 1$, but the former series converges on $[-1, 1]$, whereas its derived series converges (only) on $[-1, 1)$.

Now we can prove that the sum of a power series is a differentiable function.

THEOREM 12.5. Suppose $\Sigma a_k(x - a)^k$ has radius of convergence R and $f(x) = \Sigma_{k=0}^{\infty} a_k(x - a)^k$ for each x in $(a - R, a + R)$. Then f is differentiable there and term-by-term differentiation is valid:

$$[\Sigma_{k=0}^{\infty} a_k(x - a)^k]' = \Sigma_{k=0}^{\infty} [a_k(x - a)^k]' \tag{4}$$
$$= \Sigma_{k=1}^{\infty} k a_k(x - a)^{k-1}.$$

Proof. Theorem 12.3 and Lemma 12.2 imply that, on each closed subinterval of $(a - R, a + R)$, $\Sigma k a_k(x - a)^{k-1}$ is uniformly convergent. Also, each term of $\Sigma a_k(x - a)^k$ is obviously differentiable, and the series converges at each point of $(a - R, a + R)$. Thus the hypotheses of Theorem 11.8 are satisfied on each closed subinterval of $(a - R, a + R)$, so we may apply that result to conclude that the derivative of the sum equals the sum of the derivatives. Since this conclusion is valid on every closed subinterval of $(a - R, a + R)$, and every x in $(a - R, a + R)$ is in such a subinterval, we conclude that (4) holds throughout $(a - R, a + R)$.

COROLLARY 12.5. If

$$f(x) = \Sigma_{k=0}^{\infty} a_k(x - a)^k$$

for each x in $(a - R, a + R)$, then f is continuous there.

Proof. This is an immediate consequence of Theorem 12.5. It could also be proved directly from Theorem 12.3 (see Exercise 12.2.7).

As with Theorem 12.4, we can use Theorem 12.5 to evaluate the sums of certain series that are related to well-known series.

EXAMPLE 12.6. Evaluate $\sum_{k=1}^{\infty} k/2^k$. We start with the geometric series,

$$\sum_{k=0}^{\infty} x^k = \frac{1}{1-x} \quad \text{if} \quad -1 < x < 1,$$

differentiate term by term,

$$\sum_{k=1}^{\infty} kx^{k-1} = \frac{1}{(1-x)^2},$$

and multiply through by x,

$$\sum_{k=1}^{\infty} kx^k = \frac{x}{(1-x)^2}. \tag{5}$$

Since (5) is valid for every x in $(-1, 1)$, we can substitute $1/2$ for x to obtain

$$\sum_{k=1}^{\infty} k/2^k = 2.$$

Exercises 12.2

1. Evaluate the sum: $\sum_{k=0}^{\infty}(k + 1)x^k$.

2. Evaluate the sum: $1 - 2x^2 + 3x^4 - 4x^6 + \cdots$.

3. Evaluate the sum: $\sum_{k=1}^{\infty} k^2/2^k$.

4. Prove:

$$\pi/4 = \sum_{k=0}^{\infty} \frac{(-1)^k}{2k + 1} = 1 - \frac{1}{3} + \frac{1}{5} - \frac{1}{7} + \cdots.$$

(Hint: $\dfrac{1}{1 + x^2} = \sum_{k=0}^{\infty}(-x^2)^k$.)

5. Prove:

$$\log 2 = \sum_{k=1}^{\infty} \frac{(-1)^{k-1}}{k} = 1 - \frac{1}{2} + \frac{1}{3} - \frac{1}{4} + \cdots.$$

(Hint: $\dfrac{1}{1 + x} = \sum_{k=0}^{\infty}(-x)^k$.)

6. Prove: If g is a rational function on $\mathbb{N}$, then the series $\Sigma a_k(x - a)^k$ and $\Sigma a_k g(k)(x - a)^k$ have the same radius of convergence.

7. Give a direct proof of Corollary 12.5; that is, use Theorem 12.3 but not Theorem 12.5.

12.3. Taylor Series

The topic of this section is also familiar to the student of elementary calculus. The treatment here differs somewhat from the elementary treatments because we are primarily interested in questions of convergence and properties of the limit functions rather than the Taylor series representation of particular functions. First we make an observation about Theorem 12.5 that allows us to extend its conclusion.

THEOREM 12.6. Suppose that $\Sigma a_k(x - a)^k$ has radius of convergence $R > 0$ and $f(x) = \Sigma_{k=0}^{\infty} a_k(x - a)^k$ for each x in $(a - R, a + R)$; then f has derivatives of *all* orders throughout $(a - R, a + R)$, and for each k

$$a_k = \frac{f^{(k)}(a)}{k!};$$ (1)

thus for each x in $(a - R, a + R)$,

$$f(x) = \Sigma_{k=0}^{\infty} \frac{f^{(k)}(a)}{k!}(x - a)^k.$$ (2)

Proof. Since Lemma 12.2 guarantees that the radius of convergence remains unchanged for the derived series, we see that it, too, converges to a differentiable function on $(a - R, a + R)$. So the derived series also can be differentiated term by term to give

$$g''(x) = \Sigma_{k=2}^{\infty} k(k - 1)a_k(x - a)^{k-2} \quad \text{if} \quad x \in (a - R, a + R).$$

This power series also can be differentiated term by term, and by repeating this process we have

$$f^{(n)}(x) = \Sigma_{k=n}^{\infty} k(k - 1) \ldots (k - n + 1)a_k(x - a)^{k-n}.$$ (3)

This formula is valid throughout $(a - R, a + R)$; in particular, when $x = a$, the series in (3) has only one nonzero term ($k = n$):

$$f^{(n)}(a) = n!a_n.$$

Hence Equations (1) and (2) hold, and the theorem is proved.

The symbol $C^{(n)}(I)$ is used to denote the collection of all functions that have a continuous nth-order derivative on the interval I. A function that has derivatives of all orders on an interval I is said to belong to class $C^\infty(I)$. This symbol is read "C infinity of I." For any member of $C^\infty(a - R, a + R)$, we can determine coefficients $\{a_k\}$ from (1) and form the power series centered at a. This power series is called the *Taylor series* of f about a. For the special case in which $a = 0$, the power series $\Sigma[f^{(k)}(0)/k!]x^k$ is sometimes called the *MacLaurin series* of f.

It is important to note that we have not said that Equation (2) holds whenever f is in $C^\infty(a - R, a + R)$. We have merely established the *existence* of the Taylor series of f about a; we have not proved that it converges (except at a, where convergence is trivial); neither can we say anything conclusive about the sum of that Taylor series even if it is known to be convergent (see Example 12.7). If the function f is equal to the sum of a power series throughout some nondegenerate interval, say, $(a - R, a + R)$ where $R > 0$, then f is said to be *analytic* in $(a - R, a + R)$. This is precisely the class of functions that satisfy the hypothesis of Theorem 12.6, so we conclude from that theorem that the coefficients of the power series must be given by (1), and (2) holds for x in $(a - R, a + R)$. Hence we have proved that the Taylor series of f about a is the *only* power series whose sum equals $f(x)$ throughout an open interval centered at a. This is stated formally in the following proposition:

PROPOSITION 12.1. If f is analytic on $(a - R, a + R)$, where $R > 0$, then the power series whose sum is f must be the Taylor series of f about a.

Now we consider the converse relationship and show by an example that a function can be in $C^\infty(\mathbb{R})$ without being analytic.

EXAMPLE 12.7. If the function f is given by

$$f(x) = \begin{cases} e^{-1/x^2}, & \text{if } x \neq 0, \\ 0, & \text{if } x = 0, \end{cases}$$

then f is in $C^\infty(\mathbb{R})$, but f is not equal to the sum of its MacLaurin series. The differentiability of f is clear at all values of

x other than zero. We assert that for each k, $f^{(k)}(0) = 0$, which implies that the coefficients of the MacLaurin series are identically zero. Although this series is certainly convergent (to the identically zero function), its sum is not f. To show that $f^{(k)}(0) = 0$, we first use L'Hôpital's Rule repeatedly to do the following calculation:

$$\lim_{x \to 0} \frac{e^{-1/x^2}}{x^k} = \lim_0 \frac{x^{-k}}{e^{1/x^2}} = \lim_0 \frac{kx^{-k-1}}{2x^{-3}e^{1/x^2}}$$

$$= \lim_0 \frac{kx^{-k+2}}{2e^{1/x^2}} = \cdots = \lim_0 K \frac{x^i}{e^{1/x^2}}, \tag{4}$$

where $i = 0$ or 1 and K is a constant. Thus

$$\lim_0 \frac{e^{-1/x^2}}{x^k} = 0 \quad \text{for every } k \text{ in } \mathbb{N}.$$

Therefore

$$f'(0) = \lim_0 \frac{f(x) - f(0)}{x - 0} = \lim_0 \frac{e^{-1/x^2}}{x} = 0. \tag{5}$$

For the kth-order derivative, we have

$$f^{(k)}(0) = \lim_0 \frac{f^{(k-1)}(x) - f^{(k-1)}(0)}{x}. \tag{6}$$

To prove the assertion by mathematical induction, we may assume that $f^{(k-1)}(0) = 0$, which allows us to write (6) as

$$f^{(k)}(0) = \lim_0 \frac{f^{(k-1)}(x)}{x}. \tag{7}$$

When $x \neq 0$, $f^{(k-1)}(x)$ is obtained by repeated differentiation of e^{-1/x^2}, which yields at most 2^k terms of the form $x^{-m}e^{-1/x^2}$. Hence, by the above calculation in (4),

$$\lim_0 \frac{f^{(k-1)}(x)}{x} = 0,$$

and our proof is complete.

It is also possible for a function in C^∞ ($\mathbb{R}$) to have a Taylor series that is nonconvergent (except at a, where convergence is trivial). Such a function is much more complicated than the one we have just seen in Example 12.7, so we do not present one here. The interested reader is referred to Gelbaum and Olmstead, *Counter-Examples in Analysis* (San Francisco: Holden-Day, 1964), p. 68.

Exercises 12.3

1. Show that each of the following functions does not have a MacLaurin series:

   (a) $f(x) = |x|$;

   (b) $f(x) = \begin{cases} x^2, & \text{if } x > 0, \\ 0, & \text{if } x \le 0; \end{cases}$

   (c) $f(x) = \begin{cases} x^3, & \text{if } x > 0, \\ 0, & \text{if } x \le 0; \end{cases}$

   (d) $f(x) = \begin{cases} x^p, & \text{if } x > 0, \\ 0, & \text{if } x \le 0. \end{cases}$ where $p \in \mathbb{N}$,

2. Use Equation (1) to find the MacLaurin series for the following functions:

   (a) $f(x) = e^x$;

   (b) $f(x) = \cos x$;

   (c) $f(x) = \sin x$;

   (d) $f(x) = \log(1 + x)$;

   (e) $f(x) = \cosh x = (e^x + e^{-x})/2$;

   (f) $f(x) = \sinh x = (e^x - e^{-x})/2$.

12.4. The Remainder Term

Now that we have established that the class of analytic functions is a proper subset of C^∞, we wish to develop some means of determining whether a function is equal to the sum of its Taylor series. Let f be a function in $C^\infty(a - R, a + R)$ and define

$$S_n(x) = \sum_{k=0}^{n-1} \frac{f^{(k)}(a)}{k!} (x - a)^k$$

and

$$R_n(x) = f(x) - S_n(x).$$

Then $S_n(x)$ is the sum of the first n terms of the Taylor series of f about a, and $R_n(x)$ is called the nth *remainder term*. It is clear that

$$f(x) = \sum_{k=0}^{\infty} \frac{f^{(k)}(a)}{k!} (x - a)^k$$

if and only if

$$\lim_n R_n(x) = 0.$$

Thus f is analytic on $(a - R, a + R)$ if and only if $\lim_n R_n(x) = 0$ for every x in $(a - R, a + R)$.

We have seen the sum $S_n(x)$ in Chapter 6, where Theorem 6.6 was called Taylor's Formula with Remainder; it asserted that

$$f(x) = S_n(x) + \frac{f^{(n)}(\mu_n)}{n!} (x - a)^n, \tag{1}$$

where $f^{(n)}$ exists between a and x, and μ is between a and x. In general, Taylor's Formula asserts that if f is in $C^{(n)}(a - R, a + R)$, then for each x in $(a - R, a + R)$,

$$f(x) = S_n(x) + R_n(x),$$

where $R_n(x)$ is expressed by some explicit formula. In this section we develop three different formulas for the remainder term. The first one is the result that was proved in Chapter 6 as Taylor's Formula with Remainder. We state it here (without proof) using our present terminology and notation.

THEOREM 12.7: LAGRANGE FORM OF R_n. If f is in $C^{(n)}(a - R, a + R)$ and x is in $(a - R, a + R)$, then there is a number μ_n between a and x such that

$$R_n(x) = \frac{f^{(n)}(\mu_n)}{n!} (x - a)^n. \tag{2}$$

Our second formula for the remainder expresses R_n as a definite integral.

THEOREM 12.8: INTEGRAL FORM OF R_n. If f is in $C^{(n)}(a - R, a + R)$ and x is in $(a - R, a + R)$, then

4/23/91

$f(N) = P_n(N) + R_n(N)$ ← Actual error

235

← Polynomial approximation

→ N is value at which funct. is being evaluated

$$R_n(x) = \frac{1}{(n-1)!} \int_a^x (x-t)^{n-1} f^{(n)}(t)\,dt. \qquad (3)$$

Proof. The continuity of f' allows us to use the Fundamental Theorem of Calculus to write

$$f(x) - f(a) = \int_a^x f'(t)\,dt,$$

and making the substitution $u = x - t$, we get

$R_2(N) = \int_1^N (N-t) f^2(t)\,dt$

$$f(x) - f(a) = \int_{t=a}^{t=x} f'(x-u)(-du)$$

$= 2 \int_1^N \frac{(N-t)}{t^3}\,dt$

$$= -\int_{u=x-a}^{u=0} f'(x-u)\,du$$

$$= \int_0^{x-a} f'(x-u)\,du.$$

$R_2(N) = 2 \int_1^N \frac{(N-t)}{t^3}\,dt$

Now we repeatedly use integration by parts to obtain

$$f(x) - f(a) = [u \cdot f'(x-u)]_{u=0}^{u=x-a} + \int_0^{x-a} u \cdot f''(x-u)\,du$$

$$= f'(a)(x-a) + \int_0^{x-a} u \cdot f''(x-u)\,du \qquad \text{Recall}$$

$\left| \int_a^b f(t)\,dt \right| \leq$

$$= f'(a)(x-a) + \left[\frac{u^2}{2} \cdot f''(x-u) \right]_0^{x-a} + \int_0^{x-a} \frac{u^2}{2} \cdot f'''(x-u)\,du$$

$$= f'(a)(x-a) + \frac{f''(a)}{2}(x-a)^2 + \int_0^{x-a} \frac{u^2}{2} \cdot f'''(x-u)\,du$$

$\int_a^b |f(u)|\,du$

$$\vdots$$

$|R_n(N)| \leq \frac{1}{(n-1)!} \int_a^N |N-t|^n \cdot |f^{(n)}(t)|\,dt$

$$= f'(a)(x-a) + \cdots + \frac{f^{(n-1)}(a)}{(n-1)!}(x-a)^{n-1}$$

$\left| R_2\left(\tfrac{1}{2}\right) \right| = \left| 2 \int_1^{\frac{1}{2}} \frac{\frac{1}{2}-t}{t^3}\,dt \right|$

$$+ \int_0^{x-a} \frac{u^{n-1}}{(n-1)!} f^{(n)}(x-u)\,du$$

$= 2 \int \left(\frac{1}{2} t^{-3} - t^{-2} \right) dt$

$$= S_n(x) - f(a) + \frac{1}{(n-1)!} \int_0^{x-a} u^{n-1} f^{(n)}(x-u)\,du.$$

$= 2 \left[\frac{1}{2} \frac{t^{-2}}{-2} - \frac{t^{-1}}{-1} \right]_1^{1/2}$

Hence

$$f(x) = S_n(x) + \frac{1}{(n-1)!} \int_a^x (x-t)^{n-1} f^{(n)}(t)\,dt,$$

$= 2 \left(-\frac{1}{4t^2} + \frac{1}{t} \right)\Big|_1^{1/2}$

which is equivalent to (3), and the proof is complete.

$= 2 \left\{ (-1+2) - \left(-\frac{1}{4}+1 \right) \right\}$

$= 2 \left\{ 1 - \frac{3}{4} \right\} = \frac{1}{2}$

* $13\frac{1}{2}A$ (in note of 4/23/91) because this ↓ showed = 2, N=3 is correct (mistake was wrong) there is a mistake somewhere here!

The third formula for the remainder term can be derived from the integral form by using the Mean Value Theorem for Integrals (Exercise 7.2.7). Since the integrand function $(x - t)^{n-1} f^{(n)}(t)$ is continuous on $(a - R, a + R)$, there is a number c_n between a and x such that

$$\int_a^x (x - t)^{n-1} f^{(n)}(t) dt = (x - c_n)^{n-1} f^{(n)}(c_n)(x - a).$$

Substituting this into (3), we get the next result.

Theorem 12.9: Cauchy Form of R_n. If f is in $C^{(n)}(a - R, a + R)$ and x is in $(a - R, a + R)$, then there is a number c_n between a and x such that

$$R_n(x) = \frac{f^{(n)}(c_n)}{(n - 1)!} (x - c_n)^{n-1}(x - a). \tag{4}$$

12.5. Taylor Series of Some Elementary Functions

The final section of this chapter consists of several examples in which we prove that some well-known elementary functions are equal to the sums of their MacLaurin series (compare Exercise 12.3.2).

Example 12.8. The polynomial P is analytic on $\mathbb{R}$. If

$$P(x) = a_N x^N + \cdots + a_1 x + a_0,$$

then it is an easy calculation to show that

$$P^{(k)}(0) = (k!)a_k \quad \text{for} \quad k \le N.$$

Therefore the polynomial coefficients are the first $N + 1$ Taylor coefficients, and since $P^{(k)}(x) = 0$ when $k > N$, we see that $R_n(x) = 0$ when $n > N$. Hence it is trivial that $\lim_n R_n(x) = 0$. Note that the MacLaurin series for P is P itself.

EXAMPLE 12.9. The exponential function is analytic on $\mathbb{R}$. By repeated differentiation of e^x, we find that the kth MacLaurin coefficient is $a_k = 1/k!$ (see Exercise 12.3.2a). Also, the Lagrange form of R_n is

$$R_n(x) = \frac{e^{\mu_n} x^n}{n!},$$

where μ_n is between zero and x. Since $|\mu_n| < |x|$, we have $e^{\mu_n} < e^{|x|}$ for every n. Therefore $|R_n(x)| < e^{|x|}|x|^n/n!$. Also, $\lim_n x^n/n! = 0$, because the series $\Sigma x^n/n!$ is convergent for every x. Hence

$$|\lim_n R_n(x)| \le e^{|x|} \lim_n \frac{|x^n|}{n!} = 0,$$

and therefore

$$e^x = \Sigma_{k=0}^{\infty} \frac{1}{k!} x^k \quad \text{for every } x \text{ in } \mathbb{R}. \tag{1}$$

EXAMPLE 12.10. The sine function is analytic on $\mathbb{R}$. By direct calculation of the successive derivatives of $\sin x$, we find that $a_{2k} = 0$ for each k in $\mathbb{N}$, and $a_{2k+1} = (-1)^{k+1}/(2k + 1)!$ (see Exercise 12.3.2c). Also, the nth-order derivative of $\sin x$ is either $\pm\sin x$ or $\pm\cos x$, so its absolute value never exceeds 1. Therefore the Lagrange form of $R_n(x)$ yields

$$|R_n(x)| \le \frac{1}{n!} |x^n|,$$

and we know that $\lim_n |x^n|/n! = 0$, so $\lim_n R_n(x) = 0$. Hence

$$\sin x = \Sigma_{k=0}^{\infty} \frac{(-1)^k}{(2k + 1)!} x^{2k+1} \quad \text{for every } x \text{ in } \mathbb{R}. \tag{2}$$

EXAMPLE 12.11. The function given by $\log(1 + x)$ is analytic on the interval $(-1, 1)$. Repeated differentiation gives

$$f^{(k)}(x) = (-1)^{k-1}(k-1)!(1+x)^{-k} \quad \text{if } x > -1,$$

so the MacLaurin coefficients are given by

$$\frac{f^{(k)}(0)}{k!} = \frac{(-1)^{k-1}}{k} \quad \text{if } k > 0.$$

(See Exercise 12.3.2d.) To show that $\lim_n R_n(x) = 0$, we derive a special formula for $R_n(x)$ (compare Exercise 12.2.5). Consider the geometric sum

$$1 - t + t^2 - \cdots = \frac{1 - (-t)^{n-1}}{1 - (-t)} = \frac{1}{1 + t} - \frac{(-t)^{n-1}}{1 + t}.$$

If x is in $[-1, 1)$, then we can integrate from zero to x to obtain

$$\int_0^x \frac{dt}{1 + t} = \int_0^x [1 - t + t^2 + \cdots + (-t)^{n-2}] dt$$

$$+ (-1)^{n-1} \int_0^x \frac{t^{n-1}}{1 + t} dt,$$

which is equivalent to

$$\log(1 + x) = x - \frac{x^2}{2} + \frac{x^3}{3} - \cdots - \frac{(-x)^{n-1}}{k - 1}$$

$$+ (-1)^{n-1} \int_0^x \frac{t^{n-1}}{1 + t} dt$$

$$= S_n(x) + (-1)^{n-1} \int_0^x \frac{t^{n-1}}{1 + t} dt.$$

Hence

$$R_n(x) = (-1)^{n-1} \int_0^x \frac{t^{n-1}}{1 + t} dt. \tag{3}$$

If $0 \le x \le 1$, then $0 \le t \le x$ implies that $\dfrac{t^{n-1}}{1 + t} \le t^{n-1}$, so (3) yields

$$|R_n(x)| \le \int_0^x t^{n-1} dt = \frac{x^n}{n},$$

which shows that $\lim_n R_n(x) = 0$. If $-1 < x < 0$, then $x \leq t \leq 0$ implies that

$$\left| \frac{t^{n-1}}{1+t} \right| \leq \frac{|t^{n-1}|}{1+x},$$

so (3) yields

$$|R_n(x)| \leq \frac{1}{1+x} \left| \int_0^x t^{n-1} dt \right| = \frac{|x^n|}{(1+x)n} < \frac{1}{(1+x)n}.$$

Therefore $\lim_n R_n(x) = 0$ for every x in $(-1, 1]$, and

$$\log(1 + x) = \sum_{k=1}^{\infty} \frac{(-1)^{k+1}}{k} x^k \quad \text{if } -1 < x \leq 1. \tag{4}$$

Note that we have proved that (4) holds at $x = 1$ as well as on $(-1, 1)$. Although we do not say that $\log(1 + x)$ is analytic at an endpoint such as $x = 1$, we can substitute 1 for x in (4) and get the interesting sum

$$\log 2 = \sum_{k=1}^{\infty} \frac{(-1)^{k+1}}{k}.$$

Thus we see that the sum of the alternating harmonic series is the natural logarithm of 2 (this is the assertion of Exercise 12.2.5).

Exercises 12.5

1. Prove that the cosine function is analytic on $\mathbb{R}$ and for each real number x

$$\cos x = \sum_{k=0}^{\infty} \frac{(-1)^k}{(2k)!} x^{2k}.$$

2. Prove that the inverse tangent function is analytic on $(-1, 1)$ and for each x in $[-1, 1]$

$$\arctan x = \sum_{k=0}^{\infty} \frac{(-1)^k}{2k+1} x^{2k+1}.$$

(*Hint:* Expand $1/(1 + t^2)$ as a geometric series and integrate; compare Exercise 12.2.4.)

3. Find the MacLaurin series for $f(x) = \dfrac{1}{x} \sin x$ and show that f is analytic on $\mathbb{R}$.

4. Find the MacLaurin series for the function f given by
$$f(x) = \int_0^x e^{-t^2} dt.$$
(*Hint:* Substitute $-t^2$ for x in Example 12.9.)

5. Find the MacLaurin series for the function f given by
$$f(x) = \int_0^x \sin t^2 \, dt.$$

6. Find the MacLaurin series for $f(x) = e^{-x}$ and show that f is analytic on $\mathbb{R}$.

7. Find the MacLaurin series for $f(x) = e^{x^2}$ and show that f is analytic on $\mathbb{R}$.

8. Find the Cauchy product of the MacLaurin series for $\sin x$ and $\cos x$, and use it to prove that for every x in $\mathbb{R}$,
$\sin 2x = 2 \sin x \cos x$.

13

METRIC SPACES AND EUCLIDEAN SPACES

13.1. Metric Spaces

Our objective in this chapter is twofold: we wish to establish a basis for developing a multidimensional calculus theory, and we want this basis to be general enough to acquaint ourselves with some limit concepts in an abstract setting. In this case the word *abstract* implies that the objects we deal with do not have to be numbers or even have arithmetic properties like numbers. The one property that is essential to our idea of limits is the concept of distance between elements. Each limit concept that we have seen thus far is based on the question of whether two numbers are close enough to each other whenever certain conditions hold. So our abstraction is developed in a system whose elements need possess only the property of "distance between pairs of elements." In the following, the symbol $X \times X$ denotes the collection of all ordered pairs of elements of X, that is,

$$X \times X = \{(x, y): x \in X \text{ and } y \in X\}.$$

DEFINITION 13.1. A *metric space* is a system that consists of a set X whose elements are called "points" and a function d whose domain is all of $X \times X$ and whose range is in $[0, \infty)$ such that the following properties hold:

(i) $d(x, y) = 0$ if and only if $x = y$;
(ii) $d(x, y) = d(y, x)$ for every x and y in X;
(iii) $d(x, z) \le d(x, y) + d(y, z)$ for every x, y, and z in X.

These three properties are very natural to our intuitive idea of distance. Property (i) merely asserts that any point is zero distance from itself and no two distinct points can be zero distance apart. Item (ii) is the *symmetry property*, which asserts that the distance from point x to point y must equal the distance from y to x. The sense of the third property is less obvious; it is called the *Triangle Inequality* because in certain very familiar settings it asserts that no leg of a triangle can exceed the sum of the other two legs. In our abstract space, according to (iii) the distance function d always gives the "shortest distance" between two points; for if we go from x to z by way of some third point y, then the total distance $d(x, y) + d(y, z)$ is at least as great as the "direct distance" $d(x, z)$.

Before developing any of the theory of metric spaces we give some examples to illustrate the definition. The first example is perhaps the most natural of all. It is based on the Cartesian coordinate system, the setting of Euclidean plane geometry, and the system in which we most easily visualize objects and distances.

EXAMPLE 13.1. Let X be $\mathbb{R} \times \mathbb{R}$, the set of all ordered pairs of real numbers, and define d as follows:

if $x = (x_1, x_2)$ and $y = (y_1, y_2)$, then

$$d(x, y) = [(x_1 - y_1)^2 + (x_2 - y_2)^2]^{1/2}. \tag{1}$$

Be careful of the notation; the subscripts do not indicate a first and second point, which is common usage in many ele-

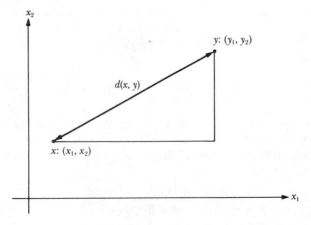

Figure 13.1

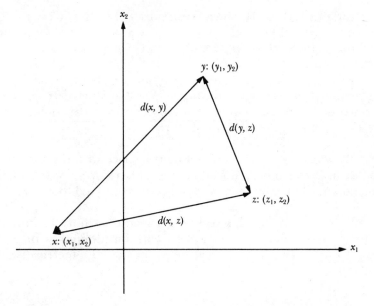

Figure 13.2

mentary texts. Rather, each point p has a "first coordinate" p_1 and a "second coordinate" p_2. As we see in Figure 13.1, the distance $d(x, y)$ is given by the Pythagorean Theorem. Figure 13.2 illustrates the Triangle Inequality. Properties (i) and (ii) are obvious.

EXAMPLE 13.2. Let X be $\mathbb{R}$ and define d by

$$d(x, y) = |x - y| \tag{2}$$

for all x and y in $\mathbb{R}$. Since the number $|x - y|$ represents the distance between the points on the number line corresponding to the numbers x and y, this is the natural definition of distance in $\mathbb{R}$. Properties (i), (ii), and (iii) are clearly satisfied.

EXAMPLE 13.3. Let $X = \mathbb{Q}$, the set of rational numbers, and define d by

$$d(x, y) = |x - y| \tag{3}$$

for all x and y in $\mathbb{Q}$. Since $\mathbb{Q}$ is a subset of $\mathbb{R}$, the fact that properties (i), (ii), and (iii) hold for all x, y, and z in $\mathbb{R}$ guarantees that they hold for all x, y, and z in $\mathbb{Q}$.

Example 13.3 illustrates a general fact about metric spaces that is sometimes useful in examples. We state it here as a proposition whose truth is an immediate consequence of the definition of metric space and the concept of subset:

PROPOSITION 13.1. If X is a metric space with distance function d and Y is a subset of X, then Y is a metric space with distance function d, where the domain of d is now restricted to $Y \times Y$.

Before finishing this section we give one more example of a metric space. This shows a very extreme definition of a distance function that still satisfies the properties of a metric distance function.

EXAMPLE 13.4. Let X be any (nonempty) set and define the *discrete metric* d by

$$d(x, y) = \begin{cases} 1, & \text{if } x \neq y, \\ 0, & \text{if } x = y, \end{cases} \tag{4}$$

for all x and y in X. It is obvious from (4) that (i) and (ii) are satisfied. To verify (iii), consider any points x, y, and z in X. If $d(x, z) = 0$, then (iii) is trivial. Suppose $d(x, z) > 0$; then (4) implies that $x \neq z$ and $d(x, z) = 1$. Now $x \neq z$ implies that y cannot equal both x and z, so at least one of the distances $d(x, y)$ or $d(y, z)$ must equal 1. Hence the Triangle Inequality must hold.

Exercises 13.1

In Exercises 1–9, X is $\mathbb{R} \times \mathbb{R}$. For the given distance function determine which of properties (i), (ii), and (iii) are satisfied and sketch the set of all points x such that the distance from x to $(0, 0)$ equals 1.

1. $d(x, y) = |x_1 - y_1| + |x_2 - y_2|$.

2. $d(x, y) = |x_1 - y_1| + 2|x_2 - y_2|$.

3. $d(x, y) = |x_1 - y_1|$.

4. $d(x, y) = |x_2 - y_2|$.

5. $d(x, y) = \max\{|x_1 - y_1|, |x_2 - y_2|\}$.

6. $d(x, y) = \max\{|x_1 - y_1|, 2|x_2 - y_2|\}$.

7. $d(x, y) = \begin{cases} [(x_1 - y_1)^2 + (x_2 - y_2)^2]^{1/2}, & \text{if } x_2 \neq y_2, \\ (1/2)[(x_1 - y_1)^2 + (x_2 - y_2)^2]^{1/2}, & \text{if } x_2 = y_2. \end{cases}$

8. $d(x, y) = \begin{cases} [(x_1 - y_1)^2 + (x_2 - y_2)^2]^{1/2}, & \text{if } x_2 \geq y_2, \\ |x_1 - y_1| + |x_2 - y_2|, & \text{if } x_2 < y_2. \end{cases}$

9. $d(x, y) = \begin{cases} 3[(x_1 - y_1)^2 + (x_2 - y_2)^2]^{1/2}, & \text{if } x_2 \geq 0 \text{ and } y_2 \geq 0, \\ [(x_1 - y_1)^2 + (x_2 - y_2)^2]^{1/2}, & \text{if } x_1 < 0 \text{ or } y_1 < 0. \end{cases}$

10. Prove the extension of the Triangle Inequality for m points: If $x^{(1)}, x^{(2)}, \ldots, x^{(m)}$ are points in the metric space X, then

$$d(x^{(1)}, x^{(m)}) \leq \sum_{j=2}^{m} d(x^{(j)}, x^{(j-1)}).$$

13.2. Euclidean n-Space

In this section we introduce a class of metric spaces that extends the familiar two-dimensional space of Example 13.1. The name *Euclidean* is derived from the fact that the two-dimensional case is the setting for Euclidean plane geometry.

Definition 13.2. Let n be a fixed positive integer and E^n denote the set of all (finite) number sequences $\mathbf{x} = \{x_k\}_{k=1}^n$. Let d be the function defined on $E^n \times E^n$ by

$$d(\mathbf{x}, \mathbf{y}) = [\sum_{k=1}^{n} |x_k - y_k|^2]^{1/2}. \tag{1}$$

Then E^n with this distance function d is called *n-dimensional Euclidean space*, or, more briefly, *Euclidean n-space*.

Note that in the case $n = 1$ we have $E^1 = \mathbb{R}$, and (1) becomes $d(\mathbf{x}, \mathbf{y}) = |x_1 - y_1| = |x - y|$. Thus E^1 is just $\mathbb{R}$ with the usual distance function.

Each element of E^n is called a *point* and is customarily written as

$$\mathbf{x} = \{x_k\}_{k=1}^n = (x_1, \ldots, x_n).$$

For each $k = 1, \ldots, n$, the number x_k is called the kth coordinate of $\mathbf{x}$. Note that the adjective "metric" was carefully omitted in defining Euclidean space. Our immediate goal, however, is to verify that formula (1) of Definition 13.2 does indeed provide E^n with a metric distance function. It is obvious from (1) that properties (i) and (ii) of Definition 13.1 are satisfied, but (iii) would require some very tedious algebra for a direct verification (the square roots cause the messiness). Therefore we prove the Triangle Inequality by a different method, one that avoids most of the algebraic computation. This method makes use of the series version of the Cauchy-Bunyakovsky-Schwarz Inequality (see Theorem 7.10).

THEOREM 13.1: CAUCHY-BUNYAKOVSKY-SCHWARZ INEQUALITY. If each of $\mathbf{x}$ and $\mathbf{y}$ is in E^n, then

$$\sum_{k=1}^n x_k y_k \leq [\sum_{k=1}^n x_k^2]^{1/2} [\sum_{k=1}^n y_k^2]^{1/2}.$$

Proof. Consider the quadratic polynomial $Q(t) = At^2 + Bt + C$, where

$$A = \sum_{k=1}^n x_k^2, \quad B = \sum_{k=1}^n 2x_k y_k, \quad \text{and} \quad C = \sum_{k=1}^n y_k^2.$$

Then

$$Q(t) = \sum_{k=1}^n (x_k^2 t^2 + 2x_k y_k t + y_k^2) = \sum_{k=1}^n (x_k t + y_k)^2 \geq 0.$$

Since $Q(t)$ is never negative, it cannot have two distinct zeros. Therefore, by the quadratic formula, its discriminant $B^2 - 4AC$ cannot be positive. For this particular quadratic, $B^2 - 4AC \leq 0$ is the same as

$$4[\sum_{k=1}^n x_k y_k]^2 - 4[\sum_{k=1}^n x_k^2][\sum_{k=1}^n y_k^2] \leq 0,$$

which immediately yields the Cauchy-Bunyakovsky-Schwarz Inequality. Note that this proof is essentially the same as the proof of Theorem 7.10, the integral version of this inequality.

COROLLARY 13.1a: TRIANGLE INEQUALITY FOR E^n. If each of $\mathbf{x}$, $\mathbf{y}$, and $\mathbf{z}$ is a point in E^n, then

$$d(\mathbf{x}, \mathbf{z}) \leq d(\mathbf{x}, \mathbf{y}) + d(\mathbf{y}, \mathbf{z}).$$

Proof. We have

$$[d(\mathbf{x}, \mathbf{y}) + d(\mathbf{y}, \mathbf{z})]^2 = d(\mathbf{x}, \mathbf{y})^2 + 2d(\mathbf{x}, \mathbf{y})d(\mathbf{y}, \mathbf{z}) + d(\mathbf{y}, \mathbf{z})^2$$

$$= \sum_{k=1}^{n}(x_k - y_k)^2 + 2[\sum_{k=1}^{n}(x_k - y_k)^2]^{1/2} [\sum_{k=1}^{n}(y_k - z_k)^2]^{1/2}$$

$$+ \sum_{k=1}^{n}(y_k - z_k)^2$$

$$\geq \sum_{k=1}^{n}(x_k - y_k)^2 + 2\sum_{k=1}^{n}(x_k - y_k)(y_k - z_k)$$

$$+ \sum_{k=1}^{n}(y_k - z_k)^2$$

$$= \sum_{k=1}^{n}[(x_k - y_k) + (y_k - z_k)]^2$$

$$= \sum_{k=1}^{n}(x_k - z_k)^2$$

$$= d(\mathbf{x}, \mathbf{z})^2.$$

Taking the square root of the first and last members, we get the Triangle Inequality.

COROLLARY 13.1b. The space E^n with the Euclidean distance function d, given in Definition 13.2, is a metric space.

Proof. This corollary follows immediately from Corollary 13.1a.

Now that we have established that $\{E^n, d\}$ is a metric space, we adopt the following convention, which is customary usage. Whenever we refer to E^n without specifying the distance function, it is assumed that E^n has the usual Euclidean distance function as in Definition 13.2. Also, when we are working with E^n, the symbol d will always denote the Euclidean distance function.

In examples and exercises it is convenient to have a short symbol for the point $(0, 0, \ldots, 0)$ in E^n; we denote it by $\mathbf{0}$. The number of coordinates will be clear from the context: in E^2, $\mathbf{0} = (0, 0)$; in E^3, $\mathbf{0} = (0, 0, 0)$; etc.

Exercises 13.2

1. Prove the Triangle Inequality for m points: If $\mathbf{x}^{(1)}, \mathbf{x}^{(2)}, \ldots, \mathbf{x}^{(m)}$ are m points in E^n, then

$$d(\mathbf{x}^{(1)}, \mathbf{x}^{(m)}) \leq \sum_{j=2}^{m} d(\mathbf{x}^{(j-1)}, \mathbf{x}^{(j)}).$$

2. Another metric distance function for E^n is the so-called taxi-cab metric:

$$d^*(\mathbf{x}, \mathbf{y}) = \sum_{k=1}^{n} |x_k - y_k|.$$

Prove that d^* satisfies properties (i), (ii), and (iii) of Definition 13.1.

3. Prove that for any $\mathbf{x}$ and $\mathbf{y}$ in E^n, $d(\mathbf{x}, \mathbf{y}) \leq d^*(\mathbf{x}, \mathbf{y})$, where d^* is defined as in Exercise 2. (*Hint:* In case $n = 2$, we could write $d^*(\mathbf{x}, \mathbf{y}) = d(\mathbf{x}, \mathbf{x}') + d(\mathbf{x}', \mathbf{y})$, where $\mathbf{x}' = (x_1, y_2)$.)

13.3. Metric Space Topology

The *topology* for a space of points provides a theory of limit points and convergence for sets of points in the space. In a metric space, this convergence theory is based on the concept of distance between points. For more abstract spaces it is based on a system of "open sets," and there may be no distance concept associated with the notion of limit. We here take the former approach and use the metric distance function to define certain "open sets," which are generalizations of the open intervals in $\mathbb{R}$. Throughout this section we assume that X is a metric space with distance function d.

DEFINITION 13.3. If x is a point in X and r is a nonnegative number, then

$$\mathbf{N}_r(x) = \{y \in X: d(x, y) < r\}$$

is called the *open sphere* of radius r about x.

$$\bar{\mathbf{N}}_r(x) = \{y \in X: d(x, y) \leq r\}$$

is called the *closed sphere* of radius r about x. A *neighborhood* of x is any subset of X that contains some open sphere about x.

Note that for any point x in X, $\mathbf{N}_0(x) = \varnothing$; that is, an open sphere of radius zero is empty. Also, $\bar{\mathbf{N}}_0(x) = \{x\}$, where $\{x\}$ denotes the singleton set consisting of the single point x.

EXAMPLE 13.5. If $X = E^1 = \mathbb{R}$, then for any x in $\mathbb{R}$ and $r > 0$,

$$\mathbf{N}_r(x) = (x - r, x + r);$$

that is, an open sphere about x is an open interval centered at x. Similarly, a closed sphere is a closed interval.

EXAMPLE 13.6. If $X = E^2$, then for any $\mathbf{x}$ in E^2 and $r > 0$, $\mathbf{N}_r(\mathbf{x})$ consists of all points lying within a circle of radius r centered at $\mathbf{x}$. The closed sphere $\bar{\mathbf{N}}_r(\mathbf{x})$ consists of those points lying within or on that circle.

EXAMPLE 13.7. If $X = E^3$, then the open and closed spheres are spherical in the usual sense, the latter consisting of the points within or on the sphere of radius r centered at $\mathbf{x}$, while the former consists of just the points within the sphere.

DEFINITION 13.4. Let A be a subset of X; the point p is a *limit point* of A provided that every open sphere about p contains a point of A other than p itself.

EXAMPLE 13.8. In E^2, let A be the set $\mathbf{N}_2(\mathbf{0})$ and let $\mathbf{p} = (\sqrt{2}, \sqrt{2})$; then $\mathbf{p}$ is a limit point of A. For, if $\mathbf{N}_\varepsilon(\mathbf{p})$ is an arbitrary

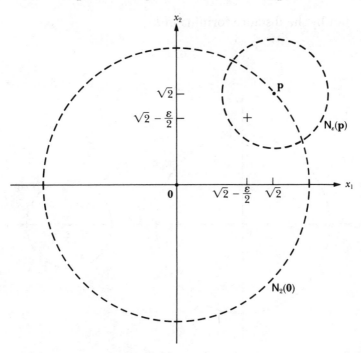

Figure 13.3

open sphere about **p**, then $N_\varepsilon(p)$ contains points that are inside the sphere $N_2(\mathbf{0})$. For example, the point $(\sqrt{2} - \varepsilon/2, \sqrt{2} - \varepsilon/2)$ is in $N_2(\mathbf{0}) \cap N_\varepsilon(\mathbf{p})$. (See Figure 13.3.) Check the inequalities to verify this. Note that **p** is not in $N_2(\mathbf{0})$, so we see that a point can be a limit point of a set without being an element of the set.

EXAMPLE 13.9. In E^2, let A be the set $N_1(\mathbf{0}) \cup \{\mathbf{p}\}$, where $\mathbf{p} = (2, 0)$. Then **p** is *not* a limit point of A, because the sphere $N_{1/2}(\mathbf{p})$ contains no point of A except **p** itself. Although this assertion appears obvious in Figure 13.4, its proof is not trivial. We argue by contradiction: Suppose there exists a point **y** in $N_{1/2}(\mathbf{p}) \cap A$ such that $\mathbf{y} \neq \mathbf{p}$. Then **y** must be in $N_1(\mathbf{0})$, so $d(\mathbf{0}, \mathbf{y}) < 1$. Also, **y** is in $N_{1/2}(\mathbf{p})$, so $d(\mathbf{y}, \mathbf{p}) < 1/2$. Therefore the Triangle Inequality gives us

$$d(\mathbf{0}, \mathbf{p}) \leq d(\mathbf{0}, \mathbf{y}) + d(\mathbf{y}, \mathbf{p})$$
$$< 1 + 1/2;$$

but by the distance formula in E^2,

$$d(\mathbf{0}, \mathbf{p}) = 2.$$

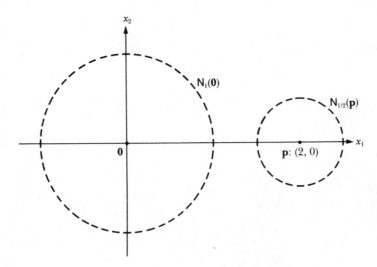

Figure 13.4

This contradiction means there can be no such point **y**, so $N_{1/2}(\mathbf{p})$ contains no point of A other than **p**.

The stipulation that each sphere about p must contain a point of A *other than* p may seem a little contrived, but it is necessary to make this concept of limit point agree with the definition as given in Chapter 2. More important, this stipulation is needed to make a distinction between the property of being "close to the set A" (that is, a limit point) and the property of being "in the set A" (that is, an element).

DEFINITION 13.5. The set A (in X) is said to be *open* if for every point x in A there is an open sphere about x that is contained in A.

Even if the set A is not open it may be true that some of its points are centers of open spheres that are contained entirely in A. Such a point is called an *interior point* of A, and the subset of A consisting of all such interior points is called the *interior* of A, denoted $A°$. It is obvious that for any set A, $A° \subseteq A$. It is also true that A is open if and only if $A° \supseteq A$ (see Exercise 13.3.5).

The concept of open set is a generalization of the notion of open interval that we have in $\mathbb{R}$. Most of the examples of open sets are developed in the exercises, but we mention a few here. An open sphere $N_r(x)$ is an open set, the proof of which is requested in Exercise 13.3.3. Also, X is obviously open because it contains *every* sphere about *every* point. And the empty set $\varnothing$ is open because it contains no point that fails the test of being an interior point.

DEFINITION 13.6. The set F in X is *closed* if F contains all of its limit points.

An easy example of a closed set is a singleton set $\{x\}$. It has no limit point, so it cannot fail to contain any of its limit points. As we see in Exercise 13.3.2, this can be extended to any finite set. The whole space X is a closed set because it certainly contains all of its limit points. Therefore X is both open and closed. Similarly, $\varnothing$ is both open and closed. In E^n these two sets are the only sets that are both open and closed. This is proved in Section 13.4. We should also note that there may be sets that are neither open nor closed, as in the following example, whose details are to be verified in Exercise 13.3.6:

EXAMPLE 13.10. Let F be the subset of E^2 consisting of

$$\left\{(1, 0), \left(\frac{1}{2}, 0\right), \left(\frac{1}{3}, 0\right), \ldots, \left(\frac{1}{k}, 0\right), \ldots\right\}.$$

Then 0 is a limit point of F, and since 0 is not in F we infer that F is not closed. Also, the point $(1, 0)$ is not an interior point of F, so F is not open.

The next result is the fundamental relationship between the notions of open set and closed set. We denote the complement of a set A by $\sim A$, which is short for $X \sim A$, that is, all points in X that are not in A.

THEOREM 13.2. The set A is open if and only if its complement $\sim A$ is closed.

Proof. Suppose A is open. To show that $\sim A$ is closed, we observe that any point x that is not in $\sim A$ is in A. Since A is open, there is a sphere $\mathbf{N}_r(x)$ contained entirely in A. Therefore $\mathbf{N}_r(x)$ contains no point of $\sim A$, so x is not a limit point of $\sim A$. Thus $\sim A$ cannot fail to contain all of its limit points.

Conversely, if A is not open, then A contains some point y that is not an interior point. Thus every sphere $\mathbf{N}_r(y)$ fails to be contained in A, so each $\mathbf{N}_r(y)$ contains a point of $\sim A$. Therefore y is a limit point of $\sim A$, and since y is in A rather than $\sim A$, we conclude that $\sim A$ is not closed.

Perhaps the greatest value of Theorem 13.2 is that it gives us an alternative method for proving that a set is open (or closed). It is sometimes easier to work with the complement of a set than with the set itself. For example, we know immediately that $X \sim \{p\}$ is open because we know that the singleton $\{p\}$ is closed.

THEOREM 13.3. The intersection of a finite number of open sets is itself an open set.

Proof. Suppose each of $A_1, \ldots, A_n$ is open, and let x be an arbitrary point in $A = \cap_{k=1}^n A_k$. Then x is in each of the open sets $A_1, \ldots, A_n$, so for each $k \leq n$ there is a positive radius r_k such that $\mathbf{N}_{r_k}(x) \subseteq A_k$. Define $r = \min\{r_1, \ldots, r_n\}$. Then for each $k \leq n$,

$$\mathbf{N}_r(x) \subseteq \mathbf{N}_{r_k}(x) \subseteq A_k \subseteq A.$$

Hence x is in A°, and we have proved that A is open.

COROLLARY 13.3. The union of a finite number of closed sets is itself a closed set.

Proof. Suppose each of $F_1, \ldots, F_n$ is a closed set. It is easy to verify that the following set equation holds:

$$\sim \bigcup_{k=1}^{n} F_k = \bigcap_{k=1}^{n} (\sim F_k), \tag{1}$$

and by Theorem 13.2, each $\sim F_k$ on the right side is open. Thus, according to Theorem 13.3 their intersection, which is the entire right member of (1), is open. Hence, $\cup_{k=1}^{n} F_k$ is closed because its complement is open.

In Theorem 13.3 and its corollary it is necessary to restrict the assertion to finite collections of sets. This is demonstrated by the next example:

EXAMPLE 13.11. The union of an infinite collection of closed sets need not be closed. For example, let F_k be the singleton set consisting of the point $(1/k, 0)$ in E^2. The distance between this point and the origin **0** is $1/k$, so it follows that **0** is a limit point of $\cup_{k=1}^{\infty} F_k$. But **0** is not in $\cup_{k=1}^{\infty} F_k$, so the union is not closed.

It is easy to give an example of an infinite collection of open sets whose intersection is not open. For example, let $A_k = \mathbf{N}_{1/k}(\mathbf{0})$. It is clear that $\cap_{k=1}^{\infty} A_k = \{\mathbf{0}\}$, and the singleton set $\{\mathbf{0}\}$ is not open.

THEOREM 13.4. The union of any collection of open sets is an open set.

Proof. Suppose that for each μ in some set M, A_μ is an open set. (The set M is called an "index set" for the collection of sets $\{A_\mu : \mu \in M\}$.) Let x be a point in $\cup_{\mu \in M} A_\mu$. Then x is in at least one of the sets, say, $x \in A_{\mu*}$. Since $A_{\mu*}$ is open, there is a sphere $\mathbf{N}_r(x)$ such that $\mathbf{N}_r(x) \subset A_{\mu*} \subseteq \cup_{\mu \in M} A_\mu$. Hence $\cup_{\mu \in M} A_\mu$ is an open set.

COROLLARY 13.4. The intersection of any collection of closed sets is a closed set.

Proof. This proof is similar to that of Corollary 13.3 and is left as Exercise 13.3.10.

We should take particular notice of the fact that the collections of sets in Theorem 13.4 and Corollary 13.4 may be arbitrarily large. In particular, the collections may contain so many sets that it is not possible to describe them as an infinite sequence of sets. (See Appendix B for a discussion of such "uncountable sets.") This is why we use the device of an index set M to describe the collection. Had we written $\{A_n\}_{n=1}^{\infty}$ for the collection of sets, this would have indicated a sequence of sets and therefore limited our discussion to countable collections.

DEFINITION 13.7. If $A \subseteq X$, then the closure of A, denoted $\bar{A}$, is the set consisting of all points x such that every open sphere about x intersects A. Thus

$$\bar{A} = \{x \in X: \text{ for every } r > 0,\ \mathbf{N}_r(x) \cap A \neq \varnothing\}.$$

It is obvious that for every set, $A \subseteq \bar{A}$.

Exercises 13.3

In these exercises, A represents a point set in some arbitrary metric space X.

1. Prove: If p is a limit point of A, then every open sphere $\mathbf{N}_\varepsilon(p)$ contains infinitely many points of A.

2. Prove: If A is a finite set, that is, A contains only a finite number of points, then A has no limit point.

3. Prove: For each x in X and $r > 0$, $\mathbf{N}_r(x)$ is an open set.

4. Prove: A° is the largest open subset of A; that is, if B is an open set such that $B \subseteq A$, then $B \subseteq A^\circ$.

5. Prove: A is open if and only if $A = A^\circ$.

6. Provide the details to prove the two assertions in Example 13.10.

7. Prove the set equality (1) used in proving Corollary 13.3.

8. Prove the set equality

$$\sim \bigcap_{\lambda \in \Lambda} A_\lambda = \bigcup_{\lambda \in \Lambda} (\sim A_\lambda).$$

9. Prove: A can be written as the union of a collection of closed sets.

10. Prove Corollary 13.4.

11. In E^2, show that the closure of $\mathbf{N}_1(\mathbf{0})$ is $\bar{\mathbf{N}}_1(\mathbf{0})$.

12. Let $\mathbb{Q}^2$ be the subset of E^2 consisting of all points $\mathbf{x} = (x_1, x_2)$ such that both x_1 and x_2 are rational numbers. What is $\bar{\mathbb{Q}^2}$? What is $(\mathbb{Q}^2)°$?

13. Prove: $\bar{A}$ is a closed set.

14. Prove: $\bar{A}$ is the smallest closed set containing A; that is, if F is a closed set such that $A \subseteq F$, then $\bar{A} \subseteq F$.

15. Prove: If L_A denotes the set of all limit points of A, then L_A is a closed set.

16. Prove: If L_A is given as in Exercise 15, then $\bar{A} = A \cup L_A$.

13.4. Connectedness

In this section we study the phenomenon of a set that cannot be "naturally separated" into two or more subsets. The vagueness of the phrase "naturally separated" suggests that it is not easy to give a precise definition of the property we want. Indeed, we define it by first stating when a set does *not* possess this property.

DEFINITION 13.8. The set S (in the metric space X) is *disconnected* if there exist nonempty sets S_1 and S_2 such that $S_1 \cup S_2 = S$,

$$\bar{S}_1 \cap S_2 = \varnothing \quad \text{and} \quad S_1 \cap \bar{S}_2 = \varnothing. \tag{1}$$

If S is not disconnected, then S is said to be *connected*.

We say that the sets A and B are *disjoint* if $A \cap B = \varnothing$. It is important to note that the sets S_1 and S_2 in (1) satisfy a property that is stronger than being disjoint. Not only must S_1 and S_2 fail to contain any point of the other set, they must also fail to contain any limit point of the other set. It is this extra stipulation that gives Definition 13.8 its significance, because any set containing two or more points can be written as the union of two nonempty disjoint subsets: If x is in S, take $S_1 = \{x\}$ and $S_2 = S \sim \{x\}$. But as we see in the next theorem, when we are dealing with open sets, disjointness is sufficient to imply that their union is disconnected.

THEOREM 13.5. The set S is connected if and only if there do not exist disjoint open sets A and B such that $S \subseteq A \cup B$, $A \cap S \neq \varnothing$, and $B \cap S \neq \varnothing$.

Proof. Suppose A and B exist as in the statement of the theorem, and let $S_1 = A \cap S$ and $S_2 = B \cap S$. Clearly S_1 and S_2 are nonempty, and $S_1 \cup S_2 = S$. Consider an arbitrary point x in S_1; then x is in the open set A, so for some positive number r, $\mathbf{N}_r(x) \subset A$. Thus $\mathbf{N}_r(x) \cap B = \varnothing$ because A and B are disjoint, so $\mathbf{N}_r(x) \cap S_2 = \varnothing$. Therefore $x \notin \bar{S}_2$; hence $S_1 \cap \bar{S}_2 = \varnothing$. Similarly, $\bar{S}_1 \cap S_2 = \varnothing$, and we conclude that S is disconnected.

Now assume that S is disconnected, say, $S = S_1 \cup S_2$ and $S_1 \cap \bar{S}_2 = \bar{S}_1 \cap S_2 = \varnothing$. Since $S_1 \cap \bar{S}_2 = \varnothing$, no point of S_1 is a limit point of S_2. For each x in S_1, choose $\mathbf{N}_r(x)$ such that $\mathbf{N}_r(x) \cap S_2 = \varnothing$. Then $A = \bigcup_{x \in S_1} \mathbf{N}_{r/2}(x)$ is an open set that contains no point of S_2. (Although our notation does not indicate it, the radius r of the sphere $\mathbf{N}_r(x)$ varies with x.) Similarly, we can choose a sphere $\mathbf{N}_{r'}(y)$ for each y in S_2 that satisfies $\mathbf{N}_{r'}(y) \cap S_1 = \varnothing$ and define $B = \bigcup_{y \in S_2} \mathbf{N}_{r'/2}(y)$, which is an open set by Theorem 13.4. It remains only to show that A and B are disjoint. Suppose not, and let p be a point in both A and B. Then for some x in S_1 and y in S_2, $p \in \mathbf{N}_{r/2}(x) \cap \mathbf{N}_{r'/2}(y)$. But this implies that

$$d(x, y) \le d(x, p) + d(p, y) < r/2 + r'/2 \le \max\{r, r'\}. \qquad (2)$$

Therefore either $y \in \mathbf{N}_r(x)$ or $x \in \mathbf{N}_{r'}(y)$, which contradicts the choice of either r or r', respectively.

THEOREM 13.6. If S is connected, then $\bar{S}$ is connected.

Proof. Suppose $\bar{S}$ is disconnected, and let A and B be disjoint open sets as in Theorem 13.5: $\bar{S} \subset A \cup B$, $\bar{S} \cap A \ne \varnothing$, and $S \cap B \ne \varnothing$. Since S is obviously contained in $A \cup B$, we need to show only that $S \cap A \ne \varnothing$ and $S \cap B \ne \varnothing$. Let x be a point in $\bar{S} \cap A$. Since A is open, there is a sphere about x such that $\mathbf{N}_r(x) \subset A$. But x is also in $\bar{S}$, so $\mathbf{N}_r(x)$ must contain some point y of S. Therefore $y \in \mathbf{N}_r(x) \cap S \subset S \cap A$, so $S \cap A \ne \varnothing$. Similarly, $S \cap B \ne \varnothing$, so by Theorem 13.5, S is disconnected.

In Exercises 13.4 we develop some examples and properties of connected and disconnected sets in the Euclidean spaces E^1 and E^2.

Exercises 13.4

1. Prove: If S is a finite subset of the metric space X, then S is disconnected.

2. In E^2, let $S = N_1(0) \cup \{p\}$, where $p = (2, 0)$. Show that S is disconnected.

3. Prove: In an arbitrary metric space X, if $S \sim L_S \neq \emptyset$, then S is disconnected.

4. Let f and g be strictly positive functions on $\mathbb{R}$ so that the graphs $G_1 = \{x \in E^2 : x_2 = f(x_1)\}$ and $G_2 = \{x \in E^2 : x_2 = -g(x_1)\}$ are sets in E^2. Prove that $S = G_1 \cup G_2$ is disconnected.

5. Give an example (with explanations) of a connected set S in E^2 that contains exactly one point p such that $S \sim \{p\}$ is disconnected.

6. Let S be a connected set in E^2 that contains points x and y such that $x_1 < y_1$. Prove that for every number μ such that $x_1 < \mu < y_1$, there is a point z in S such that $z_1 = \mu$.

7. Prove that E^1 is a connected subset of itself. (*Hint:* Use the Dedekind Cut Theorem, Exercise 1.3.14.)

8. Prove: If S is a connected set in E^1, then S satisfies the following "intermediate value property": If a and b are numbers in S, then S contains every number between a and b.

9. Prove: If S is a connected subset of E^1, then S is an interval.

10. Prove or disprove: In E^2 the intersection of connected sets is a connected set.

11. Prove or disprove: In E^1 the intersection of connected sets is a connected set.

12. Let $\mathbb{Q}^2$ be the subset of E^2 consisting of those points $x = (x_1, x_2)$ such that both x_1 and x_2 are rational numbers. Is $\mathbb{Q}^2$ a connected set?

13.5. Point Sequences

A point sequence is a function from the positive integers into a metric space X, that is, a sequence whose terms are points in X. Since our principal examples of metric spaces are the Euclidean spaces and we use subscripts to designate the coordinates of points in E^n, we use superscripts to index the terms of a point sequence: $\{x^{(k)}\}_{k=1}^{\infty}$.

DEFINITION 13.9. The point sequence $\{x^{(k)}\}_{k=1}^{\infty}$ is said to *converge* to the point p in X provided that for every positive number ε there is a number N such that

$$x^{(k)} \in N_\varepsilon(p) \quad \text{whenever} \quad k > N.$$

This is denoted by $\lim_k x^{(k)} = p$.

Our first task is to establish a connection between convergent point sequences and the open sphere limit points of Section 13.3.

THEOREM 13.7. The point p is a limit point of the point set A if and only if there is a nonrepeating point sequence in A that converges to p.

Proof. First assume p is a limit point of A. Then the sphere $N_1(p)$ contains a point $x^{(1)}$ of A other than p. Let $r(2) = \min\{1/2, d(x^{(1)}, p)\}$, and choose a point $x^{(2)}$ of $A \sim \{p\}$ in the sphere $N_{r(2)}(p)$. Continue this process: After $x^{(1)}, \ldots, x^{(k)}$ have been defined, let $r(k + 1) = \min\{1/(k + 1), d(x^{(k)}, p)\}$ and choose $x^{(k+1)}$ as a point of $A \sim \{p\}$ in the sphere $N_{r(k+1)}(p)$. Since $r(k) \le 1/k$, it is clear that $\lim_k x^{(k)} = p$; and since $r(k) \le d(x^{(k-1)}, p)$, it follows that no two $x^{(k)}$'s are the same. To prove the converse, we simply note that if A contains such a point sequence that converges to p, then it is obvious from Definition 13.9 that every open sphere about p contains infinitely many points of A.

Note that it is necessary to stipulate that the point sequence in Theorem 13.7 is nonrepeating. For example, the constant sequence in which $x^{(k)} = p$ is certainly convergent to p, but the singleton set $\{p\}$ does not have p as a limit point.

In the next theorem we prove a very strong connection between convergent point sequences in E^n and convergent number sequences, namely, that the convergence of $\{x^{(k)}\}_{k=1}^{\infty}$ is equivalent to having each of the coordinate (number) sequences converge.

THEOREM 13.8. In E^n, the point sequence $\{x^{(k)}\}_{k=1}^{\infty}$ converges to p if and only if

$$\lim_k x_i^{(k)} = p_i \quad \text{for each} \quad i = 1, \ldots, n. \tag{1}$$

Proof. First we note that the statement $\lim_k x^{(k)} = p$ is equivalent to $\lim_k d(x^{(k)}, p) = 0$, because $x^{(k)} \in N_\varepsilon(p)$ if and only if $d(x^{(k)}, p) < \varepsilon$. Since

$$d(\mathbf{x}^{(k)}, \mathbf{p}) = [\textstyle\sum_{j=1}^{n}|x_j^{(k)} - p_j|^2]^{1/2}$$

$$\geq |x_i^{(k)} - p_i| \quad \text{for} \quad i = 1, \ldots, n, \tag{2}$$

it is clear that $\lim_k d(\mathbf{x}^{(k)}, \mathbf{p}) = 0$ implies (1). Conversely, if (1) holds, then Equation (2) and Theorems 2.3 and 2.4 on the algebraic combinations of convergent number sequences allow us to conclude that $\lim_k d(\mathbf{x}^{(k)}, \mathbf{p}) = 0$.

DEFINITION 13.10. The point sequence $\{x^{(k)}\}_{k=1}^{\infty}$ is a *Cauchy sequence* in the metric space X provided that for every positive number ε there is a number N such that

$$d(x^{(k)}, x^{(m)}) < \varepsilon \quad \text{whenever} \quad k > m > N.$$

This definition is obviously a generalization of the concept of Cauchy number sequence. Therefore, the Cauchy point sequences in E^n can be identified very precisely, which we do in the next theorem.

THEOREM 13.9. In E^n a point sequence is a Cauchy sequence if and only if it converges to a point in E^n.

Proof. Suppose $\{\mathbf{x}^{(k)}\}_{k=1}^{\infty}$ is a Cauchy sequence. Since

$$d(\mathbf{x}^{(k)}, \mathbf{x}^{(m)}) = [\textstyle\sum_{j=1}^{n} |x_j^{(k)} - x_j^{(m)}|^2]^{1/2}$$

$$\geq |x_i^{(k)} - x_i^{(m)}| \quad \text{for} \quad i = 1, \ldots, n,$$

it follows that for each $i = 1, \ldots, n$, the ith coordinates $\{x_i^{(k)}\}_{k=1}^{\infty}$ form a Cauchy number sequence. From Theorem 3.2 we know that such sequences converge to a limit in $\mathbb{R}$, say, $\lim_k x_i^{(k)} = p_i$. Now Theorem 13.8 ensures that the point sequence $\{\mathbf{x}^{(k)}\}_{k=1}^{\infty}$ converges to $\mathbf{p} = \{p_1, \ldots, p_n\}$ in E^n. The converse assertion is true in *any* metric space, and therefore we give it as the next theorem.

THEOREM 13.10. If $\{x^{(k)}\}_{k=1}^{\infty}$ is a convergent sequence in the metric space X, then $\{x^{(k)}\}_{k=1}^{\infty}$ is a Cauchy sequence.

Proof. Assume that $\lim_k x^{(k)} = p$, and suppose $\varepsilon > 0$. Choose N so that

$$d(x^{(k)}, p) < \frac{\varepsilon}{2} \quad \text{whenever} \quad k > N. \tag{3}$$

Therefore when k and m are *both* greater than N, we use the Triangle Inequality and (3) to get

$$d(x^{(k)}, x^{(m)}) \leq d(x^{(k)}, p) + d(p, x^{(m)})$$

$$< \frac{\varepsilon}{2} + \frac{\varepsilon}{2}$$

$$= \varepsilon.$$

Hence $\{x^{(k)}\}_{k=1}^{\infty}$ is a Cauchy sequence.

At this point one may ask, "Why not prove *both* parts of Theorem 13.9 for a general metric space X, and then Theorem 13.9 would be only an example of a more general result?" The reason is that for general metric spaces it is not necessarily true that every Cauchy sequence converges. The subtlety is that when we say that a point sequence converges, there is an unspoken requirement that *there is a point in the space to which the sequence converges.* It is this required "limit point" that can cause the difficulty in some metric spaces. As we saw in Proposition 13.1, we can form a new metric space by removing a point p from a given space, say,

$$Y = X \sim \{p\}.$$

Now suppose that the point p had been the limit of a sequence $\{x^{(k)}\}_{k=1}^{\infty}$ of points not equal to p. Then $\{x^{(k)}\}_{k=1}^{\infty}$ is a sequence in both X and Y, and it converges in X; so by Theorem 13.10 it is a Cauchy sequence in X. The distance function d is the same in Y as in X, so $\{x^{(k)}\}_{k=1}^{\infty}$ is also a Cauchy sequence in Y. But in Y there is no point to which $\{x^{(k)}\}_{k=1}^{\infty}$ can converge, because p has been removed and limits of sequences are unique (see Exercise 13.5.6). This idea is the general principle behind the following two examples.

EXAMPLE 13.12. Let X be the sphere $\mathsf{N}_1(\mathbf{0})$ in E^2 with the usual Euclidean distance function d. Consider a sequence in $\mathsf{N}_1(\mathbf{0})$ that converges to a point on its boundary, say, $\mathbf{x}^{(k)} = (1 - 1/k, 0)$, so that $\lim_k \mathbf{x}^{(k)} = (1, 0)$. Then $\{\mathbf{x}^{(k)}\}_{k=1}^{\infty}$ is a Cauchy sequence in X (as well as in E^2), but it has no limit in X, so it does not converge in X.

EXAMPLE 13.13. Let $\mathbb{Q}^n$ be the subset of E^n consisting of all points $\mathbf{x}$ for which all coordinates $x_1, \ldots, x_n$ are rational numbers (compare Exercise 13.3.12). The point $\mathbf{p} = (\sqrt{2},$ $0, \ldots, 0)$ is in $E^n \sim \mathbb{Q}^n$, but if $\{r_k\}_{n=1}^{\infty}$ is a sequence of rational numbers such that $\lim_k r_k = \sqrt{2}$, and $\mathbf{x}^{(k)} = (r_k, 0, \ldots, 0)$, then $\lim_k \mathbf{x}_k = \mathbf{p}$. Therefore $\{\mathbf{x}^{(k)}\}_{k=1}^{\infty}$ is a Cauchy sequence but does not converge in $\mathbb{Q}^n$.

In Section 13.6 we encounter some properties that imply that Cauchy sequences converge. But for now we prove only one property that is true of Cauchy sequences in all metric spaces.

DEFINITION 13.11. The sequence $\{x^{(k)}\}_{k=1}^{\infty}$ in X is *bounded* if there exists a sphere $N_r(x)$ that contains every point $x^{(k)}$ of the sequence.

THEOREM 13.11. If $\{x^{(k)}\}_{k=1}^{\infty}$ is a Cauchy sequence in X, then it is bounded.

Proof. Assume $\{x^{(k)}\}_{k=1}^{\infty}$ is a Cauchy sequence and apply Definition 13.10 with $\varepsilon = 1$. Thus there is a number N such that

$$d(x^{(k)}, x^{(m)}) < 1 \quad \text{whenever} \quad k > m > N.$$

In particular, if $m = N + 1$ this property becomes

$$d(x^{(k)}, x^{(N+1)}) < 1 \quad \text{whenever} \quad k > N.$$

Therefore all the points $x^{(k)}$ for $k > N$ lie in the sphere $N_1(x^{(N+1)})$. Now we simply enlarge the radius of the sphere until it also includes the first N points of the sequence. Define

$$r = 1 + \max\{1, d(x^{(1)}, x^{(N+1)}), \ldots, d(x^{(N)}, x^{(N+1)})\}.$$

Thus r is at least one unit more than the distance between $x^{(N+1)}$ and $x^{(k)}$ for every $k = 1, 2, \ldots$; so $N_r(x^{(N+1)})$ contains every point $x^{(k)}$ of the sequence.

Exercises 13.5

In Exercises 1–5, the point sequence $\{\mathbf{x}^{(k)}\}_{k=1}^{\infty}$ is in E^2. Determine whether it converges and, if so, find its limit.

1. $\mathbf{x}^{(k)} = \left(\dfrac{k-1}{k}, \dfrac{k+1}{k} \right)$.

2. $\mathbf{x}^{(k)} = \left(\dfrac{1}{k}, \dfrac{k}{2^k} \right).$

3. $\mathbf{x}^{(k)} = \left(\dfrac{1}{k}, \sin \pi k \right).$

4. $\mathbf{x}^{(k)} = \left(\dfrac{1}{k}, k \right).$

5. $\mathbf{x}^{(k)} = \left(k \sin \dfrac{1}{k}, \dfrac{k^2 - 1}{2k^2 - 1} \right).$

6. Prove that the limit of a convergent sequence in the metric space X is unique by showing that if $\lim_k x^{(k)} = p$ and $\lim_k x^{(k)} = q$, then $p = q$.

7. Let $X = E^n$ and $d^*(\mathbf{x}, \mathbf{y}) = \sum_{k=1}^{n} |x_i - y_i|$; prove that $\lim_k \mathbf{x}^{(k)} = \mathbf{p}$ in $\{E^n, d^*\}$ if and only if $\lim_k x_i^{(k)} = p_i$ for $i = 1, \ldots, n$.

8. Let X be the interval $(0, 1)$ in E^1 with the usual distance: $d(x, y) = |x - y|$. Find a Cauchy sequence in X that does not converge in X.

9. Let X be the subset of E^1 consisting of the union of closed intervals

$$\bigcup_{k=1}^{\infty} \left[\frac{1}{2k}, \frac{1}{2k - 1} \right].$$

Find a Cauchy sequence in X that does not converge in X.

10. Prove that in E^n the sequence $\{\mathbf{x}^{(k)}\}_{k=1}^{\infty}$ is bounded if and only if each of its coordinate sequences $\{x_i^{(k)}\}_{k=1}^{\infty}$ is a bounded number sequence.

13.6. Completeness of E^n

In Chapter 3 we studied in detail the relationship of convergent sequences and Cauchy sequences in $\mathbb{R}$. Since $\mathbb{R}$ is merely the one-dimensional case of E^n, it is a plausible conjecture that a completeness theory holds in E^n that is analogous to that of $\mathbb{R}$. First we define the completeness of E^n.

DEFINITION 13.12. The metric space X is said to be *complete* if every Cauchy sequence in X converges to a point in X.

In the terminology of Definition 13.12, E^n is a complete metric space by Theorem 13.9. The next theorem gives four alternative properties that are equivalent to completeness in E^n. Each is an important result in its own right, and collectively they are the most important tools of analysis. They are immediately recognizable as multidimensional extensions of the four theorems of Chapter 3.

THEOREM 13.12. Each of the following five statements is true, and they are equivalent to one another:

(i) E^n is complete.

(ii) *Heine-Borel Property.* If F is a closed and bounded set in E^n and $\{A_\mu\}_{\mu \in M}$ is a collection of open sets whose union contains F, then there is a finite subcollection $A_{\mu(1)}, A_{\mu(2)}, \ldots, A_{\mu(m)}$ such that $F \subset \bigcup_{i=1}^m A_{\mu(i)}$.

(iii) *Bolzano-Weierstrass Property.* If S is an infinite bounded set in E^n, then S has a limit point in E^n.

(iv) *Bounded Sequence Property.* If $\{\mathbf{x}^{(k)}\}_{k=1}^\infty$ is a bounded point sequence in E^n, then it has a convergent subsequence.

(v) *Nested Sets Property.* If $\{F_k\}_{k=1}^\infty$ is a sequence of closed bounded nonempty sets such that for each k, $F_{k+1} \subseteq F_k \subseteq E^n$, then $\bigcap_{k=1}^\infty F_k \neq \varnothing$.

Proof. To establish their equivalence, we prove that each of the first four statements implies the one following it and (v) implies (i). As we remarked above, (i) has already been proved as Theorem 13.9, so the equivalence of the five statements will imply that they are all true.

Statements (ii–v) refer to bounded sets. It is convenient to use another form of boundedness that is equivalent to Definition 13.11. We note that a set S is bounded in E^n if and only if it is contained in an "n-cube":

$$S \subseteq C = \{\mathbf{x} \in E^n : |x_i| \leq r \quad \text{for} \quad i = 1, \ldots, n\}.$$

The proof of this assertion is requested in Exercise 13.6.1. Since the first implication requires the most complicated proof, it may be helpful to visualize the construction in E^2 as shown in Figure 13.5.

(i) *implies* (ii): Let F be a closed and bounded set and suppose we are given that $\{A_\mu\}_{\mu \in M}$ is an "open cover of F" as in (ii). Since F is bounded, there is an n-cube C_1 containing F. Suppose that F *cannot* be covered by any finite number of the A_μ's. Subdivide C_1 into 2^n n-cubes by halving each of the coordinate intervals of C_1; that is, the ith coordinate of a point $\mathbf{x}$ in one of the n-cubes is restricted to either $[0, r]$ or $[-r, 0]$ instead of $[-r, r]$. If it were possible to cover those parts of F that lie in each one of the 2^n n-cubes with finitely many A_μ's, then the union of all of these A_μ's would still be only

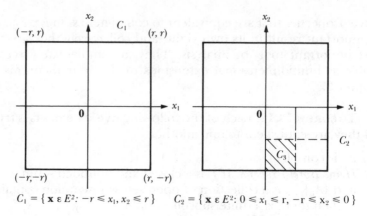

$$C_1 = \{ \mathbf{x} \; \varepsilon \; E^2 \colon -r \leqslant x_1, x_2 \leqslant r \} \qquad C_2 = \{ \mathbf{x} \; \varepsilon \; E^2 \colon 0 \leqslant x_1 \leqslant r, \; -r \leqslant x_2 \leqslant 0 \}$$

Figure 13.5

a finite number and would cover F. Therefore one of these 2^n n-cubes must contain a subset of F that cannot be covered by any finite number of the A_μ's. Call this n-cube C_2. Now subdivide C_2 into 2^n n-cubes and repeat the process ad infinitum. The result is a sequence of n-cubes such that $C_{k+1} \subset C_k$, and the ith coordinate of each point in C_k is restricted to an interval whose length is $r/2^{k-2}$. Choose a point sequence $\{\mathbf{x}^{(k)}\}_{k=1}^\infty$ such that $\mathbf{x}^{(k)}$ is in $F \cap C_k$. Then the sequence of ith coordinates forms a Cauchy number sequence, which implies, as in Theorems 13.8 and 13.9, that $\{\mathbf{x}^{(k)}\}_{k=1}^\infty$ is a Cauchy sequence. Assuming (i), we conclude that there is a point $\mathbf{p}$ such that $\lim_k \mathbf{x}^{(k)} = \mathbf{p}$. Since each $\mathbf{x}^{(k)}$ is in F and F is closed, we have $\mathbf{p} \in F$. Therefore $\mathbf{p}$ is in one of the A_μ's, say, $\mathbf{p} \in A_\mathbf{p}$. But $A_\mathbf{p}$ is open, so for some r', $\mathbf{p} \in \mathsf{N}_{r'}(\mathbf{p}) \subset A_\mathbf{p}$, which means that $\mathsf{N}_{r'}(\mathbf{p})$ contains all but a finite number of the n-cubes C_k. But this means that the points of F in all these C_k's are contained in *one* of the open sets, namely $A_\mathbf{p}$, and this contradicts the choice of C_k.

(ii) *implies* (iii): Suppose S is a bounded set that has no limit point. (We show that S must be finite, which establishes (iii).) Let $\bar{\mathsf{N}}$ be a closed (and bounded) sphere containing S. Since no point of $\bar{\mathsf{N}}$ is a limit point of S, we can choose a sphere $\mathsf{N}_{r(\mathbf{p})}(\mathbf{p})$ for each $\mathbf{p}$ in $\bar{\mathsf{N}}$ such that $\mathsf{N}_{r(\mathbf{p})}(\mathbf{p})$ contains no point of S except possibly $\mathbf{p}$ itself. Now $\bar{\mathsf{N}} \subset \cup_{\mathbf{p} \in S} \mathsf{N}_{r(\mathbf{p})}(\mathbf{p})$, so $\{\mathsf{N}_{r(\mathbf{p})}(\mathbf{p}) \colon \mathbf{p} \in S\}$ is a collection of open sets whose union contains $\bar{\mathsf{N}}$. Therefore from (ii) we conclude that there is a finite subcollection of these spheres that covers $\bar{\mathsf{N}}$. But this subcollection must also cover S (since $S \subseteq \bar{\mathsf{N}}$), and since each sphere contains at most one point of S, we conclude that S is finite.

(iii) *implies* (iv): Suppose $\{\mathbf{x}^{(k)}\}_{k=1}^\infty$ is a bounded point sequence, and let S be the range of this sequence; that is, $S = \{\mathbf{p} \in E^n \colon \mathbf{p} = \mathbf{x}^{(k)}$

for some k}. If S is a finite set, then at least one of its points must appear infinitely many times as a term of the sequence, in which case $\{\mathbf{x}^{(k)}\}_{k=1}^{\infty}$ has a *constant* subsequence. If S has infinitely many points, then (iii) implies that S has a limit point, and in Theorem 13.7 we saw how to construct a sequence in S that converges to that limit point.

(iv) *implies* (v): Let $\{F_k\}_{k=1}^{\infty}$ be a nested sequence of closed bounded sets as in the hypothesis of (v). For each k, choose a point $\mathbf{x}^{(k)}$ in F_k. Every such point is in F_1, so $\{\mathbf{x}^{(k)}\}_{k=1}^{\infty}$ is a bounded sequence. By (iv), $\{\mathbf{x}^{(k)}\}_{k=1}^{\infty}$ has a convergent subsequence whose limit we here call $\mathbf{p}$. We see that the nested property of the sets guarantees that each F_k contains all but possibly a finite number of the points $\mathbf{x}^{(k)}$. Therefore for each k there is a sequence of points in F_k that converges to $\mathbf{p}$. Since these are closed sets, $\mathbf{p}$ must belong to each F_k. Hence $\cap_{k=1}^{\infty} F_k$ contains $\mathbf{p}$, so the intersection is nonempty.

(v) *implies* (i): Let $\{\mathbf{x}^{(k)}\}_{k=1}^{\infty}$ be a Cauchy sequence, and for each m, define F_m to be the closure of the set of points $\{\mathbf{x}^{(k)}: k \geq m\}$. Thus $\{F_m\}_{m=1}^{\infty}$ is a nested sequence of closed sets. By Theorem 13.11 $\{\mathbf{x}^{(k)}\}_{k=1}^{\infty}$ is bounded, and therefore F_1 is bounded. By (v) there is a point $\mathbf{p}$ that is in every set F_m. Let $\mathsf{N}_\varepsilon(\mathbf{p})$ be any sphere about $\mathbf{p}$, and choose a number N such that $k > m > N$ implies that $d(\mathbf{x}^{(k)}, \mathbf{x}^{(m)}) < \varepsilon/2$. Since $\mathbf{p}$ is in F_N, which is the closure of the subsequence $\{\mathbf{x}^{(k)}\}_{k=N}^{\infty}$, there is some point $\mathbf{x}^{(m)}$, where $m > N$, such that $\mathbf{x}^{(m)}$ is in $\mathsf{N}_{\varepsilon/2}(\mathbf{p})$. Now if $k > N$, then

$$d(\mathbf{x}^{(k)}, \mathbf{p}) < d(\mathbf{x}^{(k)}, \mathbf{x}^{(m)}) + d(\mathbf{x}^{(m)}, \mathbf{p})$$
$$< \varepsilon/2 + \varepsilon/2$$
$$< \varepsilon.$$

Hence $\lim_k \mathbf{x}^{(k)} = \mathbf{p}$.

In examining the proof of Theorem 13.12, we find that the Euclidean distance formula was used only in proving that (i) implies (ii). Therefore we conclude that the other four implications hold in *any* metric space. Of course, the fact that all five statements are *true* in E^n is quite dependent upon the distance formula. This was needed to prove Theorem 13.9, which gave us the truth of (i). The use of the distance formula in proving that (i) implies (ii) is very subtle. It is inherent in the notions of boundedness and n-cubes that were used. A full discussion of the circumstances in which completeness implies the Heine-Borel property is not necessary to the study of E^n, so we do not attempt it here. We give a brief discussion, however, of two metric spaces that do not satisfy the Heine-Borel property. The first is the space of Example 13.4:

EXAMPLE 13.14. Let X be an infinite set and define d by

$$d(x, y) = \begin{cases} 1, & \text{if } x \neq y, \\ 0, & \text{if } x = y. \end{cases}$$

Since each point in X can be written as

$$\{x\} = \mathsf{N}_{1/2}(x),$$

we see that every singleton set is an open set in X. Also $X = \mathsf{N}_2(x)$ for any $x \in X$, so X is bounded. Therefore X is a bounded closed set, and $\cup_{x \in X}\{x\}$ is an open cover of X that obviously cannot be reduced to a finite number of open sets. We now assert that X is complete, because it is not hard to show that if $\{x^{(k)}\}_{k=1}^{\infty}$ is a Cauchy sequence, then there is an N such that $x^{(k)} = x^{(N)}$ whenever $k \geq N$ (Exercise 13.6.3). Therefore $\lim_k x^{(k)} = x^{(N)}$, so X is complete even though the Heine-Borel property does not hold.

It is not hard to give an example of a metric space in which all five statements of Theorem 13.12 are false. Rather than changing the distance formula, it is simpler to change the set of points by removing one.

EXAMPLE 13.15. Let X denote $E^n \sim \{\mathbf{p}\}$, the complement of the singleton set $\{\mathbf{p}\}$, and let the distance between points in X be the same Euclidean distance as in E^n. Now if $\{\mathbf{x}^{(k)}\}_{k=1}^{\infty}$ is a sequence that converges to $\mathbf{p}$ in E^n and $\mathbf{x}^{(k)} \neq \mathbf{p}$, then it is still a Cauchy sequence in X. But because there is no point (in X) to which it converges, the sequence $\{\mathbf{x}^{(k)}\}_{k=1}^{\infty}$ is not convergent in X. Thus X is not complete. Since completeness is implied by the other four statements, it follows that all five are false.

Exercises 13.6

1. Prove: A is a bounded set in E^n if and only if there exists an n-cube containing A.

2. Prove: In the metric space of Example 13.14, every subset of X is closed.

3. Prove: In the metric space of Example 13.14, every Cauchy sequence is eventually constant; that is, there exists an N such that $x^{(k)} = x^{(N)}$ whenever $k \geq N$.

4. Give an example in E^2 of a nested sequence of bounded nonempty sets that have empty intersection. (Of course, the sets cannot be closed; see Theorem 13.12.)

5. Give an example in E^2 of a nested sequence of nonempty closed sets that have empty intersection.

6. Give an example in a general metric space X of the type requested in Exercise 5.

7. If S is an unbounded set in the metric space X, show that S cannot have the Heine-Borel property; that is, construct a collection $\mathcal{G}$ of open sets that covers S such that no finite subcollection of $\mathcal{G}$ covers S.

8. For an arbitrary metric space, prove that if the set S is not closed then it cannot have the Heine-Borel property (see Exercise 7).

13.7. Dense Subsets of E^n

This section depends rather heavily on the notion of *countable set*, which is discussed in Appendix B. It would be helpful to review that topic before proceeding.

DEFINITION 13.13. The point set D is said to be *dense* in the metric space X if every open sphere in X contains a point of D.

For examples, we turn immediately to E^n (see also Exercises 13.7.1–13.7.3).

EXAMPLE 13.16. Let $\mathbb{Q}^n$ be the subset of E^n consisting of those points $\mathbf{q}$ such that each coordinate q_i is a rational number; then $\mathbb{Q}^n$ is a countable dense subset of E^n. The fact that $\mathbb{Q}^n$ is countable can be deduced from the countability of the set $\mathbb{Q}$ of rational numbers (see Theorem B1 of Appendix B). To show that $\mathbb{Q}^n$ is dense in E^n, let $\mathsf{N}_\varepsilon(\mathbf{p})$ be any sphere and use the density of $\mathbb{Q}$ in $\mathbb{R}$ to choose rational numbers $q_1, \dots, q_n$ such that

$$|p_i - q_i| < \frac{\varepsilon}{\sqrt{n}} \quad \text{for} \quad i = 1, \dots, n.$$

Then

$$d(\mathbf{p}, \mathbf{q}) = [\textstyle\sum_{i=1}^{n} |p_i - q_i|^2]^{1/2}$$

$$< \left[n\left(\frac{\varepsilon^2}{n} \right) \right]^{1/2}$$

$$= \varepsilon.$$

Therefore the point $\mathbf{q}$ is in $\mathbb{Q}^n \cap N_\varepsilon(\mathbf{p})$.

THEOREM 13.13. There is a countable collection $\mathscr{B}$ of open spheres in E^n such that any open set in E^n can be expressed as the union of some subcollection of $\mathscr{B}$.

Proof. Let $\mathscr{B}$ be the collection of spheres that consists of all spheres $N_r(\mathbf{q})$ such that r is a rational number and $\mathbf{q}$ is in $\mathbb{Q}^n$. For a given open set A consider the subcollection $\mathscr{B}_A$ consisting of those spheres such that $N_r(\mathbf{q}) \subseteq A$. Their union is obviously contained in A, and the collection is countable because both $\mathbb{Q}^n$ and $\mathbb{Q}$ are countable. Therefore it remains only to show that A is contained in the union of this collection. If $\mathbf{x}$ is a point in the open set A, then there is a sphere about $\mathbf{x}$ with rational radius r such that $N_r(\mathbf{x}) \subset A$. Since $\mathbb{Q}^n$ is dense, there is a point $\mathbf{q}$ of $\mathbb{Q}$ in $N_{r/2}(\mathbf{x})$. Then $\mathbf{x}$ is in $N_{r/2}(\mathbf{q})$, and $N_{r/2}(\mathbf{q})$ is in $\mathscr{B}_A$ because it is contained in $N_r(\mathbf{x})$, which lies in A. This last assertion is verified by observing that if $\mathbf{y}$ is in $N_{r/2}(\mathbf{q})$, then

$$d(\mathbf{x}, \mathbf{y}) \le d(\mathbf{x}, \mathbf{q}) + d(\mathbf{q}, \mathbf{y}) < r/2 + r/2 = r.$$

We have shown that an arbitrary point $\mathbf{x}$ in A is contained in some sphere in $\mathscr{B}_A$, so A is contained in the union of all spheres in $\mathscr{B}_A$.

We conclude this chapter with a theorem of great generality that combines the concepts of open sets, density, and countability. This theorem should be compared to the Heine-Borel property, for each statement makes an assertion about reducing an open covering of a set.

THEOREM 13.14: LEBESGUE PROPERTY. If D is any set in E^n and $\{A_\mu\}_{\mu \in M}$ is a collection of open sets whose union contains D, then there is a countable subcollection $\{A_{\mu(k)}\}_{k=1}^{\infty}$ whose union contains D.

Proof. By Theorem 13.13, each of the open sets A_μ can be written as the union of all the spheres about points of $\mathbb{Q}^n$ that have rational radii and are contained in A_μ, say, $A_\mu = \bigcup_{k=1}^\infty \mathbf{N}_{r(k)}^\mu$. Each point $\mathbf{x}$ in D is in some $\mathbf{N}_{r(k)}^\mu$, which is a subset of A_μ. For each of the spheres $\mathbf{N}_{r(k)}^\mu$, choose one A_μ that contains $\mathbf{N}_{r(k)}^\mu$ and call it $A_{r(k)}^\mu$. The subcollection of all such $A_{r(k)}^\mu$ is countable because there is (at most) one $A_{r(k)}^\mu$ for each $\mathbf{N}_{r(k)}^\mu$. Also, the union of this subcollection contains the union of the $\mathbf{N}_{r(k)}^\mu$'s, which in turn contains D. Hence there is a countable subcollection of $\{A_\mu\}_{\mu \in M}$ whose union contains D.

Note that the hypotheses of Theorem 13.14 make no restrictive assumption about the set D. That is the strength and generality of this result; it applies to *any* subset D of E^n. If it is given, in addition, that D is closed and bounded, then the countable subcollection obtained from the Lebesgue property can be further reduced to a finite subcollection whose union still contains D.

Exercises 13.7

1. Prove that D is dense in E^n, where
$$D = \{\mathbf{x} \in E^n \colon x_i \in \mathbb{R} \sim \mathbb{Q}, i = 1, \ldots, n\}.$$

2. Prove that D is dense in E^2, where
$$D = \{\mathbf{x} \in E^2 \colon x_1 \in \mathbb{Q} \text{ and } x_2 \in \mathbb{R} \sim \mathbb{Q}\}.$$

3. Prove that D is dense in the 2-cube
$$C = \{\mathbf{x} \in E^2 \colon -1 \le x_1, x_2 \le 1\},$$
where
$$D = \{\mathbf{x} \in C \colon x_1 \in \mathbb{Q}\}.$$

4. Prove: If D is an uncountable point set in E^n, then D has a limit point in E^n. (*Note:* This assertion should be compared to the Bolzano-Weierstrass property—Theorem 13.12 (iii). Here D need not be bounded, yet we are guaranteed a limit point of D simply because D has so many elements.)

5. Prove: If D is an uncountable point set in E^n, then D *contains* one of its limit points.

6. Prove: If D is an uncountable point set in E^n, then all but a countable number of the points of D are limit points of D.

14

CONTINUOUS
TRANSFORMATIONS

14.1. Transformations and Functions

The continuity of functions from E^1 to E^1 was studied in Chapter 4. In this chapter, in order to avoid confusion, we use the term *function* only when the range is E^1. We use the terms *transformation* or *mapping* to indicate a "function" whose domain is a metric space and whose range is either E^m ($m > 1$) or a more general metric space. In the many situations in which we consider more than one Euclidean space in the same discussion, we denote the Euclidean distance in E^n by d_n, in E^m by d_m, and so on.

DEFINITION 14.1. Let T be a transformation from a domain D in E^n into E^m; then T is said to be *continuous* at the point $\mathbf{p}$ in its domain D provided that if $\varepsilon > 0$, then there is a positive number δ such that

$$\mathbf{x} \in \mathbf{N}_\delta(\mathbf{p}) \cap D \quad \text{implies} \quad T(\mathbf{x}) \in \mathbf{N}_\varepsilon(T(\mathbf{p})); \qquad (1a)$$

that is,

$$\mathbf{x} \in D \quad \text{and} \quad d_n(\mathbf{x}, \mathbf{p}) < \delta \quad \text{imply} \quad d_m(T(\mathbf{x}), T(\mathbf{p})) < \varepsilon. \quad (1b)$$

If T is continuous at every point of D, then we say that T is *continuous* on D.

A graph is a familiar device for illustrating a function from E^1 into E^1. Also, the representation of a function from E^2 into E^1

by a surface in three-dimensional space can be seen in multivariable calculus. The following example helps to illustrate our current notation and terminology.

EXAMPLE 14.1. The function f on E^2 given by

$$f(\mathbf{x}) = f(x_1, x_2) = \sqrt{x_1^2 + x_2^2}$$

is represented as a surface in three dimensions by a cone with vertex at the origin. (Sketch this to understand it better.) The continuity of f is easily verified by noting that $f(\mathbf{x}) = d_2(\mathbf{x}, \mathbf{0})$; then

$$d_1(f(\mathbf{x}), f(\mathbf{p})) = |f(\mathbf{x}) - f(\mathbf{p})|$$
$$= |d_2(\mathbf{x}, \mathbf{0}) - d_2(\mathbf{p}, \mathbf{0})|$$
$$\leq d_2(\mathbf{x}, \mathbf{p}),$$

by the Triangle Inequality. Thus the definition of continuity is satisfied at every $\mathbf{p}$ in E^2.

Note that Definition 14.1 need not be restricted to transformations on E^n. Line (1a) makes perfectly good sense even for mappings whose domain is $\{X, d\}$ and whose range is $\{Y, d^*\}$. In this case, though, line (1b) would be replaced by

$$x \in D \quad \text{and} \quad d(x, p) < \delta \quad \text{imply} \quad d^*(T(x), T(p)) < \varepsilon. \quad (1c)$$

One way of producing examples of mappings is to use the same set of points for X and Y but then give them different distance functions.

EXAMPLE 14.2. Let X be E^2 with the usual distance function d_2, and let Y be E^2 with the "taxicab metric"

$$d^*(\mathbf{x}, \mathbf{y}) = |x_1 - y_1| + |x_2 - y_2|.$$

We assert that the identity mapping $T(\mathbf{x}) = \mathbf{x}$, as a mapping from X to Y, is continuous at every point in X. For, if $\varepsilon > 0$ is given, we choose $\delta = \varepsilon/2$. Then for any $\mathbf{p}$ in X, $d_2(\mathbf{x}, \mathbf{p}) < \delta$ implies that

$$d^*(T(\mathbf{x}), T(\mathbf{p})) = |T(\mathbf{x})_1 - T(\mathbf{p})_1| + |T(\mathbf{x})_2 - T(\mathbf{p})_2|$$
$$= |x_1 - p_1| + |x_2 - p_2|$$
$$= [|x_1 - p_1|^2]^{1/2} + [|x_2 - p_2|^2]^{1/2}$$
$$\leq [|x_1 - p_1|^2 + |x_2 - p_2|^2]^{1/2}$$
$$\quad + [|x_1 - p_1|^2 + |x_2 - p_2|^2]^{1/2}$$
$$= d_2(\mathbf{x}, \mathbf{p}) + d_2(\mathbf{x}, \mathbf{p})$$
$$< 2\,\delta$$
$$= \varepsilon.$$

Hence T is continuous at $\mathbf{p}$.

One of the curious consequences of Definition 14.1 occurs at a point in the domain of a mapping that is not a limit point of the domain. Such a point is called an *isolated point* of the domain, and the curious fact is that the mapping is always continuous at such a point. For if $\mathbf{p}$ is an isolated point of the domain D of a mapping T, then there is a sphere $N_\delta(\mathbf{p})$ that contains no point of D other than $\mathbf{p}$. Using this in (1a) we see that the only point $\mathbf{x}$ that satisfies both $\mathbf{x} \in N_\delta(\mathbf{p})$ and $\mathbf{x} \in D$ is the point $\mathbf{p}$, and with $\mathbf{x} = \mathbf{p}$ the conclusion of (1a) is true for any $\varepsilon > 0$. Thus T is continuous at $\mathbf{p}$.

The phenomenon discussed in the preceding paragraph can be seen in an extreme case by considering a domain consisting entirely of isolated points. On such a domain, *any* mapping is continuous at every point.

EXAMPLE 14.3. Let $X = \{\mathbf{x} \in E^2 : x_1 \text{ and } x_2 \text{ are integers}\}$, and let d_2 be the usual Euclidean distance function. Then every point of X is an isolated point because $N_{1/2}(\mathbf{p}) = \{\mathbf{p}\}$ for every $\mathbf{p}$ in X. Now define a mapping on X in a completely arbitrary way, say,

$$T(\mathbf{x}) = \begin{cases} (-1)^{x_1}\pi, & \text{if } x_2 \text{ is even,} \\[2mm] \sin\dfrac{\pi}{x_2}, & \text{if } x_2 \text{ is odd;} \end{cases}$$

then even this mapping is continuous at every point in X.

Exercises 14.1

1. Let f be the function on E^2 given by $f(\mathbf{x}) = |x_1 - x_2|$. Show that f is continuous on E^2.

2. Let f be the function on E^2 given by $f(\mathbf{x}) = x_1$. Show that f is continuous on E^2.

3. Let T be the transformation from E^3 to E^2 given by $T(\mathbf{x}) = (x_1, x_2)$. Show that T is continuous on E^3.

4. Let T be the transformation from E^n into E^n given by

$$T(\mathbf{x}) = (3x_1, 3x_2, \ldots, 3x_n).$$

   Prove that T is continuous on E^n.

5. Let $\{X, d_2\}$ and $\{Y, d^*\}$ be the metric spaces of Example 14.2, and let T be the mapping from Y to X given by $T(\mathbf{x}) = \mathbf{x}$. Show that T is continuous on Y.

6. Define $D = \{\mathbf{x} \in E^2 : (x_1^2 + x_2^2)(x_1 - 1) \geq 0\}$ and let T be the transformation from D to E^1 given by

$$T(\mathbf{x}) = \sqrt{(x_1^2 + x_2^2)(x_1 - 1)}.$$

   Show that T is continuous at $\mathbf{0}$.

7. Given $f(\mathbf{x}) = d_2(\mathbf{x}, \mathbf{0})$ for every $\mathbf{x}$ in $E^2 \sim \{\mathbf{0}\}$, define $f(\mathbf{0})$ so that f is continuous at $\mathbf{0}$.

8. Let T be a continuous mapping from $\{X, d\}$ into $\{Y, d'\}$ and let T' be a continuous mapping from $\{Y, d'\}$ into $\{Z, d''\}$. Prove that the composition given by $T'(T(\mathbf{x}))$ is a continuous mapping from $\{X, d\}$ into $\{Z, d''\}$.

9. Prove: If the transformation T from $\{X, d\}$ into $\{Y, d^*\}$ is continuous at p, then for every $\varepsilon > 0$ there is a number $\delta > 0$ such that

$$d(f(x), f(y)) < \varepsilon \quad \text{whenever} \quad x, y \in N_\delta(p).$$

14.2. Criteria for Continuity

In this section we prove two theorems that can be used to determine whether a given transformation is continuous. The first of these two results applies to a transformation that is defined on an open set. If the domain of a given transformation is not open, this theorem can still be used to determine whether the transformation is continuous on the interior of its domain or any other open set in its domain. First we introduce some terminology and notation.

DEFINITION 14.2. Let T be a transformation from the metric space X to the metric space Y. If $A \subseteq Y$, then the *inverse image* of A under the mapping T is the set denoted and given by

$$T^{-1}[A] = \{x \in X : T(x) \in A\}.$$

An easy way to describe the inverse image $T^{-1}[A]$ is the set of all points that are mapped into A by T.

EXAMPLE 14.4. If T maps E^2 into E^2 by the formula

$$T(\mathbf{x}) = (x_1 - 4, x_2)$$

and A is the set $\mathbf{N}_1(\mathbf{0})$, then

$$T^{-1}[\mathbf{N}_1(\mathbf{0})] = \mathbf{N}_1(\mathbf{p})$$

where $\mathbf{p} = (4, 0)$, because T simply shifts each point four units to the left (see Figure 14.1).

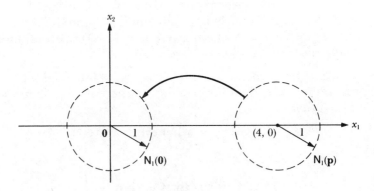

Figure 14.1

THEOREM 14.1. Let T be a transformation from the metric space X into the metric space Y. Then T is continuous on the open set D in X if and only if, for each open set A in Y, $T^{-1}[A]$ is an open set in X.

Proof. First assume T is continuous on D and let A be an open set in Y. If $T^{-1}[A] = \varnothing$, then it is open. If $T^{-1}[A] \neq \varnothing$, then there is at least one point p in D such that $T(p) \in A$. We must show that p is an interior point of $T^{-1}[A]$. Since A is open, there is a sphere $N_\varepsilon(T(p))$ contained entirely in A. By continuity there exists a sphere $N_\delta(p)$ such that if $x \in N_\delta(p) \cap D$ then $T(x) \in N_\varepsilon(T(p)) \subset A$. Therefore $N_\delta(p) \cap D \subseteq T^{-1}[A]$. Since D is open, we can choose $\delta' \leq \delta$ so that

$$N_{\delta'}(p) \subset N_\delta(p) \cap D \subseteq T^{-1}[A].$$

Thus p is an interior point of $T^{-1}[A]$, so $T^{-1}[A]$ is an open set.

Conversely, suppose $T^{-1}[A]$ is open whenever A is open, and let p be an arbitrary point in D and ε be a positive number. Since $N_\varepsilon(T(p))$ is an open set in Y, $T^{-1}[N_\varepsilon(T(p))]$ is open in X. Therefore p is an interior point, so for some positive δ,

$$N_\delta(p) \subset T^{-1}[N_\varepsilon(T(p))].$$

This means that if $x \in N_\delta(p)$, then $T(x) \in N_\varepsilon(T(p))$, that is, T is continuous at p.

The next result is exactly analogous to Theorem 4.2 for functions from $\mathbb{R}$ to $\mathbb{R}$. Although the proof is the same as before, we can gain familiarity with the notation and terminology of transformations by repeating the details of the argument.

THEOREM 14.2: SEQUENTIAL CRITERION FOR CONTINUITY. The transformation T is continuous at p if and only if for each sequence $\{x^{(k)}\}_{k=1}^\infty$ in D that converges to p, the sequence $\{T(x^{(k)})\}_{k=1}^\infty$ converges to $T(p)$.

Proof. First assume that T is continuous at p, and let $\{x^{(k)}\}_{k=1}^\infty$ be a sequence that converges to p. If $\varepsilon > 0$, then there is a positive number δ such that $x \in N_\delta(p)$ implies $T(x) \in N_\varepsilon(T(p))$. Since $\lim_k x^{(k)} = p$, there exists a number N such that $k > N$ implies that $x^{(k)} \in N_\delta(p)$, which in turn implies that $T(x^{(k)}) \in N_\varepsilon(T(p))$. Hence $\lim_k T(x^{(k)}) = T(p)$.

Now suppose T is not continuous at p. Then there is a positive number ε^* such that for every positive integer k, $N_{1/k}(p)$ contains some x for which $T(x) \notin N_{\varepsilon^*}(T(p))$. For each k, choose one such x, say, $x^{(k)}$, that satisfies

$$x^{(k)} \in N_{1/k}(p) \quad \text{and} \quad T(x^{(k)}) \notin N_{\varepsilon^*}(T(p)).$$

Since $d_n(x^{(k)}, p) < 1/k$, we have $\lim_k x^{(k)} = p$; but $d_m(T(x^{(k)}), T(p)) \geq \varepsilon^*$, so $\{T(x^{(k)})\}_{k=1}^{\infty}$ cannot converge to $T(p)$.

As in Chapter 4, the SCC is most useful in showing that a given function *fails* to be continuous at a given point. This is good to keep in mind for the following exercise set.

Exercises 14.2

1. Let T be the transformation of E^2 into E^2 given by $T(\mathbf{x}) = (3x_1, 3x_2)$, and let $A = N_6(\mathbf{0}) \subset E^2$. Find $T^{-1}[A]$.

2. Let T be the transformation of E^2 into E^2 given by

$$T(\mathbf{x}) = \left(\frac{x_1}{x_1^2 + x_2^2}, \frac{x_2}{x_1^2 + x_2^2} \right),$$

   and let $A = N_{1/4}(\mathbf{0}) \subset E^2$. Find $T^{-1}[A]$.

3. Let f be the function from E^n into E^1 given by $f(\mathbf{x}) = [x_1^2 + \cdots + x_n^2]^{1/2}$, and let A be the interval $(-3, -2)$. Find $f^{-1}[A]$.

4. Let T be the transformation from E^2 into E^2 given by $T(\mathbf{x}) = (x_1 + 2, x_2 + 2)$. Show that T is continuous on E^2.

5. Let f be the function on E^2 given by

$$f(\mathbf{x}) = \begin{cases} \dfrac{1}{x_1^2 + x_2^2}, & \text{if } \mathbf{x} \neq \mathbf{0}, \\ 0, & \text{if } \mathbf{x} = \mathbf{0}. \end{cases}$$

   Prove that f is discontinuous at $\mathbf{0}$.

6. Let f be the function on E^2 given by

$$f(\mathbf{x}) = \begin{cases} \dfrac{x_1 x_2}{|x_1 x_2|}, & \text{if } x_1 x_2 \neq 0, \\ 0, & \text{if } x_1 x_2 = 0. \end{cases}$$

   Prove that f is discontinuous at $\mathbf{0}$.

7. Prove that the function given in Exercise 6 is discontinuous at every point $\mathbf{p}$ of the form $(0, p_2)$ or $(p_1, 0)$.

8. Let T be the transformation of E^2 into E^2 given by

$$T(\mathbf{x}) = \left(\sin \frac{1}{x_1}, x_2 \right), \quad \text{if } x_1 \neq 0.$$

   Prove that T is discontinuous at every point of the form $(0, p_2)$ no matter how $T(\mathbf{x})$ is defined when $x_1 = 0$.

14.3. The Range of a Continuous Transformation

In this section we consider subsets of the range of a continuous transformation. If S is a subset of the domain of T, we follow the notation of Definition 14.2 and let $T[S]$ denote the *image* of S under T:

$$T[S] = \{T(p): p \in S\}.$$

If T is continuous and S has some particular property as a subset of the domain of T, then we ask whether its image $T[S]$ exhibits the same property as a subset of the range of T. The first theorem of this section asserts that the property of connectedness is preserved under a continuous mapping. This is the metric space generalization of the Intermediate Value Theorem (Theorem 5.3).

THEOREM 14.3. If T is a continuous transformation on the open set D and S is a connected subset of D, then T transforms S into a connected set $T[S]$.

Proof. Assume T is continuous on the open set D and $S \subseteq D$. We suppose further that $T[S]$ is *disconnected* and we show that S must therefore be disconnected. By Theorem 13.5, there exist open sets A and B such that $T[S] \subseteq A \cup B$ and both A and B intersect S. By Theorem 14.1, $T^{-1}[A]$ and $T^{-1}[B]$ are open. Also, $S \subseteq T^{-1}[A] \cup T^{-1}[B]$, because every point of S is mapped into either A or B. Next, $T^{-1}[A] \cap S \neq \varnothing$, because some point of $T[S]$ is in A. Similarly, $T^{-1}[B] \neq \varnothing$. Moreover, $T^{-1}[A]$ and $T^{-1}[B]$ are disjoint, because if some point x were in both sets, then $T(x)$ would be in both A and B. Hence, by Theorem 13.5, S is disconnected.

The next theorem is the metric space generalization of Theorems 5.1 and 5.2. In these theorems we considered the range of a continuous function on a closed interval. In the setting of an abstract metric space we consider a closed and bounded set and ask whether its image under a continuous transformation is also closed and bounded. In order to ensure an affirmative answer, we must assume that the metric space has the Heine-Borel Property (Theorem 13.12(ii)). This also implies the Bounded Sequence Property (Theorem 13.12(iv)), and we use both of them in the proof.

THEOREM 14.4. Let X be a metric space in which the Heine-Borel Property holds, and let T be a continuous transformation on $D \subseteq X$. If D is closed and bounded, then its image $T[D]$ is also closed and bounded.

Proof. We first show that $T[D]$ is bounded. Let p be a point in D so that $T(p) \in T[D]$. For brevity we write $\mathbf{N}_k$ for the open sphere $\mathbf{N}_k(T(p))$ and $A_k = T^{-1}[\mathbf{N}_k]$. By Theorem 14.1 each A_k is open in X, and since $\bigcup_{k=1}^{\infty} \mathbf{N}_k$ contains the entire range of T, it follows that $\bigcup_{k=1}^{\infty} A_k$ is an open cover of D. Now by the Heine-Borel Property there exists a finite subcollection of the A_k's that covers D. Since $A_k \subseteq A_{k+1}$, this means that there is one set, say, A_{k*}, that contains D. Thus we have $D \subseteq A_{k*}$, which implies that $T[D] \subseteq T[A_{k*}] = \mathbf{N}_{k*}$, so $T[D]$ is contained in a sphere; that is, $T[D]$ is bounded.

To show that $T[D]$ is closed, we consider a limit point q of $T[D]$. Then there is a point sequence $\{q^{(k)}\}_{k=1}^{\infty}$ in $T[D]$ such that $\lim_k q^{(k)} = q$. For each $q^{(k)}$ in $T[D]$, we can choose an $x^{(k)}$ in D such that $q^{(k)} = T(x^{(k)})$. Since D is bounded, $\{x^{(k)}\}_{k=1}^{\infty}$ is a bounded sequence, so by Theorem 13.12(iv) it has a convergent subsequence, and its limit p must be in D because D is closed, say, $\lim_m x^{(k_m)} = p$. By the SCC (Theorem 14.2), $\lim_m T(x^{(k_m)}) = T(p)$; but $\{T(x^{(k_m)})\}_{m=1}^{\infty}$ must have the same limit point as $\{T(x^{(k)})\}_{k=1}^{\infty} = \{q^{(k)}\}_{k=1}^{\infty}$. Therefore $T(p) = q$, so q is in $T[D]$. Hence $T[D]$ contains all of its limit points.

Exercises 14.3

1. Let T be a transformation from E^2 into E^2 such that for every $\mathbf{p}$ in E^2 the second coordinate of $T(\mathbf{p})$ is nonzero. If T maps the points $(0, -1)$ and $(0, 1)$ into themselves, show that T is not continuous on E^2.

2. Let T be a continuous transformation from E^2 into E^2 such that $T(\mathbf{0}) = \mathbf{0}$ and $T[E^2]$ is unbounded. Prove that there exists a point $\mathbf{p} \in E^2$ such that $d(\mathbf{0}, T(\mathbf{p})) = 1$.

3. Find a function f that is continuous on a bounded set D in E^2 and such that $T[D]$ is unbounded.

4. Find a function that is continuous on E^1 and maps the union of the intervals $[0, 1]$ and $[2, 3]$ onto the two-point set $\{0\} \cup \{1\}$.

5. Find a transformation whose domain is the sphere $\mathbf{N}_1(\mathbf{0})$ in E^2 that maps $\mathbf{N}_1(\mathbf{0})$ continuously onto the line $L = \{\mathbf{x} \in E^2 : x_2 = 0\}$.

6. Find a function f that is continuous on E^2 and maps a closed set D onto a nonclosed set $T[D]$.

7. Let D be a closed and bounded set in E^2 and let $\mathbf{p}$ be a point in $E^2 \sim D$. Prove that there is a point $\mathbf{q}$ in D that is closest to $\mathbf{p}$, that is, for each $\mathbf{x}$ in D, $d(\mathbf{x}, \mathbf{p}) \geq d(\mathbf{q}, \mathbf{p})$.

14.4. Continuity in E^n

In order to check the continuity of a function from E^n into E^1 it is perhaps natural to consider the effects of sequential limits taken one coordinate at a time. For example, if f is given by the formula

$$f(x_1, x_2) = \begin{cases} x_1/x_2, & \text{if } x_2 \neq 0, \\ 0, & \text{if } x_2 = 0, \end{cases}$$

then for any fixed nonzero value of x_2, f is continuous as a function of x_1. But for any fixed value of x_1 other than zero, f is discontinuous at $x_2 = 0$. (More precisely, f is discontinuous at $(x_1, 0)$ in E^2.) This type of reasoning can provide a very simple way of establishing a point of discontinuity, but it is important to keep in mind that continuity in each coordinate does *not* guarantee that f is continuous in E^n. Continuity at a point $\mathbf{p}$ in E^n requires that $f(\mathbf{x})$ approaches $f(\mathbf{p})$ as $\mathbf{x}$ approaches $\mathbf{p}$ *along any path*. The following two examples illustrate this fact.

EXAMPLE 14.5. Let f be defined on E^2 by $f(\mathbf{0}) = 0$ and

$$f(\mathbf{x}) = f(x_1, x_2) = \frac{x_1 x_2}{x_1^2 + x_2^2}, \quad \text{if } \mathbf{x} \neq \mathbf{0}.$$

Then $f(x_1, 0) = f(0, x_2) = 0$ for any x_1 or x_2, so as $\{\mathbf{x}^{(k)}\}_{k=1}^{\infty}$ approaches $\mathbf{0}$ along the axes, we see that $\{f(\mathbf{x}^{(k)})\}_{k=1}^{\infty}$ converges to $f(\mathbf{0})$. But consider a different sequence that approaches $\mathbf{0}$ via another route, say, $\mathbf{x}^{(k)} = (1/k, 1/k)$. Then

$$f(\mathbf{x}^{(k)}) = \frac{\left(\dfrac{1}{k}\right)\left(\dfrac{1}{k}\right)}{\left(\dfrac{1}{k}\right)^2 + \left(\dfrac{1}{k}\right)^2} = \frac{1}{2}$$

for every k, so $\{f(\mathbf{x}^{(k)})\}_{k=1}^{\infty}$ converges to $1/2$. Hence, by Theorem 14.2, f is not continuous at $\mathbf{0}$.

EXAMPLE 14.6. Let F denote the set $\{\mathbf{x} \in E^2 : 0 < x_2 < x_1^2\}$, which consists of all points lying between the x_1-axis and the

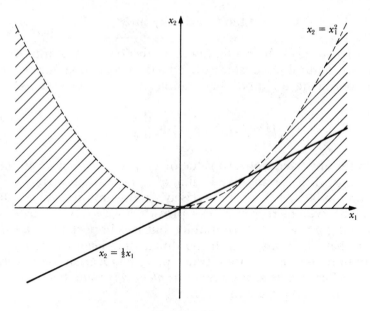

Figure 14.2

parabola given by $x_2 = x_1^2$ (see Figure 14.2). Define f to be the so-called "characteristic function" of F:

$$f(\mathbf{x}) = \begin{cases} 1, & \text{if } \mathbf{x} \in F, \\ 0, & \text{if } \mathbf{x} \notin F. \end{cases}$$

Then $f(\mathbf{x})$ approaches 0 ($= f(\mathbf{0})$) as $\mathbf{x}$ approaches $\mathbf{0}$ along any straight-line path, say, $\mathbf{x}^{(k)} = (t_k, mt_k)$, where $\lim_k t_k = 0$, because when t_k is sufficiently small (between 0 and m) $\mathbf{x}^{(k)}$ is not in F. Nevertheless, f is *not* continuous at $\mathbf{0}$, because if $\mathbf{y}^{(k)} = (1/k, k^2/2)$, then $\mathbf{y}^{(k)} \in F$, so $\{f(\mathbf{y}^{(k)})\}_{k=1}^{\infty}$ converges to 1 ($\neq f(\mathbf{0})$) as k increases without bound.

The preceding examples show that coordinate-wise limits are not sufficient to imply continuity of a function with a multi-dimensional domain. They are, however, a necessary condition for such continuity. In Examples 14.5 and 14.6 we relied on the SCC to draw our conclusions of discontinuity, but this necessary condition can be stated without reference to a sequential limit. This is achieved by a device familiar to the student of multivariable calculus. We let the point $\mathbf{x}$ approach a given point $\mathbf{p}$ along a path in which all but one of the coordinates of **x** have the fixed value of the corresponding coordinate of **p**. This "single coordinate limit" can be examined for each of the n coordinates in E^n, and in order for a function f to be continuous at **p**, each of these n limits must have the value $f(\mathbf{p})$. This is summarized in the following theorem.

THEOREM 14.5. If f is continuous at the point **p** in E^n, then for each $i = 1, \ldots, n$,

$$\lim_{x_i \to p_i} f(p_1, \ldots, x_i, \ldots, p_n) = f(\mathbf{p}), \qquad (1)$$

where $(p_1, \ldots, x_i, \ldots, p_n)$ denotes the point **y** given by $y_i = x_i$ and $y_j = p_j$ for $j \neq i$.

It should be noted that the limit in (1) is a limit of a function from $\mathbb{R}$ into $\mathbb{R}$, as studied in Chapter 4; therefore it needs no further definition. We have not given a formal definition of a function limit for functions on E^n; for $n > 1$ we have defined only the continuity of such functions.

Exercises 14.4

In Exercises 1–3, show that f is discontinuous at **0** even though $f(0, x_2) = 0 = f(x_1, 0)$ for all x_1 and x_2.

1. $f(\mathbf{0}) = 0$, and if $\mathbf{x} \neq \mathbf{0}$, then $f(\mathbf{x}) = \dfrac{x_1 x_2^3}{x_1^4 + x_2^4}$.

2. $f(\mathbf{x}) = 0$ if $x_1 x_2 = 0$, otherwise $f(\mathbf{x}) = \dfrac{x_1 x_2}{|x_1 x_2|}$.

3. $f(\mathbf{x}) = 0$ if $x_2 = 0$, otherwise $f(\mathbf{x}) = \dfrac{\sin x_1}{x_2}$.

4. Let f be the function from E^3 defined by

$$f(\mathbf{x}) = \begin{cases} \dfrac{x_1 x_2 x_3^2}{x_1^4 + x_2^4 + x_3^4}, & \text{if } \mathbf{x} \neq \mathbf{0}, \\[2mm] 0, & \text{if } \mathbf{x} = \mathbf{0}. \end{cases}$$

Show that f is discontinuous at **0** although $f(\mathbf{x}^{(k)})$ tends to 0 as $\mathbf{x}^{(k)}$ approaches **0** along any one of the coordinate axes (compare Example 14.5).

5. Define $F = \{\mathbf{x} \in E^3 : 0 < x_3 < x_1^2 + x_2^2\}$ and let f be the function from E^3 given by

$$f(\mathbf{x}) = \begin{cases} 1, & \text{if } \mathbf{x} \in F, \\ 0, & \text{if } \mathbf{x} \notin F. \end{cases}$$

Show that $f(\mathbf{x}^{(k)})$ tends to $f(\mathbf{0})$ as $\mathbf{x}^{(k)}$ approaches $\mathbf{0}$ along any "linear path," but f is discontinuous at $\mathbf{0}$.

6. The transformation T from E^n into E^m is said to be "distance reducing" if there is a number r such that $0 < r < 1$ and

$$d_m(T(\mathbf{x}), T(\mathbf{y})) \leq r \cdot d_n(\mathbf{x}, \mathbf{y}) \quad \text{for all } \mathbf{x}, \mathbf{y} \text{ in } E^n.$$

Prove that if T is distance reducing from all of E^n into E^m, then there is exactly one "fixed point" $\mathbf{p}^*$ such that $T(\mathbf{p}^*) = \mathbf{p}^*$. (*Hint:* For some $\mathbf{x}$ in E^n, define $\mathbf{x}^{(1)} = T(\mathbf{x})$, $\mathbf{x}^{(2)} = T(T(\mathbf{x})), \ldots, \mathbf{x}^{(k)} = T(\mathbf{x}^{(k-1)}), \ldots$. Show that $\{\mathbf{x}^{(k)}\}_{k=1}^{\infty}$ is a Cauchy sequence, then take $\mathbf{p}^* = \lim_k \mathbf{x}^{(k)}$.)

14.5. Linear Transformations

The space E^n can be endowed with an algebraic structure by defining addition of points in E^n using coordinate addition:

$$\begin{aligned} \mathbf{x} + \mathbf{y} &= (x_1, \ldots, x_n) + (y_1, \ldots, y_n) \\ &= (x_1 + y_1, \ldots, x_n + y_n). \end{aligned} \tag{1}$$

"Multiplication by scalars" is defined as follows: If $\mathbf{x}$ is in E^n and a is in $\mathbb{R} \, (= E^1)$, then

$$a\mathbf{x} = a(x_1, \ldots, x_n) = (ax_1, \ldots, ax_n). \tag{2}$$

We also introduce the *basis vectors:*

$$\begin{aligned} \mathbf{e}^{(1)} &= (1, 0, \ldots, 0), \\ \mathbf{e}^{(2)} &= (0, 1, 0, \ldots, 0), \\ &\quad \vdots \\ \mathbf{e}^{(n)} &= (0, \ldots, 0, 1). \end{aligned} \tag{3}$$

Using (1) and (2), we see that any point $\mathbf{x}$ can be expressed as a linear combination of these n points:

$$\mathbf{x} = (x_1, x_2, \ldots, x_n)$$
$$= x_1(1, 0, \ldots, 0) + x_2(0, 1, 0, \ldots, 0)$$
$$+ \cdots + x_n(0, \ldots, 0, 1) \tag{4}$$
$$= x_1\mathbf{e}^{(1)} + x_2\mathbf{e}^{(2)} + \cdots + x_n\mathbf{e}^n.$$

Finally, the *norm* of the point $\mathbf{x}$ is defined as

$$\|\mathbf{x}\| = [x_1^2 + x_2^2 + \cdots + x_n^2]^{1/2}. \tag{5}$$

Note that $\|\mathbf{x}\| = d_n(\mathbf{x}, \mathbf{0})$, the Euclidean distance between $\mathbf{x}$ and $\mathbf{0}$, and $\|\mathbf{x} - \mathbf{y}\| = d_n(\mathbf{x}, \mathbf{y})$.

The result of the preceding definitions is the familiar n-dimensional vector space that is studied in detail in linear algebra courses. It is not within our purview to reproduce the theory of finite dimensional vector spaces, but it is appropriate for us to examine briefly some linear transformations, because they are important—as well as simple—examples of continuous transformations from E^n into E^m.

DEFINITION 14.3. The transformation T from E^n into E^m is said to be *linear* provided that for all $\mathbf{x}$ and $\mathbf{y}$ in E^n and all a and b in E^1

$$T(a\mathbf{x} + b\mathbf{y}) = aT(\mathbf{x}) + bT(\mathbf{y}).$$

The main result of this section is that any such linear transformation can be represented by a matrix product; that is, there is an $m \times n$ matrix $[a_{i,j}]$ such that if $\mathbf{y} = T(\mathbf{x})$, then

$$T(\mathbf{x}) = \begin{bmatrix} a_{1,1} & a_{1,2} & \cdots & a_{1,n} \\ a_{2,1} & a_{2,2} & \cdots & a_{2,n} \\ \cdot & \cdot & & \\ \cdot & \cdot & & \\ \cdot & \cdot & & \\ a_{m,1} & a_{m,2} & \cdots & a_{m,n} \end{bmatrix} \begin{bmatrix} x_1 \\ x_2 \\ \cdot \\ \cdot \\ \cdot \\ x_n \end{bmatrix} = \begin{bmatrix} y_1 \\ y_2 \\ \cdot \\ \cdot \\ \cdot \\ y_m \end{bmatrix} = \mathbf{y}.$$

This is the meaning of the next theorem.

THEOREM 14.6. If T is a linear transformation from E^n into E^m, then T is given by

$$T(\mathbf{x}) = T(x_1, \ldots, x_n) = (y_1, \ldots, y_m),$$

where

$$y_1 = a_{1,1}x_1 + a_{1,2}x_2 + \cdots + a_{1,n}x_n$$
$$y_2 = a_{2,1}x_1 + a_{2,2}x_2 + \cdots + a_{2,n}x_n$$
$$\cdot$$
$$\cdot \qquad\qquad\qquad\qquad\qquad\qquad (6)$$
$$\cdot$$
$$y_m = a_{m,1}x_1 + a_{m,2}x_2 + \cdots + a_{m,n}x_n,$$

and each coefficient $a_{i,j}$ is a real number.

Proof. First consider the case in which $m = 1$, so T is a function from E^n into E^1. We focus our attention on n particular points, the basis vectors $\{\mathbf{e}^{(i)}\}$. The images of these n points are numbers, say, $T(\mathbf{e}^{(1)}) = A_1$, $T(\mathbf{e}^{(2)}) = A_2, \ldots, T(\mathbf{e}^{(n)}) = A_n$. By (4) and the linearity of T we have

$$T(\mathbf{x}) = T(x_1\mathbf{e}^{(1)} + x_2\mathbf{e}^{(2)} + \cdots + x_n\mathbf{e}^{(n)})$$
$$= x_1 T(\mathbf{e}^{(1)}) + x_2 T(\mathbf{e}^{(2)}) + \cdots + x_n T(\mathbf{e}^{(n)})$$
$$= A_1 x_1 + A_2 x_2 + \cdots + A_n x_n.$$

Thus T is represented by the row matrix $[A_1\, A_2 \ldots A_n]$.

Now suppose $m > 1$. Then $T(x_1, \ldots, x_n) = (y_1, \ldots, y_m)$, where each coordinate y_i is a function of $\mathbf{x}$, say $y_i = f_i(x_1, \ldots, x_n)$. The linearity of T implies that T is linear in each coordinate; that is, each f_i is a linear function from E^n into E^1. Therefore, by the first case, each f_i is given by a row matrix:

$$y_i = f_i(x_1, \ldots, x_n) = a_{i,1}x_1 + a_{i,2}x_2 + \ldots + a_{i,n}.$$

This set of m linear functions $f_1, \ldots, f_m$ determines the m rows of the matrix $[a_{i,j}]$ that represents T as in (6).

The idea behind the proof of Theorem 14.6 can be used to find the matrix representation of a linear transformation if the images of n points are known.

EXAMPLE 14.7. Let T be a linear transformation that maps E^3 into E^3 such that

$$T(1, 0, 0) = (1, 2, -1),$$
$$T(0, 1, 0) = (4, 0, 3),$$

and

$$T(0, 0, 1) = (-3, 1, 1).$$

Then T is represented by the matrix

$$\begin{bmatrix} 1 & 4 & -3 \\ 2 & 0 & 1 \\ -1 & 3 & 1 \end{bmatrix}$$

whose columns are the images of $T(1, 0, 0)$, $T(0, 1, 0)$, and $T(0, 0, 1)$, respectively.

EXAMPLE 14.8. Let T be the linear transformation that maps E^2 into E^2 such that $T(1, 1) = (1, 2)$ and $T(2, -1) = (-1, 1)$. Suppose T is represented by the 2×2 matrix

$$\begin{bmatrix} a_{11} & a_{12} \\ a_{21} & a_{22} \end{bmatrix}.$$

To get $T(1, 1)$ we multiply this matrix times the "point" $\begin{bmatrix} 1 \\ 1 \end{bmatrix}$ and equate the result to $\begin{bmatrix} 1 \\ 2 \end{bmatrix}$:

$$\begin{bmatrix} a_{11} & a_{12} \\ a_{21} & a_{22} \end{bmatrix} \begin{bmatrix} 1 \\ 1 \end{bmatrix} = \begin{bmatrix} 1 \\ 2 \end{bmatrix},$$

thus

$$a_{11} + a_{12} = 1 \quad \text{and} \quad a_{21} + a_{22} = 2. \tag{7}$$

Similarly, for $T(2, -1)$ we multiply the matrix times $\begin{bmatrix} 2 \\ -1 \end{bmatrix}$ and equate the result to $\begin{bmatrix} -1 \\ 1 \end{bmatrix}$:

$$2a_{11} - a_{12} = -1 \quad \text{and} \quad 2a_{21} - a_{22} = 1. \tag{8}$$

The left-hand equations in (7) and (8) form a 2×2 system that is easily solved to get $a_{11} = 0$ and $a_{12} = 1$. Similarly, the right-hand equations of (7) and (8) can be used to find $a_{21} = 1$ and $a_{22} = 1$. Hence T is given by the matrix

$$\begin{bmatrix} 0 & 1 \\ 1 & 1 \end{bmatrix}.$$

The final result of this section establishes the continuity of linear transformations. In proving that fact we use the following general property of linear transformations between Euclidean spaces.

LEMMA 14.1. If T is a linear transformation from E^m into E^n, then there is a number M_T such that for every $\mathbf{x}$ in E^m,

$$\| T(\mathbf{x}) \| \leq M_T \| \mathbf{x} \|. \tag{9}$$

Proof. Let T be represented by the $m \times n$ matrix $[a_{ij}]$ and define

$$M_T = \left\{ \sum_{i=1}^{m} \sum_{j=1}^{n} a_{i,j}^2 \right\}^{1/2} \tag{10}$$

where the double sum is taken over all mn entries of the matrix. Then

$$\| T(\mathbf{x}) \| = \left\{ \left(\sum_{j=1}^{n} a_{i,j} x_j \right)^2 + \cdots + \left(\sum_{j=1}^{n} a_{m,j} x_j \right)^2 \right\}^{1/2}. \tag{11}$$

Each row of $[a_{i,j}]$ can be treated as a point in E^n, and therefore each sum on the right-hand side of (11) can be estimated using the Cauchy-Bunyakovsky-Schwartz Inequality (Theorem 13.1):

$$\left(\sum_{j=1}^{n} a_{i,j} x_j \right)^2 \leq \left(\sum_{j=1}^{n} a_{i,j}^2 \right) \left(\sum_{j=1}^{n} x_j^2 \right).$$

Thus (11) leads to

$$\| T(\mathbf{x}) \| \leq \left\{ \sum_{j=1}^{n} a_{1,j}^2 + \cdots + \sum_{j=1}^{n} a_{m,j}^2 \right\}^{1/2} \left\{ \sum_{j=1}^{n} x_j^2 \right\}^{1/2}$$

$$\leq \left\{ \sum_{i=1}^{m} \sum_{j=1}^{n} a_{i,j}^2 \right\}^{1/2} \| \mathbf{x} \|$$

$$= M_T \| \mathbf{x} \|.$$

THEOREM 14.7. If T is a linear transformation from E^n into E^m, then T is continuous on E^n.

Proof. Let $\mathbf{p}$ be an arbitrary point in E^n and let M_T be the number from Lemma 14.1. For any $\mathbf{x}$ in E^n, we have

$$
\begin{aligned}
d_m(T(\mathbf{x}), T(\mathbf{p})) &= \| T(\mathbf{x}) - T(\mathbf{p}) \| \\
&= \| T(\mathbf{x} - \mathbf{p}) \| \\
&\leq M_T \| \mathbf{x} - \mathbf{p} \| \\
&= M_T d_n(\mathbf{x}, \mathbf{p}).
\end{aligned}
$$

If $\varepsilon > 0$ is given, we can define $\delta = \varepsilon / M_T$, and this implies that

$$
d_m(T(\mathbf{x}), T(\mathbf{p})) < \varepsilon \quad \text{whenever} \quad d_n(\mathbf{x}, \mathbf{p}) < \delta.
$$

We note that the choice of δ in the preceding proof is independent of the point $\mathbf{p}$ where continuity is established. Thus we conclude that linear transformations, like linear functions on $\mathbb{R}$, are *uniformly* continuous on their domains.

Exercises 14.5

1. Let T be the linear transformation from E^4 into E^3 given by the matrix

$$
\begin{bmatrix}
2 & -1 & 1 & 2 \\
0 & 2 & -1 & 3 \\
5 & 1 & -1 & 2
\end{bmatrix}.
$$

   Find the images of the points $(1, 2, -1, 3)$ and $(-2, 1, 3, 2)$.

2. Let T be the linear transformation from E^4 into E^4 such that
   $$T(1, 0, 0) = (2, 0, -1, 4), \quad T(0, 1, 0, 0) = (0, 1, 2, 3),$$
   $$T(0, 0, 1, 0) = (3, 2, -3, 1), \quad \text{and} \quad T(0, 0, 0, 1) = (-6, 2, 4, 1).$$

   Find the matrix representation of T.

3. Let T be the linear transformation from E^2 into E^2 such that
   $$T(1, 1) = (3, -1) \quad \text{and} \quad T(1, -1) = (1, 7).$$

   Find the matrix representation of T.

4. Let T be the linear transformation from E^2 into E^2 such that

$$T(2, 1) = (4, -5) \quad \text{and} \quad T(1, -3) = (-5, 1).$$

Find the matrix representation of T.

5. Show that the number M_T as chosen in (10) is not necessarily the smallest number that satisfies (9). (*Hint:* Consider the identity transformation $T(\mathbf{x}) = \mathbf{x}$.)

15

DIFFERENTIAL CALCULUS IN EUCLIDEAN SPACES

15.1. Partial Derivatives and Directional Derivatives

Throughout this section we let f be a function from a domain D in E^n into E^1 and $\mathbf{x}$ be a point in $D°$, the interior of D. Consider the "difference quotient"

$$\frac{[f(x_1, \ldots, x_i + t, x_{i+1}, \ldots, x_n) - f(x_1, \ldots, x_n)]}{t}.$$

Using the basis vectors, we can write the difference quotient as

$$\frac{[f(\mathbf{x} + t\mathbf{e}^{(i)}) - f(\mathbf{x})]}{t}.$$

For a given point $\mathbf{x}$ in $D°$, this quotient describes a function (of t) from some set $(-\delta, 0) \cup (0, \delta)$ into E^1. The requirement that $\mathbf{x}$ be an interior point of D guarantees that this difference quotient is defined whenever t is sufficiently close to zero. Since zero is therefore an interior point of the domain of the quotient, we can consider the possibility that the difference quotient approaches a limit as t tends to zero.

DEFINITION 15.1. If the following limit exists, we write

$$f_i(\mathbf{x}) = \lim_{t \to 0} \frac{f(\mathbf{x} + t\mathbf{e}^{(i)}) - f(\mathbf{x})}{t}.$$

This limit is the *partial derivative* of f at $\mathbf{x}$ in the ith coordinate. If this limit exists for each $\mathbf{x}$ in a subset D^* of D, then it determines a function f_i on D^*.

EXAMPLE 15.1. If

$$f(\mathbf{x}) = x_1^2 x_2 + x_2^3 \sin(x_3^2),$$

then

$$f_1(\mathbf{x}) = 2x_1 x_2,$$
$$f_2(\mathbf{x}) = x_1^2 + 3x_2^2 \sin(x_3^2),$$
$$f_3(\mathbf{x}) = 2x_2^3 x_3 \cos(x_3^2).$$

Note that each of f_1, f_2, and f_3 is a function on E^3, so we can "differentiate" them, that is, apply Definition 15.1, to get derivatives of the second order:

$$f_{11}(\mathbf{x}) = 2x_2,$$
$$f_{22}(\mathbf{x}) = 6x_2 \sin(x_3^2),$$
$$f_{33}(\mathbf{x}) = 2x_2^3 \cos(x_3^2) - 4x_2^3 x_3^2 \sin(x_3^2).$$

Also

$$f_{12}(\mathbf{x}) = [f_1(x_1, x_2, x_3)]_2 = 2x_1,$$
$$f_{13}(\mathbf{x}) = [f_1(x_1, x_2, x_3)]_3 = 0,$$
$$f_{21}(\mathbf{x}) = 2x_1,$$
$$f_{23}(\mathbf{x}) = 6x_2^2 x_3 \cos(x_3^2),$$
$$f_{31}(\mathbf{x}) = 0,$$
$$f_{32}(\mathbf{x}) = 6x_2^2 x_3 \cos(x_3^2).$$

Note that in the preceding example, $f_{12} = f_{21}$, $f_{13} = f_{31}$, and $f_{23} = f_{32}$. This suggests that these "mixed partial derivatives" yield the same result irrespective of the order of differentiation. We see later that when all the derivatives are continuous, this is indeed the case.

Suppose we want to take the limit of the difference quotient by approaching $\mathbf{x}$ along some direction other than $\mathbf{e}^{(1)}$, $\mathbf{e}^{(2)}$, . . . , or $\mathbf{e}^{(n)}$. We can approach $\mathbf{x}$ along any "straight-line" path in the following way. Let $\mathbf{u}$ be a point in E^n such that $\|\mathbf{u}\| = 1$; then $\mathbf{u}$ determines a "direction" in the sense that if t is any number, then $\mathbf{x} + t\mathbf{u}$ is a point lying on the "line" through $\mathbf{x}$ and $\mathbf{x} + t\mathbf{u}$ such that $d(\mathbf{x}, \mathbf{x} + t\mathbf{u}) = t$. (By the "line through $\mathbf{x}$ and $\mathbf{y}$" we mean the set $\{a\mathbf{x} + b\mathbf{y}: a \in \mathbb{R}, b \in \mathbb{R}\}$ consisting of all linear combinations of $\mathbf{x}$ and $\mathbf{y}$.)

DEFINITION 15.2. If the following limit exists, we write

$$D_{\mathbf{u}}f(\mathbf{x}) = \lim_{t \to 0} \frac{f(\mathbf{x} + t\mathbf{u}) - f(\mathbf{x})}{t},$$

which is called the *directional derivative* of f at $\mathbf{x}$.

Note that f_i is simply $D_{\mathbf{u}}f$ in the special case where $\mathbf{u} = \mathbf{e}^{(i)}$.

EXAMPLE 15.2. Let f be the function on E^2 given by

$$f(\mathbf{x}) = f(x_1, x_2) = x_1 x_2 + x_2^2,$$

and let $\mathbf{x} = (0, 1)$ and $\mathbf{u} = (1/\sqrt{2}, 1/\sqrt{2})$. Then

$$\begin{aligned}
\frac{f(\mathbf{x} + t\mathbf{u}) - f(\mathbf{x})}{t} &= \frac{1}{t}\left[\frac{t}{\sqrt{2}}\left(1 + \frac{t}{\sqrt{2}}\right) \right. \\
&\qquad \left. + \left(1 + \frac{t}{\sqrt{2}}\right)^2 - 1 \right] \\
&= \frac{1}{t}\left[\frac{3t}{\sqrt{2}} + \frac{t^2}{2} \right] \\
&= \frac{3}{\sqrt{2}} + \frac{t}{2}.
\end{aligned}$$

Letting $t \to 0$, we get

$$D_{\mathbf{u}}f(0, 1) = \frac{3}{\sqrt{2}}.$$

A familiar result (Theorem 6.1) in the theory of functions from $\mathbb{R}$ into $\mathbb{R}$ is that differentiability implies continuity. It is important to realize that this implication is not valid in the case of partial derivatives of functions on E^n, where $n > 1$. In Examples 14.5 and 14.6, both functions satisfy $f_1(\mathbf{0}) = f_2(\mathbf{0}) = 0$, but f is discontinuous at $\mathbf{0}$. Indeed, in the second of these examples, $D_{\mathbf{u}} f(\mathbf{0}) = 0$ for *every* directional derivative, yet f fails to be continuous at $\mathbf{0}$.

Exercises 15.1

1. Given $f(\mathbf{x}) = x_1 x_2 \log(x_1^2 + x_2^2)$, find all first- and second-order partial derivatives.

In Exercises 2–5, find $D_{\mathbf{u}} f(\mathbf{x})$.

2. $f(\mathbf{x}) = x_1^2 + x_2^2$; $\mathbf{u} = (\sqrt{2}/2, \sqrt{2}/2)$; $\mathbf{x} = (1, 2)$.

3. $f(\mathbf{x}) = x_1 x_2$; $\mathbf{u} = (\sqrt{3}/2, 1/2)$; $\mathbf{x} = (1, 1)$.

4. $f(\mathbf{x}) = x_1 x_2 + x_2 x_3$; $\mathbf{u} = (\sqrt{3}/3, \sqrt{3}/3, \sqrt{3}/3)$; $\mathbf{x} = (1, 1, 1)$.

5. $f(\mathbf{x}) = x_1^2 + x_2 x_3$; $\mathbf{u} = (\sqrt{2}/2, 1/2, 1/2)$; $\mathbf{x} = (1, 2, 1)$.

15.2. Differentials and the Approximation Property

We now go a step further by trying to develop a nondirectional derivative. Again, $\mathbf{x}$ is an interior point of the domain D of f. We let $\boldsymbol{\Delta}\mathbf{x}$ be a point that is sufficiently close to $\mathbf{0}$ so that $\mathbf{x} + \boldsymbol{\Delta}\mathbf{x}$ is in D, and then we consider the difference $f(\mathbf{x} + \boldsymbol{\Delta}\mathbf{x}) - f(\mathbf{x})$.

EXAMPLE 15.3. Let f be the function on E^3 given by

$$f(\mathbf{x}) = f(x_1, x_2, x_3) = x_1 x_2 - x_3^2.$$

Then

$$
\begin{aligned}
f(\mathbf{x} + \boldsymbol{\Delta}\mathbf{x}) &= (x_1 + \Delta x_1)(x_2 + \Delta x_2) - (x_3 + \Delta x_3)^2 \\
&= (x_1 x_2 - x_3^2) + \{ f_1(x)\Delta x_1 + f_2(x)\Delta x_2 + f_3(x)\Delta x_3 \} \\
&\quad + \{ \Delta x_1 \Delta x_2 - (\Delta x_3)^2 \} \\
&= f(\mathbf{x}) + \{ f_1(x)\Delta x_1 + f_2(x)\Delta x_2 + f_3(x)\Delta x_3 \} \\
&\quad + \{ \Delta x_1 \Delta x_2 - (\Delta x_3)^2 \}.
\end{aligned}
$$

The middle term can be interpreted as the image of the point $\Delta\mathbf{x}$ under the linear function L represented by the row matrix $[f_1(x)\ f_2(x)\ f_3(x)]$. When $\|\Delta\mathbf{x}\|$ is small (for example, $\|\Delta\mathbf{x}\| < 1$), the third term is even smaller, so we have an approximation:

$$f(\mathbf{x} + \Delta\mathbf{x}) \approx f(\mathbf{x}) + L(\Delta\mathbf{x}). \tag{1}$$

DEFINITION 15.3. If the n partial derivatives of f exist at the point $\mathbf{x}$, then the *differential* of f at $\mathbf{x}$ is the linear function $df_{\mathbf{x}}$ represented by the row matrix $df_{\mathbf{x}} = [f_1(x)\quad f_2(x)\quad \cdots \quad f_n(x)]$.

The next major result is the approximation property that was observed in Example 15.3. In order to prove this theorem we need the following preliminary result:

LEMMA 15.1. If f is a function such that its n partial derivatives exist throughout the sphere $\mathbf{N}_r(\mathbf{x})$ and $\Delta\mathbf{x}$ is a point in $\mathbf{N}_r(\mathbf{0})$, then there exist points $\mathbf{p}^{(1)}, \ldots, \mathbf{p}^{(n)}$ in $\mathbf{N}_r(\mathbf{x})$ such that

$$\begin{aligned} f(\mathbf{x} + \Delta\mathbf{x}) - f(\mathbf{x}) &= f_1(\mathbf{p}^{(1)})\Delta x_1 + f_2(\mathbf{p}^{(2)})\Delta x_2 \\ &\quad + \cdots + f_n(\mathbf{p}^{(n)})\Delta x_n. \end{aligned} \tag{2}$$

Outline of Proof. The points $\mathbf{p}^{(1)}, \ldots, \mathbf{p}^{(n)}$ are the result of applying the Law of the Mean in each of the n coordinates. For example, on the line segment $\{(1 - t)\mathbf{x} + t(\mathbf{x} + \mathbf{e}^{(1)}\Delta x_1): 0 \le t \le 1\}$ we can consider f as a number function (of t) that is differentiable on $[0, 1]$. By the Law of the Mean there is a number t_1 between 0 and 1 such that

$$f_1((1 - t_1)\mathbf{x} + t_1(\mathbf{x} + \mathbf{e}^{(1)}\Delta x_1)) = f_1(\mathbf{p}^{(1)}) = \frac{[f(\mathbf{x} + \mathbf{e}^{(1)}\Delta x_1) - f(\mathbf{x})]}{\Delta x_1}.$$

Next apply the Law of the Mean on the segment between the points $\mathbf{x} + \mathbf{e}^{(1)}\Delta x_1$ and $\mathbf{x} + \mathbf{e}^{(1)}\Delta x_1 + \mathbf{e}^{(2)}\Delta x_2$ to get $\mathbf{p}^{(2)}$, a point on that segment such that

$$f_2(\mathbf{p}^{(2)}) = \frac{[f(\mathbf{x} + \mathbf{e}^{(1)}\Delta x_1 + \mathbf{e}^{(2)}\Delta x_2) - f(\mathbf{x} + \mathbf{e}^{(1)}\Delta x_1)]}{\Delta x_2}.$$

We may think of this procedure as moving from $\mathbf{x}$ to $\mathbf{x} + \Delta\mathbf{x}$ along a polygonal path of n segments such that on each segment only one coordinate changes. Thus $f(\mathbf{x} + \Delta\mathbf{x}) - f(\mathbf{x})$ can be written as a "collapsing sum":

$$f(\mathbf{x} + \Delta\mathbf{x}) - f(\mathbf{x}) = \sum_{i=1}^{n} \left[f\left(\mathbf{x} + \sum_{j=1}^{i} \mathbf{e}^{(j)}\Delta x_j\right) - f\left(\mathbf{x} + \sum_{j=1}^{i-1} \mathbf{e}^{(j)}\Delta x_j\right) \right]$$

$$= \sum_{i=1}^{n} f_i(\mathbf{p}^{(i)})\Delta x_i.$$

Thus (2) holds. (In the first term of this sum we have written $\sum_{j=1}^{0} \mathbf{e}^{(j)}\Delta x_j$ in place of $\mathbf{0}$.)

THEOREM 15.1: APPROXIMATION PROPERTY. If f is a function on E^n whose n partial derivatives are continuous on the open set D, then for each $\mathbf{x}$ in D,

$$f(\mathbf{x} + \Delta\mathbf{x}) = f(\mathbf{x}) + df(\Delta\mathbf{x}) + R(\mathbf{x}, \Delta\mathbf{x}),$$

where R is a function such that

$$\lim_{\|\Delta\mathbf{x}\| \to 0} \left[\frac{R(\mathbf{x}, \Delta\mathbf{x})}{\|\Delta\mathbf{x}\|} \right] = 0.$$

Proof. First choose the points $\mathbf{p}^{(1)}, \ldots, \mathbf{p}^{(n)}$ as in Lemma 15.1; thus

$$\begin{aligned} f(\mathbf{x} + \Delta\mathbf{x}) &= f(\mathbf{x}) + f_1(\mathbf{p}^{(1)})\Delta x_1 + \cdots + f_n(\mathbf{p}^{(n)})\Delta x_n \\ &= f(\mathbf{x}) + f_1(\mathbf{x})\Delta x_1 + \cdots + f_n(\mathbf{x})\Delta x_n \\ &\quad + [f_1(\mathbf{p}^{(1)}) - f_1(\mathbf{x})]\Delta x_1 + \cdots + [f_n(\mathbf{p}^{(n)}) - f_n(\mathbf{x})]\Delta x_n \\ &= f(\mathbf{x}) + df_x(\Delta\mathbf{x}) + R(\mathbf{x}, \Delta\mathbf{x}). \end{aligned}$$

For each i,

$$|\Delta x_i| \leq \left\{ \sum_{i=1}^{n} |\Delta x_i|^2 \right\}^{1/2} = \|\Delta\mathbf{x}\|,$$

so it follows that

$$\frac{|R(\mathbf{x}, \Delta\mathbf{x})|}{\|\Delta\mathbf{x}\|} \leq \sum_{i=1}^{n} |f_i(\mathbf{p}^{(i)}) - f_i(\mathbf{x})| \frac{|\Delta x_i|}{\|\Delta\mathbf{x}\|}$$

$$\leq \sum_{i=1}^{n} |f_i(\mathbf{p}^{(i)}) - f(\mathbf{x})|.$$

Since each f_i is continuous at $\mathbf{x}$, the right-hand member tends to zero as $\|\Delta\mathbf{x}\|$ approaches zero, which completes the proof.

THEOREM 15.2. If f is a function whose partial derivatives are continuous in D, then all directional derivatives $D_\mathbf{u} f(\mathbf{x})$ exist at every point $\mathbf{x}$ in D. Moreover,

$$D_\mathbf{u} f(\mathbf{x}) = df_\mathbf{x}(\mathbf{u}). \tag{3}$$

Proof. By Theorem 15.1, we can write

$$f(\mathbf{x} + t\mathbf{u}) - f(\mathbf{x}) = df_x(t\mathbf{u}) + R(\mathbf{x}, t\mathbf{u}),$$

where

$$\lim_{\|\Delta\mathbf{x}\| \to 0} \frac{R(\mathbf{x}, t\mathbf{u})}{\|\Delta\mathbf{x}\|} = 0.$$

Note that $\|t\mathbf{u}\| = d_n(\mathbf{0}, t\mathbf{u}) = t$. Using the linearity of $df_\mathbf{x}$, we see that

$$\frac{f(\mathbf{x} + t\mathbf{u}) - f(\mathbf{x})}{t} = \frac{df_\mathbf{x}(t\mathbf{u}) + R(\mathbf{x}, t\mathbf{u})}{t}$$

$$= \frac{t\, df_\mathbf{x}(\mathbf{u}) + R(\mathbf{x}, t\mathbf{u})}{t}$$

$$= df_\mathbf{x}(\mathbf{u}) + \frac{R(\mathbf{x}, t\mathbf{u})}{\|t\mathbf{u}\|}.$$

As t approaches zero, the left-hand member approaches $D_\mathbf{u} f(\mathbf{x})$, and the right-hand member approaches $df_\mathbf{x}(\mathbf{u})$.

Equation (3) of Theorem 15.2 expresses the "dot product" method of calculating the directional derivative that is familiar from elementary calculus. For, writing $df_\mathbf{x}(\mathbf{u})$ in expanded form, we see that (3) becomes

$$D_\mathbf{u} f(\mathbf{x}) = f_1(\mathbf{x}) u_1 + f_2(\mathbf{x}) u_2 + \cdots + f_n(\mathbf{x}) u_n. \tag{4}$$

We illustrate this by recalculating $D_\mathbf{u} f(\mathbf{x})$ of Example 15.2.

EXAMPLE 15.4. Let f be the function on E^2 given by

$$f(\mathbf{x}) = x_1 x_2 + x_2^2,$$

and let $\mathbf{x} = (0, 1)$ and $\mathbf{u} = (1/\sqrt{2}, 1/\sqrt{2})$. Then

$$f_1(\mathbf{x}) = x_2 = 1 \quad \text{and} \quad f_2(\mathbf{x}) = x_1 + 2x_2 = 2,$$

so by (4),

$$D_{\mathbf{u}} f(\mathbf{x}) = 1 \cdot \frac{1}{\sqrt{2}} + 2 \cdot \frac{1}{\sqrt{2}} = \frac{3}{\sqrt{2}}.$$

It is obvious that the partial derivatives of a constant function are identically zero. The converse of this observation is not only nontrivial, it is a deep and useful result. In the one-dimensional case, the Law of the Mean was used to prove that if f' is identically zero, then f is a constant function. In the present setting of E^n, we can once again rely on the Law of the Mean to prove an n-dimensional analogue of the earlier result. This use of the Law of the Mean has already occurred in the proof of Lemma 15.1, and that lemma is now used again to establish the n-dimensional analogue of the Law of the Mean.

THEOREM 15.3. If f is a function on E^n whose partial derivatives are identically zero on the open connected set D, then f is constant on D.

Proof. Let $\mathbf{x}$ be an arbitrary point in D, and suppose $f(\mathbf{x}) = c$. Define the sets A and B by

$$A = \{\mathbf{p} \in D : f(\mathbf{p}) = c\} \quad \text{and} \quad B = \{\mathbf{p} \in D : f(\mathbf{p}) \neq c\}.$$

We show that $B = \varnothing$. First, we note that Lemma 15.1 guarantees that f is continuous on D. From the continuity of f we can infer that B is open, because B is the inverse image of the open set $(-\infty, c) \cup (c, \infty)$. Also, A is open, because if $\mathbf{p}$ is in A and $N_r(\mathbf{p}) \subset D$, then by Lemma 15.1 $f(\mathbf{p} + \Delta\mathbf{p}) - f(\mathbf{p}) = 0$, for every $\Delta\mathbf{p}$ in $N_r(\mathbf{0})$. Therefore $\mathbf{p} + \Delta\mathbf{p}$ is in A, so $\mathbf{p}$ is an interior point of A. It is now clear that $D = A \cup B$, where A and B are disjoint open sets. If both A and B were nonempty, D would be disconnected, so one of these sets must be empty. Since A contains the point $\mathbf{x}$, we conclude that B is empty. Hence for every point $\mathbf{p}$ in D, $f(\mathbf{p}) = c$.

Exercises 15.2

1. Find $df_{\mathbf{x}}$ for the given function and point:

   (a) $f(\mathbf{x}) = x_1^2 x_2 + x_1 x_2^2$; $\mathbf{x} = (-2, 3)$.

(b) $f(\mathbf{x}) = x_1 x_2 + x_1 x_3 + x_2 x_3^2$; $\quad \mathbf{x} = (1, -3, 2)$.

(c) $f(\mathbf{x}) = x_1 x_2 x_3^2 \sin(\pi x_4/2)$; $\quad \mathbf{x} = (3, -1, 2, 1)$.

2. Use Equation (4) to find $D_{\mathbf{u}} f(\mathbf{x})$ for Exercises 15.1.2–15.1.5.

3. Use Equation (4) to find $D_{\mathbf{u}} f(\mathbf{x})$:

   (a) $f(\mathbf{x}) = x_1 x_2^3 + \cos(\pi x_1/2)$; $\quad \mathbf{x} = (1, 2)$; $\quad \mathbf{u} = (-2, 4)$.

   (b) $f(\mathbf{x}) = x_1 x_2^2 x_3$; $\quad \mathbf{x} = (3, -1, 2)$; $\quad \mathbf{u} = (2, -2, 1)$.

   (c) $f(\mathbf{x}) = x_1 - x_2 x_3 + x_1 x_4$; $\quad \mathbf{x} = (1, 0, -1, 2)$;
   $\mathbf{u} = (1, 2, -1, -2)$.

4. Use Equation (4) and Theorem 13.1 to prove that if f is a function on E^n with continuous partial derivatives in D°, then for any $\mathbf{u}$ and any $\mathbf{x}$ in D°,

$$|df_{\mathbf{x}}(\mathbf{u})| \leq \left\{ \sum_{i=1}^{n} f_i(\mathbf{x})^2 \right\}^{1/2}. \tag{5}$$

5. Show that in Exercise 4 there exists a point $\mathbf{u}$ with $\|\mathbf{u}\| = 1$ such that equality holds in (5) and therefore $|df_{\mathbf{x}}(\mathbf{u})|$ achieves its maximum.

6. Prove: If the function f takes on a relative extreme value at the point $\mathbf{x}$ in $D^{\circ} \subseteq E^n$ and each of its partial derivatives exists in D°, then

$$f_1(\mathbf{x}) = \cdots = f_n(\mathbf{x}) = 0$$

and every directional derivative at $\mathbf{x}$ is 0:

$$D_{\mathbf{u}} f(\mathbf{x}) = 0.$$

7. Prove: If f is a function on E^2 such that f_1 and f_2 are bounded throughout some sphere $N_r(\mathbf{x})$, then f is continuous at $\mathbf{x}$. (*Hint:* Use Lemma 15.1.)

8. Prove: If f is a function on E^2 such that f_1 and f_2 are bounded throughout the open set D, then f is continuous on D.

15.3. The Chain Rule

In one-dimensional theory the idea of a composite function is a fairly simple concept; in elementary courses one often speaks of a composite function as a "function of a function." In that setting it is the Chain Rule that enables us to compute easily the derivative of such a composite. In the multidimensional theory the complications began with the initial idea of a composite function. For example, one might have n different functions $g^{(1)}, \ldots, g^{(n)}$ to determine the n coordinates of a point $\mathbf{q}$ in the domain of f; thus

$$F(\mathbf{x}) = F(x_1, \ldots, x_n) = f(g^{(1)}(\mathbf{x}), \ldots, g^{(n)}(\mathbf{x})).$$

In this case we would seek a chain rule that gives the partial derivatives of F in terms of those of f, $g^{(1)}, \ldots$, and $g^{(n)}$. Such a result is certainly obtainable; in fact, with the right continuity conditions, the ith coordinate partial derivative of F is the image of the point $(g_i^{(1)}(\mathbf{x}), \ldots, g_i^{(n)}(\mathbf{x}))$ under the linear function $df_\mathbf{q}$, where $\mathbf{q} = (g^{(1)}(\mathbf{x}), \ldots, g^{(n)}(\mathbf{x}))$. Using the matrix representation of the linear function $df_\mathbf{q}$, we can express this more efficiently as a matrix product:

$$f_i(\mathbf{x}) = [f_1(\mathbf{q}) \, f_2(\mathbf{q}) \ldots f_n(\mathbf{q})] \cdot \begin{bmatrix} g_i^{(1)}(\mathbf{x}) \\ g_i^{(2)}(\mathbf{x}) \\ \cdot \\ \cdot \\ \cdot \\ g_i^{(n)}(\mathbf{x}) \end{bmatrix}$$

In order to avoid becoming mired in notational difficulties, we do not prove this n-dimensional chain rule here. We instead prove it for the special case $n = 2$. This allows us to work without the superscripts by replacing $g^{(1)}$ and $g^{(2)}$ with g and h.

THEOREM 15.4: Chain Rule in E^2. Let each of f, g, and h be a function on E^2. Suppose that g and h have continuous partial derivatives in a sphere about the point $\mathbf{x}$, and assume f has continuous partial derivatives in an open sphere about the point $\mathbf{q}$, where $\mathbf{q} = (g(\mathbf{x}), h(\mathbf{x}))$. If F is the composite function given by $F(\mathbf{p}) = f(g(\mathbf{p}), h(\mathbf{p}))$, then F has continuous partial derivatives in an open sphere about $\mathbf{x}$, and they are given by

$$\begin{aligned} F_1(\mathbf{x}) &= f_1(\mathbf{q})g_1(\mathbf{x}) + f_2(\mathbf{q})h_1(\mathbf{x}), \\ F_2(\mathbf{x}) &= f_1(\mathbf{q})g_2(\mathbf{x}) + f_2(\mathbf{q})h_2(\mathbf{x}). \end{aligned} \qquad (1)$$

Note: In the frequently used Leibnitz notation, the formulas in (1) become

$$\frac{\partial F}{\partial x_1} = \frac{\partial f}{\partial g} \frac{\partial g}{\partial x_1} + \frac{\partial f}{\partial h} \frac{\partial h}{\partial x_1},$$

$$\frac{\partial F}{\partial x_2} = \frac{\partial f}{\partial g} \frac{\partial g}{\partial x_2} + \frac{\partial f}{\partial h} \frac{\partial h}{\partial x_2}.$$

Proof. We have $\mathbf{x} = (x_1, x_2)$ and $\mathbf{\Delta x} = (\Delta x_1, \Delta x_2)$; also we define

$$\mathbf{\Delta q} = (g(\mathbf{x} + \mathbf{\Delta x}) - g(\mathbf{x}), h(\mathbf{x} + \mathbf{\Delta x}) - h(\mathbf{x})). \qquad (2)$$

By the Approximation Property (Theorem 15.1), we have

$$g(\mathbf{x} + \mathbf{\Delta x}) - g(\mathbf{x}) = dg_\mathbf{x}(\mathbf{\Delta x}) + R_h(\mathbf{x}, \mathbf{\Delta x}),$$
$$h(\mathbf{x} + \mathbf{\Delta x}) - h(\mathbf{x}) = dh_\mathbf{x}(\mathbf{\Delta x}) + R_h(\mathbf{x}, \mathbf{\Delta x}),$$

and

$$F(\mathbf{x} + \mathbf{\Delta x}) - F(\mathbf{x}) = f(\mathbf{q} + \mathbf{\Delta q}) - f(\mathbf{q}) = df_\mathbf{q}(\mathbf{\Delta q}) + R_f(\mathbf{q}, \mathbf{\Delta q}),$$

where the remainder terms R_g, R_h, and R_f "tend to zero rapidly." By substituting these into (2), we get

$$\mathbf{\Delta q} = (dg_\mathbf{x}(\mathbf{\Delta x}), dh_\mathbf{x}(\mathbf{\Delta x})) + (R_g(\mathbf{x}, \mathbf{\Delta x}), R_h(\mathbf{x}, \mathbf{\Delta x})),$$

which gives

$$df_\mathbf{q}(\mathbf{\Delta q}) = df_\mathbf{q}(dg_\mathbf{x}(\mathbf{\Delta x}), dh_\mathbf{x}(\mathbf{\Delta x})) + df_\mathbf{q}(R_g(\mathbf{x}, \mathbf{\Delta x}), R_h(\mathbf{x}, \mathbf{\Delta x})).$$

Thus

$$F(\mathbf{x} + \mathbf{\Delta x}) - F(\mathbf{x}) = df_\mathbf{q}(dg_\mathbf{x}(\mathbf{\Delta x}), dh_\mathbf{x}(\mathbf{\Delta x})) + R(\mathbf{x}, \mathbf{\Delta x}), \qquad (3)$$

where

$$R(\mathbf{x}, \mathbf{\Delta x}) = df_\mathbf{q}(R_g(\mathbf{x}, \mathbf{\Delta x}), R_h(\mathbf{x}, \mathbf{\Delta x})) + R_f(\mathbf{q}, \mathbf{\Delta q}), \qquad (3a)$$

and we note that $\mathbf{q}$ and $\mathbf{\Delta q}$ in the last term are determined by $\mathbf{x}$ and $\mathbf{\Delta x}$. Now by the definition of $df_\mathbf{q}$, we have

$$\begin{aligned} df_\mathbf{q}(dg_\mathbf{x}(\mathbf{\Delta x}), dh_\mathbf{x}(\mathbf{\Delta x})) &= f_1(\mathbf{q})dg_\mathbf{x}(\mathbf{\Delta x}) + f_2(\mathbf{q})dh_\mathbf{x}(\mathbf{\Delta x}) \\ &= f_1(\mathbf{q})\{g_1(\mathbf{x})\Delta x_1 + g_2(\mathbf{x})\Delta x_2\} \qquad (4) \\ &\quad + f_2(\mathbf{q})\{h_1(\mathbf{x})\Delta x_1 + h_2(\mathbf{x})\Delta x_2\}. \end{aligned}$$

To get the partial derivative $F_1(\mathbf{x})$, we replace Δx_2 in (4) by zero and substitute into (3); then divide by Δx_1 to form the difference quotient. The result is

$$\frac{F(\mathbf{x} + \Delta\mathbf{x}) - F(\mathbf{x})}{\Delta x_1} = \frac{f_1(\mathbf{q})\{g_1(\mathbf{x})\Delta x_1 + 0\}}{\Delta x_1}$$

$$+ \frac{f_2(\mathbf{q})\{h_1(\mathbf{x})\Delta x_1 + 0\}}{\Delta x_1} + \frac{R(\mathbf{x}, \Delta\mathbf{x})}{\Delta x_1} \quad (5)$$

$$= f_1(\mathbf{q})g_1(\mathbf{x}) + f_2(\mathbf{q})h_1(\mathbf{x}) + \frac{R(\mathbf{x}, \Delta\mathbf{x})}{\Delta x_1}.$$

Similarly, to get $F_2(\mathbf{x})$ we use $\Delta\mathbf{x} = (0, \Delta x_2)$ and arrive at

$$\frac{F(\mathbf{x} + \Delta\mathbf{x}) - F(\mathbf{x})}{\Delta x_2} = f_1(\mathbf{q})g_2(\mathbf{x}) + f_2(\mathbf{q})h_2(\mathbf{x}) + \frac{R(\mathbf{x}, \Delta\mathbf{x})}{\Delta x_2}. \quad (6)$$

By comparing (5) and (6) with the right-hand members in (1), we see that proving (1) is equivalent to showing that the remainder term $R(\mathbf{x}, \Delta\mathbf{x})/\Delta x_i$ approaches zero as Δx_i tends to zero ($i = 1$ or 2).

Since $df_\mathbf{q}$ is a linear function, we can multiply (3a) by $1/|\Delta x_i|$ to get

$$\frac{R(\mathbf{x}, \Delta\mathbf{x})}{|\Delta x_i|} = df_\mathbf{q}\left(\frac{R_g(\mathbf{x}, \Delta\mathbf{x})}{|\Delta x_i|}, \frac{R_h(\mathbf{x}, \Delta\mathbf{x})}{|\Delta x_i|} + \frac{R_f(\mathbf{q}, \Delta\mathbf{q})}{|\Delta x_i|}\right). \quad (7)$$

As Δx_i tends to zero, both coordinates in the first term approach zero, and $df_\mathbf{q}$ is continuous at $\mathbf{0}$. Therefore the first term approaches $df_\mathbf{q}(\mathbf{0})$, which equals zero (by the linearity of $df_\mathbf{q}$). To show that the second term in (7) approaches zero, recall that

$$\Delta\mathbf{q} = (dg_\mathbf{x}(\Delta\mathbf{x}) + R_g(\mathbf{x}, \Delta\mathbf{x}), dh_\mathbf{x}(\Delta\mathbf{x}) + R_h(\mathbf{x}, \Delta\mathbf{x})).$$

Since $dg_\mathbf{x}$ and $dh_\mathbf{x}$ are linear, there is a number M such that

$$|dg_\mathbf{x}(\Delta x_i)| \le M|\Delta x_i| \quad \text{and} \quad |dh_\mathbf{x}(\Delta x_i)| \le M|\Delta x_i|;$$

and since

$$\lim_{\Delta x_i \to 0}\left[\frac{R_g(\mathbf{x}, \Delta\mathbf{x})}{\Delta x_i}\right] = 0 \quad \text{and} \quad \lim_{\Delta x_i \to 0}\left[\frac{R_h(\mathbf{x}, \Delta\mathbf{x})}{\Delta x_i}\right] = 0,$$

we can choose M large enough so that

$$|R_g(\mathbf{x}, \Delta\mathbf{x})| \le M|\Delta x_i| \quad \text{and} \quad |R_h(\mathbf{x}, \Delta\mathbf{x})| \le M|\Delta x_i|.$$

Therefore it follows that $\|\Delta\mathbf{q}\| \le 4M|\Delta x_i|$. Thus

$$\frac{|R_f(\mathbf{q}, \Delta\mathbf{q})|}{|\Delta x_i|} \leq \frac{4M|R_f(\mathbf{q}, \Delta\mathbf{q})|}{\|\Delta\mathbf{q}\|},$$

and since $\lim_{\Delta x_i \to 0} \|\Delta\mathbf{q}\| = 0$, we conclude that the second term in (7) approaches zero as Δx_i tends to zero, which completes the proof.

We demonstrate the Chain Rule in the following example. Since this type of problem is standard in elementary calculus, our primary purpose here is to familiarize ourselves with the notation.

EXAMPLE 15.5. Define the functions f, g, h, and F by

$$f(\mathbf{x}) = x_1 x_2^2, \quad g(\mathbf{x}) = x_1 \sin x_2, \quad h(\mathbf{x}) = x_1^2 - x_2,$$

and

$$F(\mathbf{x}) = f(g(\mathbf{x}), h(\mathbf{x})).$$

Then by (1),

$$\begin{aligned}
F_1(\mathbf{x}) &= f_1(g(\mathbf{x}), h(\mathbf{x}))g_1(\mathbf{x}) + f_2(g(\mathbf{x}), h(\mathbf{x}))h_1(\mathbf{x}) \\
&= [h(\mathbf{x})]^2 \sin x_2 + 2g(\mathbf{x})h(\mathbf{x}) \cdot 2x_1 \\
&= (x_1^2 - x_2)^2 \sin x_2 + 4x_1^2(\sin x_2)(x_1^2 - x_2) \\
&= (x_1^2 - x_2)[5x_1^2 - x_2]\sin x_2.
\end{aligned}$$

Also,

$$\begin{aligned}
F_2(\mathbf{x}) &= f_1(g(\mathbf{x}), h(\mathbf{x}))g_2(\mathbf{x}) + f_2(g(\mathbf{x}), h(\mathbf{x}))h_2(\mathbf{x}) \\
&= [h(\mathbf{x})]^2 x_1 \cos x_2 + 2g(\mathbf{x})h(\mathbf{x})(-1) \\
&= (x_1^2 - x_2)^2 x_1 \cos x_2 - 2x_1(\sin x_2)(x_1^2 - x_2) \\
&= (x_1^2 - x_2)x_1[(x_1^2 - x_2)\cos x_2 - 2x_1 \sin x_2].
\end{aligned}$$

In this example, it is easy to check the derivatives by writing F explicitly in terms of x:

$$F(\mathbf{x}) = x_1(\sin x_2)(x_1^2 - x_2)^2.$$

Now F_1 and F_2 can be calculated directly and compared to the above formulas.

Exercises 15.3

1. Given that $f(\mathbf{x}) = x_1^2 x_2$, $g(\mathbf{x}) = \cos(2x_1 + x_2)$, $h(\mathbf{x}) = \sin(x_1 + 3x_2)$, and $F(\mathbf{x}) = f(g(\mathbf{x}), h(\mathbf{x}))$, find $F_1(\mathbf{x})$ and $f_2(\mathbf{x})$.

2. Given that $f(\mathbf{x}) = \log x_1 x_2$, $g(\mathbf{x}) = x_1^2 x_2$, $h(\mathbf{x}) = 2x_1 - x_2$, and $F(\mathbf{x}) = f(g(\mathbf{x}), h(\mathbf{x}))$, find $F_1(\mathbf{x})$ and $F_2(\mathbf{x})$.

3. Let $F(\mathbf{x}) = f(g(\mathbf{x}), h(\mathbf{x}))$, where f_1 and f_2 exist, $g(\mathbf{x}) = x_1 \cos x_2$, and $h(\mathbf{x}) = x_1 \sin x_2$. Show that

$$F_1(\mathbf{x}) \cos x_2 - F_2(\mathbf{x}) \frac{\sin x_2}{x_1} = f_1(g(\mathbf{x}), h(\mathbf{x})),$$

$$F_1(\mathbf{x}) \sin x_2 + F_2(\mathbf{x}) \frac{\cos x_2}{x_1} = f_2(g(\mathbf{x}), h(\mathbf{x})), \quad \text{and}$$

$$[f_1(g(\mathbf{x}), h(\mathbf{x}))]^2 + [f_2 g((\mathbf{x}), h(\mathbf{x}))]^2 = [F_1(\mathbf{x})]^2 + \frac{1}{x_1^2} [F_2(\mathbf{x})]^2.$$

15.4. The Law of the Mean

The next result is a law-of-the-mean type of theorem for functions on E^n. Of course, there is no geometric interpretation for the n-dimensional case such as the familiar one for functions on an interval in E^1 (see Figure 6.1). Nevertheless, there is a true generalization in the analytic sense. Recall that the Law of the Mean says that there is a number c between a and b such that $f(b) - f(a) = f'(c)(b - a)$, where f is differentiable on $[a, b]$. Since $f'(c)(b - a)$ is the value of the differential df_c at the number $b - a$, this suggests that we should seek the existence of an intermediate point where the differential takes on a value equal to the difference of the function values.

THEOREM 15.5: Law of the Mean. If f is a function on E^n that has continuous partial derivatives in an open sphere containing the points $\mathbf{x}$ and $\mathbf{y}$, then there is a point $\mathbf{p}$ on the segment between $\mathbf{x}$ and $\mathbf{y}$ such that $f(\mathbf{y}) - f(\mathbf{x}) = df_{\mathbf{p}}(\mathbf{y} - \mathbf{x})$.

Proof. Although the result is true in E^n for every n, we prove it only for the case in which $n = 2$. Let $\Delta\mathbf{x}$ denote $\mathbf{y} - \mathbf{x} = (y_1 - x_1, y_2 - x_2)$. We must show that there is a number t^* between 0 and 1 such that if $\mathbf{p} = \mathbf{x} + t^*(\mathbf{y} - \mathbf{x})$, then

$$f(\mathbf{y}) - f(\mathbf{x}) = f_1(\mathbf{p})\Delta x_1 + f_2(\mathbf{p})\Delta x_2.$$

Define the function F from $[0, 1]$ into E^1 as the composite function

$$F(t) = f(\mathbf{x} + t[\mathbf{y} - \mathbf{x}]) = f(x_1 + t\Delta x_1, x_2 + t\Delta x_2).$$

By the Chain Rule, F is differentiable on $[0, 1]$, so by the Law of the Mean (Theorem 6.3), there is a number t^* between 0 and 1 such that

$$F'(t^*) = \frac{F(1) - F(0)}{1 - 0} = f(\mathbf{y}) - f(\mathbf{x}).$$

If $\mathbf{p} = \mathbf{x} + t^*[\mathbf{y} - \mathbf{x}]$, then by Theorem 15.4 we have

$$f(\mathbf{y}) - f(\mathbf{x}) = F'(t^*) = F_1(t^*)$$

$$= f_1(\mathbf{p}) \frac{d}{dt}[x_1 + t\Delta x_1] + f_2(\mathbf{p}) \frac{d}{dt}[x_2 + t\Delta x_2]$$

$$= f_1(\mathbf{p})\Delta x_1 + f_2(\mathbf{p})\Delta x_2$$

$$= df_{\mathbf{p}}(\Delta \mathbf{x}).$$

15.5. Mixed Partial Derivatives

In Example 15.1 we observed the equality of the "mixed" second-order partial derivatives such as f_{12} and f_{21}. In the next theorem we prove that this equality holds so long as one of the mixed partials is continuous. As before, we avoid the general n-dimensional case and consider only the two-dimensional setting. The result is valid, however, for functions on E^n; indeed, the proof that we give for a function on E^2 would be valid in any number of dimensions. The reason is that all but two of the coordinates are held constant in the evaluation of f_{12} and f_{21}.

THEOREM 15.6: Equality of Mixed Partial Derivatives. Suppose that f is a function on E^2 such that f_1, f_2, and f_{21} exist throughout a sphere $\mathbf{N}_r(\mathbf{x})$, and f_{21} is continuous at $\mathbf{x}$; then f_{12} exists at $\mathbf{x}$, and $f_{12}(\mathbf{x}) = f_{21}(\mathbf{x})$.

Proof. Let $\Delta \mathbf{x}$ be a point close enough to $\mathbf{0}$ so that $\mathbf{x} + \Delta \mathbf{x} \in \mathbf{N}_r(\mathbf{x})$, and define the functions ϕ (on $D \subset E^2$) and g (on $(-\delta, \delta) \subset E^1$) by

$$\phi(\Delta x_1, \Delta x_2) = f(x_1 + \Delta x_1, x_2 + \Delta x_2) - f(x_1 + \Delta x_1, x_2) \qquad (1)$$
$$- f(x_1, x_2 + \Delta x_2) + f(x_1, x_2),$$

and

$$g(t) = f(x_1 + \Delta x_1, t) - f(x_1, t).$$

(Note that g is defined for a particular Δx_1, which then remains unchanged.) Then

$$\phi(\Delta \mathbf{x}) = g(x_2 + \Delta x_2) - g(x_2),$$

and since f_2 exists throughout $\mathbf{N}_r(\mathbf{x})$, h is differentiable when t is near x_2. Therefore by the Law of the Mean (Theorem 6.3), there exists some t_2 in $(0, 1)$ such that

$$\phi(\Delta \mathbf{x}) = g'(x_2 + t_2 \Delta x_2) \Delta x_2 \tag{2}$$
$$= \{f_2(x_1 + \Delta x_1, x_2 + t_2 \Delta x_2) - f_2(x_1, x_2 + t_2 \Delta x_2)\} \Delta x_2.$$

Now consider $f_2(x_1, x_2 + t_2 \Delta x_2)$ as a function of x; it is differentiable because f_{21} exists, so we can use the Law of the Mean (Theorem 6.3) to write

$$\{f_2(x_1 + \Delta x_1, x_2 + t_2 \Delta x_2) - f_2(x_1, x_2 + t_2 \Delta x_2)\}$$
$$= f_{21}(x_1 + t_1 \Delta x_1, x_2 + t_2 \Delta x_2) \Delta x_1,$$

for some t_1 in $(0, 1)$. Substituting this expression into (2), we get

$$\phi(\Delta \mathbf{x}) = \Delta x_1 \Delta x_2 f_{21}(x_1 + t_1 \Delta x_1, x_2 + t_2 \Delta x_2);$$

therefore

$$\frac{\phi(\Delta \mathbf{x})}{\Delta x_1 \Delta x_2} = f_{21}(x_1 + t_1 \Delta x_1, x_2 + t_2 \Delta x_2). \tag{3}$$

Since f_{21} is continuous at $\mathbf{x}$, the right-hand member of (3) approaches $f_{21}(\mathbf{x})$ as Δx_1 and Δx_2 both tend to zero, that is, as $\|\Delta \mathbf{x}\|$ tends to zero. Hence

$$\lim_{\substack{\Delta x_1 \to 0 \\ \Delta x_2 \to 0}} \left[\frac{\phi(\Delta \mathbf{x})}{\Delta x_1 \Delta x_2} \right] = f_{21}(\mathbf{x}). \tag{4}$$

The continuity of f_{21} assures us that the limit in (4) is the same regardless of the route by which $\Delta \mathbf{x}$ tends to $\mathbf{0}$. Therefore we may evaluate this limit in an iterated fashion by first letting Δx_1 tend to zero and then letting Δx_2 tend to zero. To do this we define another number function h by

$$h(s) = f(s, x_2 + \Delta x_2) - f(s, x_2). \tag{5}$$

From (1) we have

$$\frac{\phi(\Delta\mathbf{x})}{\Delta x_1 \Delta x_2} = \frac{1}{\Delta x_2} \frac{h(x_1 + \Delta x_1) - h(x_1)}{\Delta x_1}. \tag{6}$$

We know that h is differentiable because f_1 exists, so the difference quotient in the right-hand member of (6) converges as x_1 tends to zero. Combining this with (5), we get

$$\lim_{\Delta x_1 \to 0} \left[\frac{\phi(\Delta\mathbf{x})}{\Delta x_1 \Delta x_2} \right] = \frac{1}{\Delta x_2} h'(x_1)$$
$$= \frac{1}{\Delta x_2} \{ f_1(x_1, x_2 + \Delta x_2) - f_1(x_1, x_2) \}. \tag{7}$$

The right-hand member is the difference quotient for $f_{12}(\mathbf{x})$, and we are assuming that f_{12} exists. Thus we let Δx_2 tend to zero in (7), which yields

$$\lim_{\Delta x_2 \to 0} \left\{ \lim_{\Delta x_1 \to 0} \left[\frac{\phi(\Delta\mathbf{x})}{\Delta x_1 \Delta x_2} \right] \right\} = f_{12}(\mathbf{x}). \tag{8}$$

Hence (4) and (8) combine to give $f_{12}(\mathbf{x}) = f_{21}(\mathbf{x})$.

Exercises 15.5

1. Given $f(\mathbf{x}) = \|\mathbf{x}\|^2$, find a point $\mathbf{p}$ on the segment from $\mathbf{e}^{(1)} = (1, 0)$ to $\mathbf{e}^{(2)} = (0, 1)$ that satisfies the Law of the Mean conclusion; that is,

$$f(\mathbf{e}^{(2)}) - f(\mathbf{e}^{(1)}) = df_{\mathbf{p}}(\mathbf{e}^{(2)} - \mathbf{e}^{(1)}).$$

2. Given $f(\mathbf{x}) = 3x_1 + x_2^2$, find a point $\mathbf{p}$ as in Exercise 1.

3. Find all second-order partial derivatives and verify the equality of the mixed partial derivatives:

(a) $f(\mathbf{x}) = \dfrac{x_1}{x_1 + x_2}$.

(b) $g(\mathbf{x}) = x_1 \sin x_2$.

(c) $h(\mathbf{x}) = \dfrac{x_1^2 \cos x_2}{\|\mathbf{x}\|^2}$.

4. Define $f(\mathbf{0}) = 0$ and, if $\mathbf{x} \neq \mathbf{0}$,

$$f(\mathbf{x}) = x_1 x_2 \frac{x_1^2 - x_2^2}{\|x\|^2}.$$

Find $f_1(0, x_2)$ and $f_2(x_1, 0)$ and use them to find $f_{12}(\mathbf{0})$ and $f_{21}(\mathbf{0})$. Why does the result that $f_{12}(\mathbf{0}) \neq f_{21}(\mathbf{0})$ not contradict Theorem 15.6?

15.6. The Implicit Function Theorem

A familiar situation in calculus and analytic geometry is one in which two quantities are related by some rule or equation, and it is hoped that one quantity is determined as a function of the other. For example, a set of points in E^2 may be described as all points whose coordinates satisfy an equation

$$f(\mathbf{x}) = f(x_1, x_2) = 0. \tag{1}$$

Equation (1) gives an "implicit" relationship between x_1 and x_2, and the hope is that there exists a number function ϕ such that

$$x_2 = \phi(x_1) \tag{2}$$

describes the same point set described by (1). Equation (2) is an "explicit" relationship between x_1 and x_2. It is often preferable to have an explicit formulation rather than an implicit one, but it is not always possible to get an explicit formula. This can be seen in the following very familiar implicit relationship.

EXAMPLE 15.6. Let C be the "unit circle" in E^2; that is, C consists of all points $\mathbf{x}$ such that

$$f(\mathbf{x}) = x_1^2 + x_2^2 - 1 = 0. \tag{3}$$

It is simple enough to solve for x_2 in terms of x_1:

$$x_2 = \pm \sqrt{1 - x_1^2} \, ; \tag{4}$$

but this does not give x_2 as a *function* of x_1, because for each x_1 between -1 and 1 there are *two* values of x_2 that satisfy (4). In order to define a single value $\phi(x_1)$ for x_2, we "localize" the problem. Thus we do not attempt to give one function ϕ such that for all x_1 in $[-1, 1]$

$$x_1^2 + \phi(x_1)^2 = 1. \tag{5}$$

Instead we consider only those points of C in some neighborhood of a point $\mathbf{p}$ on C (see Figure 15.1). Then we can find a function ϕ that satisfies (5) for all x_1 in some interval $p_1 - \delta < x_1 < p_1 + \delta$. From Figure 15.1 it is clear that we could choose

$$\phi(x_1) = \sqrt{1 - x_1^2} \quad \text{for } x \text{ in } \mathsf{N}_\delta(\mathbf{p});$$

and we could choose

$$\phi^*(x_1) = -\sqrt{1 - x_1^2} \quad \text{for } x \text{ in } \mathsf{N}_\varepsilon(\mathbf{q}).$$

But in any neighborhood of $(1, 0)$ or $(-1, 0)$ there is no function ϕ satisfying (5), because every sphere about $(1, 0)$ or $(-1, 0)$ contains a pair of points with the same first coordinate.

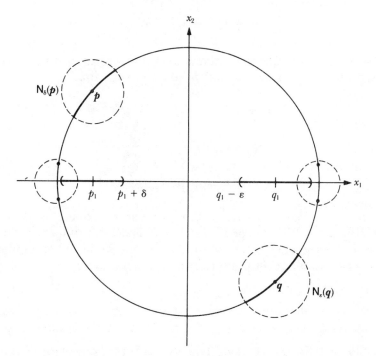

Figure 15.1

The preceding example serves both to motivate and to illustrate the next theorem. It is a pure "existence theorem" that asserts the existence of some implicit function without giving any method for finding the function. It does, however, give an explicit formula for the derivative of this elusive function.

THEOREM 15.7: Implicit Function Theorem. Let f be a function whose domain D is in E^2 and let $\mathbf{p}$ be a point in $D°$. If $f(\mathbf{p}) = 0$, $f_2(\mathbf{p}) \neq 0$, and f and f_2 are continuous in some sphere $\mathsf{N}_r(\mathbf{p})$,

then there exist open intervals I_1 about p_1 and I_2 about p_2 and a unique function ϕ from I_1 into I_2 such that

(i) for each x_1 in I_1,

$$(x_1, \phi(x_1)) \text{ is in } \mathsf{N}_r(\mathbf{p}) \quad \text{and} \quad f(x_1, \phi(x_1)) = 0;$$

(ii) $\phi(p_1) = p_2$;

(iii) ϕ is continuous on I_1;

(iv) if f_1 is continuous throughout $\mathsf{N}_r(\mathbf{p})$, then ϕ' exists and is continuous in I_1, and for each x_1 in I_1,

$$\phi_1'(x_1) = -\frac{f_1(x_1, \phi(x_1))}{f_2(x_1, \phi(x_1))}.$$

Proof. Since f_2 is continuous and nonzero at $\mathbf{p}$, there exist closed intervals J_1 about p_1 and J_2 about p_2 such that

$$J_1 \times J_2 = \{\mathbf{x} \in E^2 : x_1 \in J_1, x_2 \in J_2\} \subseteq \mathsf{N}_r(\mathbf{p})$$

and

$$\text{if} \quad \mathbf{x} \in J_1 \times J_2, \quad \text{then} \quad f_2(\mathbf{x}) \neq 0.$$

We show first that for each x_1 in J_1 there is at most one x_2 in J_2 such that $f(x_1, x_2) = 0$. Suppose x_2 and x_2' are two such numbers for the same x_1. By the Law of the Mean (Theorem 6.3) there is a number μ between x_2 and x_2' (and therefore μ is in J_2) such that

$$f_2(x_1, \mu) = \frac{f(x_1, x_2) - f(x_1, x_2')}{x_2 - x_2'} = 0;$$

but $f_2(\mathbf{x}) \neq 0$ for $\mathbf{x}$ in $J_1 \times J_2$. This contradiction shows that x_2 and

x_2' cannot be distinct. Now define

$$F(\mathbf{x}) = f(\mathbf{x})^2. \tag{6}$$

Therefore $F(\mathbf{p}) = 0$, and if $p_2 \pm c$ are the endpoints of J_2, then $F(p_1, p_2 \pm c) \neq 0$. Thus if we define

$$\varepsilon = 2 \min\{F(p_1, p_2 - c), F(p_1, p_2 + c)\},$$

then

$$F(p_1, p_2 \pm c) \geq 2\varepsilon > 0, \tag{7}$$

Let us temporarily consider F as a function of x_1 with x_2 fixed as either p_2, $p_2 - c$, or $p_2 + c$. Since F is continuous in x_1, there exists an open interval $I_1 \subset J_1$ such that for all x_1 in I_1,

$$F(x_1, p_2) < \varepsilon \quad \text{and} \quad F(x_1, p_2 \pm c) > \varepsilon. \tag{8}$$

Now consider F as a function of x_2 on J_2: For each (fixed) x_1 in J_1, F is continuous in x_2 and so it attains its minimum value on the closed interval J_2; and (8) shows that this minimum cannot occur at an endpoint. If x_2^* is the interior point of J_2 where this minimum occurs, then by Lemma 6.1,

$$F_2(x_1, x_2^*) = 0. \tag{9}$$

If (6) is differentiated with respect to x_2 and combined with (9), we get

$$2f(x_1, x_2^*) f_2(x_1, x_2^*) = 0.$$

Since f_2 cannot vanish at this point, this implies

$$f(x_1, x_2^*) = 0. \tag{10}$$

Thus we define the function ϕ on I_1 by

$$\phi(x_1) = x_2^*,$$

and choose $I_2 = (J_2)^0$. Therefore we have immediately that ϕ satisfies conclusions (i) and (ii).

To prove that ϕ is continuous, take x_1 and z_1 in I_1 and use the Law of the Mean to write

$$0 - 0 = f(x_1, \phi(x_1)) - f(z_1, \phi(z_1))$$
$$= [f(x_1, \phi(x_1)) - f(x_1, \phi(z_1))]$$
$$+ [f(x_1, \phi(z_1)) - f(z_1, \phi(z_1))]$$
$$= f_2(x_1, \mu)[\phi(x_1) - \phi(z_1)] + [f(x_1, \phi(z_1)) - f(z_1, \phi(z_1))]$$

for some μ between $\phi(x_1)$ and $\phi(z_1)$. Therefore

$$\phi(x_1) - \phi(z_1) = -\frac{f(x_1, \phi(z_1)) - f(z_1, \phi(z_1))}{f_2(x_1, \mu)}. \qquad (11)$$

Now the continuity of f and nonvanishing of f_2 on $I_1 \times I_2$ imply that

$$\lim_{x_1 \to z_1} [\phi(x_1) - \phi(z_1)] = 0;$$

that is, ϕ is continuous at z_1.

Finally, to prove (iv) we assume that f_1 exists and apply the Law of the Mean to the numerator of the right-hand side of (11). This yields a number ξ between x_1 and z_1 where

$$\phi(x_1) - \phi(z_1) = -\frac{f_1(\xi, \phi(z_1))[x_1 - z_1]}{f_2(x_1, \mu)}, \quad \text{or}$$

$$\frac{\phi(x_1) - \phi(z_1)}{x_1 - z_1} = -\frac{f_1(\xi, \phi(z_1))}{f_2(x_1, \mu)}. \qquad (12)$$

Equation (12) holds for any x_1 and z_1 in I_1, and as x_1 tends to z_1 we also have ξ approaches z_1 and $\phi(x_1)$ approaches $\phi(z_1)$, which implies that μ tends to $\phi(y_1)$. With the continuity of f_1 and f_2, this yields

$$\phi'(z_1) = \lim_{x_1 \to z_2} -\frac{f_1(\xi, \phi(z_1))}{f_2(x_1, \mu)} = -\frac{f_1(z_1, \phi(z_1))}{f_2(z_1, \phi(z_1))}. \qquad (13)$$

As Equation (13) shows, ϕ' is the quotient of continuous functions, and therefore ϕ' is itself continuous. Thus the proof is complete.

The Implicit Function Theorem has many forms and extensions. In general, if there is an equation that gives a relationship among $m + n$ variables, one wishes to know whether m of these variables can be given as a well-behaved transformation of the remaining n variables. Of the many variations of the Implicit Function Theorem we here state only one other to illustrate the possibilities.

THEOREM 15.8. Let f be a function whose domain D is in E^3 and let $\mathbf{p}$ be a point in D°. If $f(\mathbf{p}) = 0$, $f_3(\mathbf{p}) \neq 0$, and f, f_1, and f_2 are continuous in some sphere $N_r(\mathbf{p})$,

then there exist open intervals I_1 about p_1, I_2 about p_2, and I_3 about p_3, and a unique function ϕ from $I_1 \times I_2$ into I_3 such that

(i) for each $\mathbf{x}$ in $I_1 \times I_2 \times I_3$, $(x_1, x_2, \phi(x_1, x_2))$ is in $N_r(\mathbf{p})$ and $f(x_1, x_2, \phi(x_1, x_2)) = 0$;
(ii) $\phi(p_1, p_2) = p_3$;
(iii) ϕ is continuous on $I_1 \times I_2$;
(iv) ϕ_1 and ϕ_2 exist and are continuous on $N_r(\mathbf{p})$, and for each (x_1, x_2) in $I_1 \times I_2$,

$$\phi_1(x_1, x_2) = -\frac{f_1(x_1, x_2, \phi(x_1, x_2))}{f_3(x_1, x_2, \phi(x_1, x_2))}, \quad \text{and}$$

$$\phi_2(x_1, x_2) = -\frac{f_2(x_1, x_2, \phi(x_1, x_2))}{f_3(x_1, x_2, \phi(x_1, x_2))}.$$

Exercises 15.6

In Exercises 1–5, determine whether the function satisfies the hypotheses of Theorem 15.7 (or 15.8) in a neighborhood of the given point $\mathbf{p}$.

1. $f(\mathbf{x}) = x_1^2 + x_2^2 - 1$; $\mathbf{p} = (0, 1)$.

2. $f(\mathbf{x}) = x_1^2 + x_2^2 - 1$; $\mathbf{p} = (1, 0)$.

3. $f(\mathbf{x}) = (x_1^2 + x_2^2 + x_3^2)^{1/2} - \cos x_3$; $\mathbf{p} = (0, 0, 1)$.

4. $f(\mathbf{x}) = x_1 x_2 + \log x_1 x_2$; $\mathbf{p} = (1, 1)$.

5. $f(\mathbf{x}) = \{(x_1^2 + x_2^2)(1 - [x_1^2 + x_2^2])\}^{1/2}$; $\mathbf{p} = (0, 0)$.

6. Does there exist a function ϕ on E^2 that is continuous at $(1, 1)$ and such that

$$x_1^3 + x_2^3 + [\phi(x_1, x_2)]^3 - 3x_1 x_2 \phi(x_1, x_2) - 4 = 0$$

for all $\mathbf{x}$ in some neighborhood of $(1, 1)$?

16

AREA AND
INTEGRATION
IN E^2

16.1. Integration on a Bounded Set

The theory to be developed here could be done in E^n instead of E^2; but, as before, it is easier to say things and to picture examples in the plane. Therefore we present the theory in the special two-dimensional case.

In general, an integral represents an "average value" of a function f over a subset D of its domain. This "average value" is approximated by a sum of values of f weighted by the "size" of the subset of D from which that value is determined.

Let f be a function whose domain is the rectangular set $[a, b] \times [c, d]$, that is,

$$R = \{\mathbf{x} \in E^2: a \leq x_1 \leq b \quad \text{and} \quad c \leq x_2 \leq d\}. \tag{1}$$

Throughout this chapter the word *rectangle* stands for a set of the form given by (1); that is, its sides are parallel to the coordinate axes, and it is a closed set unless it is stated otherwise. A *net* $\mathcal{N}$ on R is a subset of R determined by a partition $\mathcal{P}_1$ on $[a, b]$ and a partition $\mathcal{P}_2$ on $[c, d]$; then

$$\mathcal{N} = \{\mathbf{x} \in R: \quad \text{either } x_1 \in \mathcal{P}_1 \quad \text{or} \quad x_2 \in \mathcal{P}_2\}.$$

The set $\{\mathbf{x} \in R: x_1 \in \mathcal{P}_1\}$ consists of a finite number of line segments parallel to the x_2-axis with endpoints on the boundary of R. The set $\{\mathbf{x} \in R: x_2 \in \mathcal{P}_2\}$ consists of a finite number of line segments

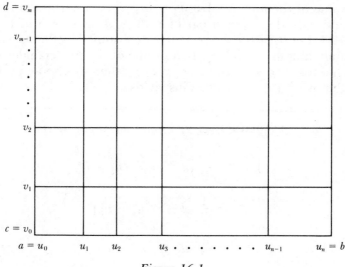

Figure 16.1

parallel to the x_1-axis with endpoints on the boundary of R. Thus the net $\mathcal{N}$ determines a finite number of closed rectangles R_i whose union is R and whose intersections are at most a line segment (see Figure 16.1). Let $A(R_i)$ denote the area of the ith rectangle R_i, and let $\|\mathcal{N}\|$ denote the *norm* of $\mathcal{N}$, which is defined to be the maximum of the lengths of the diagonals of the R_i's.

DEFINITION 16.1. The function f is said to be *integrable* on R provided that the limit

$$\lim_{\|\mathcal{N}\| \to 0} \sum_{i=1}^{n} f(\mathbf{p}^{(i)}) A(R_i)$$

exists, in which case the limit value is denoted by $\iint_R f$. This limit concept means that if $\varepsilon > 0$, there is a positive number δ such that for any net with norm less than δ and any set of points $\{\mathbf{p}^{(i)}\}_{i=1}^{n}$ such that $\mathbf{p}^{(i)} \in R_i$,

$$\left| \iint_R f - \sum_{i=1}^{n} f(\mathbf{p}^{(i)}) A(R_i) \right| < \varepsilon.$$

EXAMPLE 16.1. Let $f(\mathbf{x}) = x_1$ for $\mathbf{x}$ in the rectangle R given in (1). We assume that f is integrable on R and calculate the value of the integral. Let $\mathcal{P}_1 = \{u_k\}_{k=0}^n$ be a partition of $[a, b]$ and $\mathcal{P}_2 = \{v_k\}_{k=0}^m$ be a partition of $[c, d]$ as shown in Figure 16.1. Let R_i be the ith rectangle determined by the net corresponding to $\mathcal{P}_1$ and $\mathcal{P}_2$. If R_i is one of the m rectangles such that for each $\mathbf{x}$ in R_i, $u_{k-1} \le x_1 \le u_k$, then we choose $\mathbf{p}^{(i)}$ such that $\mathbf{p}^{(i)} = (u_{k-1} + u_k)/2$. This yields

$$\sum_{i=1}^{mn} f(\mathbf{p}^{(i)}) A(R_i) = \sum_{k=1}^{n} \frac{1}{2}(u_{k-1} + u_k)(u_k - u_{k-1})(d - c)$$

$$= \frac{1}{2}(d - c) \sum_{k=1}^{n} (u_k^2 - u_{k-1}^2)$$

$$= \frac{1}{2}(d - c)[u_n^2 - u_0^2]$$

$$= \frac{1}{2}(d - c)(b^2 - a^2).$$

Thus for any net on R an appropriate choice of $\{\mathbf{p}^{(i)}\}$ gives an approximating sum whose value is $(d - c)(b^2 - a^2)/2$. These sums must approach a limit because we are assuming that f is integrable, and therefore the only possible limit value is $(d - c)(b^2 - c^2)/2$.

DEFINITION 16.2. The function f is said to be *integrable* on the *bounded set* S provided that if R is a (closed) rectangle containing S and f_S is the function given by

$$f_S(\mathbf{p}) = \begin{cases} f(\mathbf{p}), & \text{if } \mathbf{p} \text{ is in } S, \\ 0, & \text{if } \mathbf{p} \text{ is in } R \sim S, \end{cases}$$

then f_S is integrable on R. In this case $\iint_S f \equiv \iint_R f_S$.

It should be noted that the foregoing definitions depend upon the notion of the area of the rectangle R_i. This is the "weighting factor" by which we multiply the value $f(\mathbf{p}^{(i)})$ in the approximating sum. The area $A(R_i)$ is the "size" of the subset R_i to which we alluded above. There is no difficulty with this because we can easily agree on a definition of the area $A(R_i)$, namely, the product of its length times its width. It is not so trivial, however, to define area for figures other than rectangles. Of course, we could define the area of a right triangle by subdividing a rectangle along one of its diagonals; and since any polygon can be subdivided into non-overlapping right triangles, this would lead to a notion of area for all polygons. But our goals are more ambitious than this. Indeed, we could hardly consider our definition to be complete if it did not give us the area of a circle! Thus we are led naturally to the necessity of defining area as the limit of approximating sums, which is done by using the integral defined above. But we must first resolve what appears to be an ambiguity in Definition 16.2.

LEMMA 16.1. If each of the rectangles R and R^* contains the set S and f is defined on S, then f_S is integrable on R if and only if it is integrable on R^*; moreover, $\iint_R f_S = \iint_{R^*} f_S$.

The proof of this lemma is left as Exercise 16.2.1. The result allows us to conclude that $\iint_R f_S$, and therefore $\iint_S f$, is independent of the choice of the rectangle R that contains S. This removes the ambiguity from the definition of $\iint_S f$ and also allows us to define area in terms of such an integral.

16.2. Inner and Outer Area

DEFINITION 16.3. The bounded set S *has area* provided that the function c, which is identically 1, is integrable on S. In this case,

$$A(S) = \iint_S c = \iint_R c_S, \quad \text{where } S \subset R.$$

The fact that we have very carefully stated when the bounded set S has area suggests the possibility that some bounded sets may not have area. This is not simply the case in which a set has zero area, for zero is a perfectly good limit value for nonnegative sums that may approximate the integral $\iint_R c_S$. In such a case the set has area, and its area happens to equal zero. The situation with which we are concerned is the possibility that the approximating sums

$$\sum_{i=1}^n c_S(\mathbf{p}^{(i)}) A(R_i)$$

do not approach a limit as $\|\mathcal{N}\|$ tends to zero. This is a possibility that does not arise in studying the Riemann integral in one dimension, for then a constant function such as c is always integrable over any interval. If we had attempted to integrate over arbitrary bounded sets instead of just intervals, we would have encountered the same problem that we now face: Can we define the concept of the "size" or "measure" of an arbitrary bounded set that coincides with the area of the set for simple special cases including rectangles and other polygons? This leads to the development of Lebesgue measure, which one studies in full detail in graduate analysis courses. But even in the theory of Lebesgue measure it is seen that it is not possible to extend the notion of area from simple sets to include *all* bounded sets. For the present we shall be content to work with a more elementary extension of the area concept, sometimes called the *Jordan content* of a set. This is the notion of area that has been defined above.

Let us consider this notion of area in more detail. If S is a bounded set and R is a rectangle containing S, then c_S is called the *characteristic function* of S. Let $\mathcal{N}$ be a net on R and consider the sum

$$\sum_{i=1}^{n} c_S(\mathbf{p}^{(i)}) A(R_i), \quad \text{where } \mathbf{p}^{(i)} \text{ is in } R_i.$$

This is an approximating sum for the integral $\iint_S c$, which we have defined to be the area of S. The value of the ith term $c_S(\mathbf{p}^{(i)}) A(R_i)$ is given by one of the following possibilities:

(i) if $R_i \subset S$, then $c_S(\mathbf{p}^{(i)}) = 1$, so the ith term equals $A(R_i)$;
(ii) if $R_i \subset (\sim S)$, then $c_S(\mathbf{p}^{(i)}) = 0$, so the ith term is 0;
(iii) if $R_i \cap S \neq \varnothing$ and $R_i \cap (\sim S) \neq \varnothing$, then the ith term is either $A(R_i)$ or 0 depending on whether $\mathbf{p}^{(i)}$ is in S or $\sim S$.

This is illustrated in Figure 16.2.

Corresponding to the net $\mathcal{N}$, we define upper and lower sums

$$a^-(\mathcal{N}) = \sum_{R_i \subset S} A(R_i) \quad \text{and} \quad a^+(\mathcal{N}) = \sum_{R_i \cap S \neq \varnothing} A(R_i);$$

the terms in the sum for $a^-(\mathcal{N})$ correspond to the rectangles of type (i), whereas the terms in the sum for $a^+(\mathcal{N})$ correspond to the rectangles of types (i) and (iii). It is clear that for any choice of points $\{\mathbf{p}^{(i)}\}_{i=1}^{n}$ such that $\mathbf{p}^{(i)}$ is in R_i,

$$a^-(\mathcal{N}) \leq \sum_{i=1}^{n} c_S(\mathbf{p}^{(i)}) A(R_i) \leq a^+(\mathcal{N}). \tag{1}$$

LEMMA 16.2. If S is a subset of the rectangle R, and $\mathcal{N}$ and $\mathcal{N}'$ are any two nets on R, then $a^-(\mathcal{N}) \leq a^+(\mathcal{N}')$.

Proof. As in the proof of the analogous result for integrals on E^1, we note that $\mathcal{N}$ and $\mathcal{N}'$ have a common refinement $\mathcal{N}^*$ (for ex-

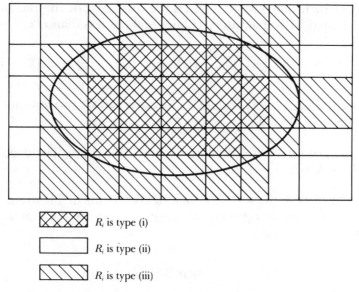

R_i is type (i)

R_i is type (ii)

R_i is type (iii)

Figure 16.2

ample, $\mathcal{N}^* = \mathcal{N} \cup \mathcal{N}'$ is one such refinement); and for any refinement, $a^-(\mathcal{N}) \leq a^-(\mathcal{N}^*)$ and $a^+(\mathcal{N}^*) \leq a^+(\mathcal{N})$. Thus the conclusion follows immediately from (1).

DEFINITION 16.4. The *inner area* of the bounded set S is denoted and given by

$$A^-(S) = \text{lub}\{a^-(\mathcal{N}): \quad \mathcal{N} \text{ is a net on } R, \text{ where } S \subset R\};$$

the *outer area* of S is denoted and given by

$$A^+(S) = \text{glb}\{a^+(\mathcal{N}): \quad \mathcal{N} \text{ is a net on } R, \text{ where } S \subset R\}.$$

It follows immediately from Lemma 16.2 that for any bounded set S, $A^-(S) \leq A^+(S)$.

THEOREM 16.1. If S is a bounded set, then

$$\lim_{\|\mathcal{N}\| \to 0} a^-(\mathcal{N}) = A^-(S) \quad \text{and} \quad \lim_{\|\mathcal{N}\| \to 0} a^+(\mathcal{N}) = A^+(S).$$

Proof. Suppose that $\varepsilon > 0$. By the definition of $A^-(S)$, there is some net $\mathcal{N}^*$ such that $a(\mathcal{N}^*) > A^-(S) - \varepsilon/2$. Choose δ so that if $\mathcal{N}$

is a net such that $\|\mathcal{N}\| < \delta$, then the total area of the rectangles determined by $\mathcal{N}$ that intersect lines of $\mathcal{N}^*$ is less than $\varepsilon/2$. Then

$$a^-(\mathcal{N}) > a^-(\mathcal{N}^*) - \sum_{R_i \cap \mathcal{N}^* \neq \varnothing} A(R_i) > a^-(\mathcal{N}^*) - \frac{\varepsilon}{2} > A^-(S) - \varepsilon.$$

Hence $\lim_{\|\mathcal{N}\| \to 0} a^-(\mathcal{N}) = A^-(S)$. The other limit is proved similarly (see Exercise 16.2.2).

THEOREM 16.2. The bounded set S has area if and only if $A^-(S) = A^+(S)$.

Proof. If $A^-(S) = A^+(S)$, then the two limits in Theorem 16.1 are equal. Combining this with inequality (1), we see that all the approximating sums

$$\sum_{i=1}^n c_S(\mathbf{p}^{(i)}) A(R_i)$$

must converge to the common value of $A^-(S)$ and $A^+(S)$. Conversely, if $A^-(S) \neq A^+(S)$, then the approximating sums cannot converge, so c is not integrable on S. Hence S does not have area.

EXAMPLE 16.2. Let ϕ and ψ be number functions that are Riemann integrable on the interval $[a, b]$, where $\phi(t) \leq \psi(t)$, and consider the point set S given by

$$S = \{\mathbf{x} \in E^2 \colon a \leq x_1 \leq b \quad \text{and} \quad \phi(x_1) \leq x_2 \leq \psi(x_1)\}. \quad (2)$$

Then S has area, because the sums $a^-(\mathcal{N})$ and $a^+(\mathcal{N})$ that approximate the inner and outer area, respectively, are also the sums that approximate the lower and upper Riemann integrals, respectively, of the function $\psi - \phi$. Since the difference of these two integrable functions is integrable, the upper and lower integrals are equal, and hence the inner and outer areas of S are equal.

Many of the familiar figures in the Euclidean plane can be described like the set S above with boundaries that are the graphs of continuous (and thus integrable) functions. In this way, the preceding example can be used to show that many common sets have area. One such set is a disk, which is considered next.

EXAMPLE 16.3. For any point $\mathbf{p}$ in E^2 and any positive number r, the closed 2-sphere $\bar{N}_r(\mathbf{p})$ has area πr^2. For we can take

$$\phi(x_1) = -\sqrt{r^2 - (x_1 - p_1)^2} \quad \text{and}$$

$$\psi(x_1) = \sqrt{r^2 - (x_1 - p_1)^2},$$

and $\bar{N}_r(\mathbf{p})$ is the set S in Equation (2). From the observation about the approximating sums in Example 16.2, we see that

$$A(N_r(\mathbf{p})) = \int_{p_1-r}^{p_1+r} (\psi - \phi)$$

$$= 2\int_{p_1-r}^{p_1+r} \sqrt{r^2 - (x_1 - p_1)^2} \, dx_1$$

$$= \pi r^2.$$

While we are citing examples, we should also give one to justify our caution in defining area; that is, we should present some set that does *not* have area. By calling on past experience, one does not have to be clairvoyant to predict that such a set can be described using the set $\mathbb{Q}^2$, which consists of those points $\mathbf{q}$ with rational coordinates.

EXAMPLE 16.4. Let $S = \mathbb{Q}^2 \cap \{\mathbf{x}: 0 \le x_1 \le 1\}$. Then we can use the rectangle $[0, 1] \times [0, 1]$ as the R in determining whether $\iint_R c_S$ exists. We know that both $\mathbb{Q}^2$ and $\sim\mathbb{Q}^2$ are dense in E^2, so for *any* rectangle R_i, the point $\mathbf{p}^i$ could be chosen in either $\mathbb{Q}^2$ or $\sim\mathbb{Q}^2$. Therefore, no matter how small the norm of the net $\mathcal{N}$ may be, every R_i intersects S, so

$$a^+(\mathcal{N}) = \sum_{R_i \cap S \ne \varnothing} A(R_i) = A(R) = 1.$$

But the density of $\sim S$ implies that no $\sim R_i$ is contained in S, so $a^-(\mathcal{N}) = 0$. Hence $A^+(S) = 1$ and $A^-(S) = 0$, so S does not have area.

The *boundary* $b[S]$ of a set S consists of those points $\mathbf{p}$ such that every neighborhood of $\mathbf{p}$ intersects both S and $\sim S$. We also note that for any net $\mathcal{N}$ on a rectangle R containing S, the difference $a^+(\mathcal{N}) - a^-(\mathcal{N})$ is the total area of those subrectangles of type

(iii) that intersect both S and $\sim S$. This suggests that the equality of $A^-(S)$ and $A^+(S)$ occurs when $b[S]$ has area zero. This is the assertion of the next theorem. First, however, we state a useful preliminary result.

LEMMA 16.3. The bounded set B has area zero if and only if $A^+(B) = 0$.

Proof. We know that $0 \le A^-(B) \le A^+(B)$, so if $A^+(B) = 0$, then it is clear that $A^-(B) = A^+(B)$. The converse is immediately true from the definition of area.

THEOREM 16.3. The bounded set S has area if and only if its boundary $b[S]$ has area zero.

Proof. First assume that S has area, and let R be a rectangle such that $\bar{S} \subset R^\circ$. Since $A^-(S) = A^+(S)$, we can choose a net $\mathcal{N}$ such that $a^+(\mathcal{N}) - a^-(\mathcal{N}) < \varepsilon$, where ε is an arbitrary positive number. Let $\{R_i^*\}$ be the collection of subrectangles that intersect both S and $\sim S$, thus $\Sigma A(R_i^*) = a^+(\mathcal{N}) - a^-(\mathcal{N}) < \varepsilon$. We want to show that every boundary point is in some R_i^*, so that $b[S] \subset \cup R_i^*$, which means that $A^+(b[S]) < \varepsilon$, and consequently $A(b[S]) = 0$. Consider a point $\mathbf{p}$ in $b[S]$: first, if $\mathbf{p}$ is in R_i°, then R_i must be one of the R_i^*'s (because R_i is then a neighborhood of $\mathbf{p}$); second, if $\mathbf{p}$ is on the common edge of two or more R_i's, then one of these R_i's must contain points of both S and $\sim S$, so it is an R_i^*. In either case, $\mathbf{p}$ is in $\cup R_i^*$; hence it follows that $A(b[S]) = 0$.

Now assume that $A(b[S]) = 0$ and $\varepsilon > 0$. Choose a net $\mathcal{N}$ such that the total area of all the subrectangles R_i' that intersect $b[S]$ is less than ε. We assert that every R_i that intersects both S and $\sim S$ is one of this set $\{R_i'\}$, so that

$$a^+(\mathcal{N}) - a^-(\mathcal{N}) \le \Sigma A(R_i') < \varepsilon;$$

for if $\mathbf{p}$ is in $R_i \cap S$ and $\mathbf{q}$ is in $R_i \cap (\sim S)$, then the segment between $\mathbf{p}$ and $\mathbf{q}$ lies entirely in R_i. Define t^* by

$$t^* = \text{lub}\{t \in [0, 1]: \quad t\mathbf{p} + (1 - t)\mathbf{q} \in S\},$$

and let $\mathbf{p}^*$ denote the point $t^*\mathbf{p} + (1 - t^*)\mathbf{q}$. Then every sphere about $\mathbf{p}^*$ contains points on this segment of both types: $t \le t^*$ and $t \ge t^*$. Since one type is in S and the other is in $\sim S$, we conclude that $\mathbf{p}^*$ is in $b[S]$. Hence R_i is one of the subrectangles R_i' that intersects $b[S]$. We have shown that for every positive number ε there is a net $\mathcal{N}$ such that $a^+(\mathcal{N}) - a^-(\mathcal{N}) < \varepsilon$; consequently,

$$A^+(S) = \lim_{\|\mathcal{N}\| \to 0} a^+(\mathcal{N}) = \lim_{\|\mathcal{N}\| \to 0} a^-(\mathcal{N}) = A^-(S).$$

Exercises 16.2

1. Prove Lemma 16.1. (*Hint: $R \cap R^*$ is a rectangle containing S, and $f(\mathbf{p}^{(i)}) = 0$ for any $\mathbf{p}^{(i)} \notin R \cap R^*$.*)

2. Prove the second limit assertion in Theorem 16.1: $\lim_{\|\mathcal{N}\| \to 0} a^+(\mathcal{N}) = A^+(S)$.

3. Let R be as in Equation (2) and define
$$f(\mathbf{x}) = \begin{cases} 1, & \text{if } x_1 = c, \text{ where } a < c < b, \\ 0, & \text{otherwise.} \end{cases}$$
Prove that $\iint_R f = 0$.

4. Let R be as in Equation (2) and $R' = [b, e] \times [c, d]$, where $e > b$. Prove: If f is integrable on R and on R', then f is integrable on $R \cup R'$ and
$$\iint_{R \cup R'} f = \iint_R f + \iint_{R'} f.$$

5. Given $S = \{\mathbf{x} \in E^2: x_1 \in \mathbb{Q} \cap [0, 1] \text{ and } x_2 \in [0, 1]\}$, find $A^-(S)$ and $A^+(S)$.

6. Given $S = \{\mathbf{x} \in E^2: x_1 = x_2 \in [0, 1]\}$, find $A^-(S)$ and $A^+(S)$.

7. Prove: If S is an open set, then $A^-(S) > 0$.

8. Prove: If S is a dense subset of the rectangle R, then $A^+(S) = A(R)$.

9. Let $S = \mathbb{Q}^2 \cap ([0, 1] \times [0, 1])$ as in Example 16.4 and find $b[S]$.

10. What is $b[\varnothing]$? Is there a nonempty set S such that $b[S] = \varnothing$?

11. What is $b[\mathbb{Q}^2]$?

12. Let R be a rectangle and $\mathcal{N}$ be a net on R; find $b[\mathcal{N}]$.

13. Prove: If p is an isolated point of S, then $p \in b[S]$.

14. Construct and explain an example of an open set that does not have area. (*Hint:* Write the points of S in Example 16.4

as a sequence $\{\mathbf{q}^{(i)}\}_{i=1}^{\infty}$ and for each i let $\mathbf{N}_{r(i)}(\mathbf{p}^{(i)})$ be a sphere of area 2^{-i-1}; then

$$A^- \left[\bigcup_{i=1}^{\infty} \mathbf{N}_{r(i)}(\mathbf{p}^{(i)}) \right] \le \sum_{i=1}^{\infty} 2^{-i-1}.$$

16.3. Properties of the Double Integral

The content of this section is a listing of properties of the integral as we have defined it on a subset of E^2. These properties will be recognizable as ones that were proved for the Riemann integral and the Stieltjes integral on E^1. For the most part, they are the fundamental properties that one finds in the theory of any type of integral. There are a few exceptions, and they are results concerning the nature of the domain S over which the integral is taken. The only reason that these results are not analogues of similar ones for the one-dimensional integrals is that we never considered the possibility of integrating over subsets of E^1 other than intervals. Our first theorem is such a result, and therefore we give a complete proof of it. In most cases the proofs are left as exercises, because they are totally similar to the proof of the corresponding result for the Riemann integral.

THEOREM 16.4. Let f be a bounded function on the bounded set S, and suppose there is a subset Z of S such that $A(Z) = 0$ and $f(\mathbf{x}) = 0$ whenever $\mathbf{x}$ is in $S \sim Z$; then f is integrable on S and $\iint_S f = 0$.

Proof. Suppose $|f(\mathbf{x})| < K$ on S, and $\varepsilon > 0$. Let R be a rectangle containing S, and choose a net $\mathcal{N}$ such that the total area of R_i^*'s that intersect Z is less than ε/K. Then for any $\mathbf{p}^{(i)}$ in an R_i that is not an R_i^*, $f(\mathbf{p}^{(i)}) = 0$; so no matter how the $\mathbf{p}^{(i)}$'s are chosen, we see that

$$\left| \sum_{i=1}^{n} f_S(\mathbf{p}^{(i)}) A(R_i) \right| = \left| \sum f_S(\mathbf{p}^{(i)*}) A(R_i^*) \right|$$

$$\le \sum |f_S(\mathbf{p}^{(i)*})| A(R_i^*)$$

$$\le K \sum A(R_i^*)$$

$$< \varepsilon.$$

Hence $\lim_{\|\mathcal{N}\| \to 0} \sum_{i=1}^{n} f_S(\mathbf{p}^{(i)}) A(R_i) = 0$; that is, $\iint_S f = 0$.

THEOREM 16.5. If each of f and g is integrable on the bounded set S, then $f + g$ and $f - g$ are also integrable on S, and

$$\iint_S (f \pm g) = \iint_S f \pm \iint_S g.$$

Proof. Left as Exercise 16.3.1.

COROLLARY 16.5. Suppose f and g are bounded on the bounded set S and there is a subset Z of S such that $f(\mathbf{x}) = g(\mathbf{x})$ whenever $\mathbf{x}$ is in S; then f is integrable on S if and only if g is integrable on S, and $\iint_S f = \iint_S g$.

Proof. By Theorem 16.4, the difference $f - g$ is integrable on S, and $\iint_S (f - g) = 0$. Therefore the integrability of one of the functions implies that of the other (by Theorem 16.5), and the equality of their values follows from the formula in Theorem 16.5.

THEOREM 16.6. If each of f and g is integrable on S and $f(\mathbf{x}) \le g(\mathbf{x})$ whenever $\mathbf{x}$ is in S, then $\iint_S f \le \iint_S g$.

Proof. Left as Exercise 16.3.2.

THEOREM 16.7. If f is integrable on S and k is a number, then kf is integrable on S, and $\iint_S kf = k \iint_S f$.

Proof. Left as Exercise 16.3.3.

COROLLARY 16.7. If S is a set that has area and f is a constant function, say, $f(\mathbf{x}) = k$, then f is integrable on S, and $\iint_S k = kA(S)$.

Proof. Left as Exercise 16.3.4.

THEOREM 16.8. Suppose f is integrable on both of the sets S and T, where $A(S \cap T) = 0$; then f is integrable on $S \cup T$, and

$$\iint_{S \cup T} f = \iint_S f + \iint_T f.$$

Proof. Let R be a rectangle containing $S \cup T$. Then f_S and f_T are both integrable on R, so by Theorem 16.5,

$$\iint_R (f_S + f_T) = \iint_R f_S + \iint_R f_T.$$

If $\mathbf{x}$ is in $R \sim (S \cap T)$, then $f_{S \cup T}(\mathbf{x}) = f_S(\mathbf{x}) + f_T(\mathbf{x})$. Therefore, since $A(S \cap T) = 0$, it follows from Corollary 16.5 that $f_{S \cup T}$ is also integrable on R, and

$$\iint_R f_{S \cup T} = \iint_R (f_S + f_T).$$

Hence f is integrable on $S \cup T$, and its integral is given by

$$\iint_{S \cup T} f = \iint_R f_{S \cup T}$$

$$= \iint_R f_S + \iint_R f_T$$

$$= \iint_S f + \iint_T f.$$

Exercises 16.3

1. Prove Theorem 16.5.

2. Prove Theorem 16.6.

3. Prove Theorem 16.7.

4. Prove Corollary 16.7.

5. Prove: If the function f is integrable on R, then it is bounded there.

6. A *step function S* on a rectangle R is a bounded function for which there is a net $\mathcal{N}$ on R such that s is constant on the interior R_i° of each subrectangle determined by $\mathcal{N}$. Prove that any such step function is integrable on R and if $s(\mathbf{x}) = k_i$ on R_i°, then

$$\iint_R s = \sum_{i=1}^{n} k_i A(R_i).$$

16.4. Line Integrals

In this section we introduce another type of single integral that is related to the double integrals we have just studied. The relationship to double integrals is twofold: This single integral involves a function whose domain is in E^2, and, as we see in Section 16.7, it can be used to evaluate certain double integrals. First we describe the sets in E^2 that serve as domains for the functions to be integrated.

DEFINITION 16.5. The set C is called a *curve* in E^2 if there exist functions g and h continuous on an interval $[a, b]$, such that

$$C = \{\mathbf{x} \in E^2: x_1 = g(t), x_2 = h(t), t \in [a, b]\}. \tag{1}$$

When $t = a$, the resulting point $(g(a), h(a))$ is called the *initial point* of C; the point $(g(b), h(b))$ is called the *terminal point* of C.

The curve C is called a *closed curve* if, for some parametrization, its initial and terminal points coincide. It is called a *simple curve* if $t, t' \in [a, b)$ implies $(g(t), h(t)) \neq (g(t'), h(t'))$. (This means that the curve does not "intersect itself," except perhaps at the terminal point.) If g and h have continuous derivatives on $[a, b]$, then C is called a *smooth curve*; and C is *sectionally smooth* if it is the union of a finite number of smooth curves that are connected "terminal point to initial point."

EXAMPLE 16.5. Let $x_1 = \sin t$ and $x_2 = \cos t$ for $0 \leq t \leq \pi$. Then C is the simple smooth curve pictured in Figure 16.3a.

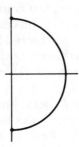

Figure 16.3a

EXAMPLE 16.6. Let $x_1 = \cos^3 t$ and $x_2 = \sin^3 t$ for $0 \leq t \leq 2\pi$. Then C is the astroid pictured in Figure 16.3b, a simple closed curve that is sectionally smooth.

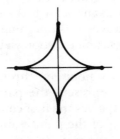

Figure 16.3b

EXAMPLE 16.7. The curve pictured in Figure 16.3c is closed and smooth but not simple.

Figure 16.3c

Recall from elementary calculus that a given curve has many parametrizations, and so the functions g and h in Definition 16.5 are not unique. There is, however, one distinction in the parametrization of C that is very useful. It is sometimes desirable to interchange the initial and terminal points of a curve C and traverse it in the opposite direction. This is achieved by replacing the parametric functions $g(t)$ and $h(t)$ with the composition functions $g(-t + a + b)$ and $h(-t + a + b)$, respectively, and we denote this by $-C$.

DEFINITION 16.6. Let F be a function whose domain D is an open connected set that includes the curve C as given by (1); let ϕ be a function on $[a, b]$; then the composition $F(g(t), h(t))$ is also a function on $[a, b]$, and we can form the Riemann-Stieltjes integral $\int_a^b F(g, h) \, d\phi$. This is called a *line integral* and is denoted by $\int_C F \, d\phi$. Thus

$$\int_C F \, d\phi = \lim_{\|\pi\| \to 0} \sum_{k=1}^n F(g(\mu_k), h(\mu_k))[\phi(t_k) - \phi(t_{k-1})]. \qquad (2)$$

As we see in Chapter 10, when ϕ has a continuous derivative, the integral (2) reduces to the Riemann integral $\int_a^b F\phi'$. It should be noted that the line integral (2) is strongly dependent on the parametrization of C. For example, if C is a simple closed curve and two parametrizations traverse C one time and two times, respectively, then the second parametrization yields an integral value that is twice that of the first.

In order to simplify notation, throughout this chapter the letters g and h always represent the parametric functions as in Equation (1); that is, $g(t)$ gives the first coordinate x_1 and $h(t)$ gives the second coordinate x_2 of the point **x** on C.

The particular line integrals that are of most frequent interest are obtained by using either of the parametric functions g or h in place of ϕ. This yields the line integrals

$$\int_C P \, dg \quad \text{and} \quad \int_C Q \, dh. \tag{3}$$

By adding these two integrals, we get a third type:

$$\int_C [P \, dg + Q \, dh]. \tag{4}$$

Exercises 16.4

In Exercises 1–5, sketch the curve C and determine whether C is simple, closed, or smooth.

1. $g(t) = 2 \cos \pi t$, $h(t) = \sin \pi t$; $\quad 0 \le t \le 2$.

2. $g(t) = t^3$, $h(t) = t^2$; $\quad 0 \le t \le 1$.

3. $g(t) = \begin{cases} \cos \pi t, & \text{if } 0 \le t < 1, \\ 2t - 3, & \text{if } 1 \le t \le 2; \end{cases}$

   $h(t) = \begin{cases} \sin \pi t, & \text{if } 0 \le t < 1, \\ 0, & \text{if } 1 \le t \le 2; \end{cases}$

   $0 \le t \le 2$.

4. $g(t) = \begin{cases} t, & \text{if } 0 \le t < 2\pi, \\ 4\pi - t, & \text{if } 2\pi \le t \le 4\pi; \end{cases}$

   $h(t) = \begin{cases} \sin \pi t, & \text{if } 0 \le t < 2\pi, \\ 0, & \text{if } 2\pi \le t \le 4\pi; \end{cases}$

   $0 \le t \le 4\pi$.

5. $g(t) = \begin{cases} 1, & \text{if } 0 \le t < 1, \\ 2 - t, & \text{if } 1 \le t \le 2; \end{cases} \quad h(t) = \begin{cases} t, & \text{if } 0 \le t < 1, \\ 1, & \text{if } 1 \le t \le 2; \end{cases}$

   $0 \le t \le 2$.

6. Evaluate $\int_C f \, d\phi$, where $f(\mathbf{x}) = x_1^2 x_2 + 3x_2$, $\phi(t) = t^2$, and $C = \{(t, t^2): 0 \le t \le 1\}$.

7. Evaluate $\int_C f \, d\phi$, where $f(\mathbf{x}) = x_1^2 x_2 + 3x_2$, $\phi(t) = t$, and $C = \{(\cos t, \sin t): 0 \le t \le \pi/2\}$.

8. Evaluate $\int_C f \, d\phi$, where $f(\mathbf{x}) = x_1^2 x_2 + 3x_2$, $\phi(t) = t$, and C is the curve given in Exercise 5.

9. Evaluate $\int_c P\, dg$, where $P(\mathbf{x}) = x_1^2 + 4x_2^2$ and C is the curve given in Exercise 1.

10. Evaluate $\int_c [P\, dg + Q\, dh]$, where $P(\mathbf{x}) = x_1^2 + 4x_2^2$, $Q(\mathbf{x}) = 2x_2$, and C is the curve given in Exercise 2.

11. Evaluate $\int_c [P\, dg + Q\, dh]$, where $P(\mathbf{x}) = \sqrt{x_1^2 + x_2^2}$, $Q(\mathbf{x}) = 2x_2$, and C is the curve given in Exercise 4.

12. Let C be a curve given by (1) and let $\mathbf{p}^*$ be a point on C other than the initial point $\mathbf{p}$ or the terminal point $\mathbf{q}$. Show that C can be written as $C = C_1 \cup C_2$, where C_1 is a curve from $\mathbf{p}$ to $\mathbf{p}^*$ and C_2 is a curve from $\mathbf{p}^*$ to $\mathbf{q}$.

16.5. Independence of Path and Exact Differentials

In this section we define two concepts related to the line integral and prove a relationship between them. Throughout this section, D denotes an open connected set in E^2.

DEFINITION 16.7. If F is a function on D, then the line integral $\int_C F\, d\phi$ is *independent of the path* in D provided that if $\mathbf{p}$ and $\mathbf{q}$ are any two points in D, and C_1 and C_2 are any two curves in D having $\mathbf{p}$ and $\mathbf{q}$ as initial and terminal points, respectively, then the two line integrals have the same value:

$$\int_{C_1} F\, d\phi = \int_{C_2} F\, d\phi.$$

The first result on line integrals is an intuitive characterization of independence of the path.

THEOREM 16.9. The line integral $\int_C F\, d\phi$ is independent of the path in D if and only if for every sectionally smooth closed curve C in D,

$$\int_C F\, d\phi = 0. \tag{1}$$

Proof. First let $\mathbf{p}$ and $\mathbf{q}$ be two points in D and assume (1) holds for every sectionally smooth closed curve in D. Let C_1 and C_2 be two such curves from $\mathbf{p}$ to $\mathbf{q}$. Define $C = C_1 \cup (-C_2)$, where $-C_2$ is the curve C_2 with the direction reversed (see Figure 16.4). More precisely, C is defined as follows: we may assume that

$$C_1 = \{(g_1(t),\, h_1(t)) \colon \quad a \le t \le b\}$$

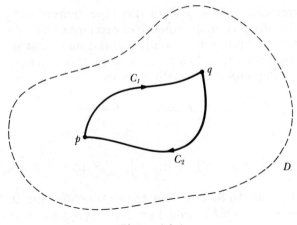

Figure 16.4

and

$$C_2 = \{(g_2(t), h_2(t)): \quad b \le t \le c\}.$$

Then we choose

$$C = \{(g(t), h(t)): \quad a \le t \le c\},$$

where

$$g(t) = \begin{cases} g_1(t), & \text{if } a \le t < b, \\ g_2(-t + b + c), & \text{if } b \le t \le c, \end{cases}$$

and

$$h(t) = \begin{cases} h_1(t), & \text{if } a \le t < b, \\ h_2(-t + b + c), & \text{if } b \le t \le c. \end{cases}$$

Then C is a sectionally smooth closed curve, so by (1)

$$\int_{C_1 \cup (-C_2)} F \, d\phi = 0.$$

From properties of the Riemann-Stieltjes integral (Theorems 10.4 and 10.5), this equation is equivalent to

$$\int_{C_1} F \, d\phi - \int_{C_2} F \, d\phi = 0.$$

Conversely, suppose $\int_C F \, d\phi$ is independent of the path in D, and let C be any sectionally smooth closed curve in D. As shown in Figure 16.4, we choose two points $\mathbf{p}$ and $\mathbf{q}$ on C and subdivide C into C_1 (from $\mathbf{p}$ to $\mathbf{q}$) and C_2 (from $\mathbf{q}$ to $\mathbf{p}$). (See Exercise 16.4.12.) Then the independence of path gives us

$$\int_{C_1} F \, d\phi = \int_{-C_2} F \, d\phi,$$

so

$$0 = \int_{C_1} F \, d\phi + \int_{C_2} F \, d\phi = \int_{C_1 \cup C_2} F \, d\phi = \int_C F \, d\phi.$$

DEFINITION 16.8. If P and Q are functions on D, then the expression $P \, dg + Q \, dh$ is called an *exact differential* in D provided there is a function f on D such that

$$f_1 = P \quad \text{and} \quad f_2 = Q.$$

The name *exact differential* describes the fact that the row matrix $[P \ Q]$ is the differential of some function f. The notation $P \, dg + Q \, dh$ is used instead of the matrix notation because of its strong connection with line integrals. Some of that connection is seen in the next theorem, which is analogous to the Fundamental Theorem of Calculus.

THEOREM 16.10. If $P \, dg + Q \, dh$ is an exact differential in D and if C is a sectionally smooth curve in D from $\mathbf{p}$ to $\mathbf{q}$, then there is a function f on D such that

$$\int_C [P \, dg + Q \, dh] = f(\mathbf{q}) - f(\mathbf{p}). \tag{2}$$

Proof. First consider the case in which C is a smooth curve; then g and h are continuously differentiable on $[a, b]$, and we have

$$\int_C [P \, dg + Q \, dh] = \int_a^b P \cdot g' + \int_a^b Q \cdot h'$$
$$= \int_a^b (f_1 g' + f_2 h').$$

By the Chain Rule in E^2 (Theorem 15.4), the integrand of the last integral is the derivative $f'(g, h)$ of the composite function $f(g, h)$. Therefore, by the Fundamental Theorem of Calculus, we have

$$\int_C [P \, dg + Q \, dh] = f(g(b), h(b)) - f(g(a), h(a))$$
$$= f(\mathbf{q}) - f(\mathbf{p}).$$

In case C is the union of a finite number of sectionally smooth curves, the integral over C is formed by integrating over the n smooth sections on which the above case can be applied. Then the n resulting equations can be added, and the right-hand members collapse to give $f(\mathbf{q}) - f(\mathbf{p})$. (The details of this case are requested in Exercise 16.5.1.)

The next theorem is the principal result of this section. It gives the relationship between the concepts of independence of path and exact differentials.

THEOREM 16.11. If P and Q are continuous in D, then $\int_C [Pdg + Qdh]$ is independent of the path in D if and only if $Pdg + Qdh$ is an exact differential in D.

Proof. One of the implications in this assertion is an immediate consequence of Theorem 16.10. For if $Pdg + Qdh$ is an exact differential, say, of f, then the value of the line integral $\int_C [Pdg + Qdh]$ is given by $f(\mathbf{q}) - f(\mathbf{p})$, which clearly depends only on the endpoints $\mathbf{p}$ and $\mathbf{q}$. Hence the integral is independent of the path from $\mathbf{p}$ to $\mathbf{q}$.

Now assume $\int_C [Pdg + Qdh]$ is independent of the path in D. We need to find a function f that satisfies $f_1 = P$ and $f_2 = Q$. Let $\mathbf{p}^*$ be a point in D and define f by the equation

$$f(\mathbf{x}) = \int_C [Pdg + Qdh],$$

where C is any sectionally smooth curve in D from $\mathbf{p}^*$ to $\mathbf{x}$. (Note that the independence of path guarantees that f is well defined.) Consider the difference quotient

$$\frac{f(x_1 + \Delta x_1, x_2) - f(x)}{\Delta x_1} = \frac{1}{\Delta x_1} \left\{ \int_{\mathbf{p}^*}^{\mathbf{x}+\Delta\mathbf{x}} [Pdg + Qdh] \right.$$

$$\left. - \int_{\mathbf{p}^*}^{\mathbf{x}} [Pdg + Qdh] \right\}. \tag{3}$$

Since the integrals on the right side are independent of the paths, we have indicated only the endpoints, and we may choose the paths to suit our own purposes. Let C_1 be a sectionally smooth curve from $\mathbf{p}^*$ to $\mathbf{x}$, and then define C_2 to be the path from $\mathbf{p}^*$ to $\mathbf{x} + \Delta\mathbf{x}$ given by $C_2 = C_1 \cup L$, where L is the line segment from $\mathbf{x} = (x_1, x_2)$ to $\mathbf{x} + \Delta\mathbf{x} = (x_1 + \Delta x_1, x_2)$. (We can assume that $L \subseteq D$ because D is open and $\|\Delta\mathbf{x}\|$ can be chosen as small as necessary.)

Then

$$\int_{C_2} [Pdg + Qdh] = \int_{C_1} [Pdg + Qdh] + \int_L [Pdg + Qdh],$$

and Equation (3) becomes

$$\frac{f(x_1 + \Delta x_1, x_2) - f(x_1, x_2)}{\Delta x_1} = \frac{1}{\Delta x_1} \int_L [Pdg + Qdh]. \qquad (4)$$

But L was chosen so that the second coordinate is constant on L. This implies that $\int_L Qdh = 0$, so (4) reduces to

$$\frac{f(x_1 + \Delta x_1, x_2) - f(x_1, x_2)}{\Delta x_1} = \frac{1}{\Delta x_1} \int_L Pdg. \qquad (5)$$

Now we can evaluate the integral over L by choosing a particularly simple parametrization of L, namely, let $g(t) = t$ on the interval $[x_1, x_1 + \Delta x_1]$. Then

$$\int_L P \, dg = \int_{x_1}^{x_1 + \Delta x_1} P(t, x_2) dt.$$

Since P is continuous, the Mean Value Theorem guarantees that there is a number μ in $[x_1, x_1 + \Delta x_1]$ such that

$$P(\mu, x_2) \cdot \Delta x_1 = \int_{x_1}^{x_1 + \Delta x_1} P.$$

Substituting this into (5), we get

$$\frac{f(x_1 + \Delta x_1, x_2) - f(x_1, x_2)}{\Delta x_1} = P(\mu, x_2).$$

As Δx_1 tends to zero, the left side approaches $f_1(\mathbf{x})$. Also, μ approaches x_1 as Δx_1 tends to zero, and the continuity of P implies that the right side approaches $P(x_1, x_2)$. Hence $f_1(\mathbf{x}) = P(\mathbf{x})$. The proof of $f_2 = Q$ is done similarly (see Exercise 16.5.2).

Exercises 16.5

1. Give the details of the proof of Theorem 16.10 for the case in which C is sectionally smooth.

2. Give the details of the proof that $f_2 = Q$ in Theorem 16.11.

3. Show that the expression $Pdg + Qdh$ is an exact differential for the given P and Q by finding a function f as in Definition 16.8:

(a) $P(\mathbf{x}) = 2x_1 x_2$, $\quad Q(\mathbf{x}) = x_1^2$;

(b) $P(\mathbf{x}) = x_2^3$, $\quad Q(\mathbf{x}) = 3x_1 x_2^2$;

(c) $P(\mathbf{x}) = x_2^2 e^{x_1 x_2}$, $\quad Q(\mathbf{x}) = (1 + x_1 x_2) e^{x_1 x_2}$.

4. Use Theorem 16.10 to evaluate the line integral $\int_C [Pdg + Qdh]$:

(a) P and Q as in Exercise (3a), and $C = \{(t, t^2): 0 \leq t \leq 2\}$;

(b) P and Q as in Exercise (3b), and $C = \{(\cos t, \sin t): 0 \leq t \leq \pi\}$;

(c) P and Q as in Exercise (3c), and $C = \{(t^2, t^3): -1 \leq t \leq 1\}$.

5. Prove: If the functions P, Q, P_2, and Q_1 are continuous in D and $Pdg + Qdh$ is an exact differential in D, then $P_2 = Q_1$ throughout D. (*Hint:* Use Theorem 15.6.)

6. Show that the expression $Pdg + Qdh$ for the given functions P and Q is not an exact differential in E^2:

(a) $P(\mathbf{x}) = \sin x_1 y_1$, $\quad Q(\mathbf{x}) = \cos x_1 y_1$;

(b) $P(\mathbf{x}) = x_1^3 - 3x_1 x_2^2$, $\quad Q(\mathbf{x}) = 3x_1^2 x_2 - x_2^3$;

(c) $P(\mathbf{x}) = x_1 \sin x_2$, $\quad Q(\mathbf{x}) = x_1 \cos x_2$.

16.6. Green's Theorem

As our final result on line integrals, we establish a relationship between the double integral over a set S and the line integral over the boundary curve of S. Of course, not every bounded set in E^2 has a boundary that is a nice enough curve to use for a line integral, so we must first describe the sets that we can consider.

DEFINITION 16.9. The set S in E^2 is an *elementary set* provided that there are continuously differentiable functions u, v, u^*, and v^* such that $u(t) \leq v(t)$ on $[a, b]$ and $u^*(t) \leq v^*(t)$ on $[c, d]$ and

$$S = \{\mathbf{x}: a \leq x_1 \leq b \quad \text{and} \quad u(x_1) \leq x_2 \leq v(x_1)\}$$

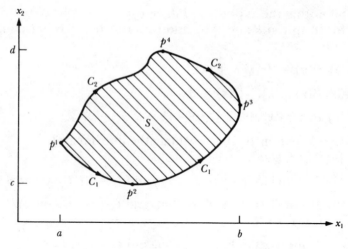

Figure 16.5

and

$$S = \{\mathbf{x}: c \leq x_2 \leq d \quad \text{and} \quad u^*(x_2) \leq x_1 \leq v^*(x_2)\}.$$

(Compare the set in Example 16.2.)

An elementary set is pictured in Figure 16.5. The graph of u is the curve C_1 from $\mathbf{p}^{(1)}$ through $\mathbf{p}^{(2)}$ to $\mathbf{p}^{(3)}$; the graph of v is the curve C_2 from $\mathbf{p}^{(1)}$ through $\mathbf{p}^{(4)}$ to $\mathbf{p}^{(3)}$. Similarly, the graphs of u^* and v^* go from $\mathbf{p}^{(2)}$ to $\mathbf{p}^{(4)}$ through $\mathbf{p}^{(1)}$ and $\mathbf{p}^{(3)}$, respectively. The line integrals that we consider here are to be taken over the boundary of S, which is a sectionally smooth simple closed curve. This curve is to be traversed exactly once, moving with an orientation so that the set S remains on the left while traversing C. For example, we could take $C = C_1 \cup (-C_2)$, where C_1 is parametrized by

$$C_1 = \{(t, u(t)): \quad a \leq t \leq b\}$$

and C_2 is parametrized by

$$C_2 = \{(t, v(t)): \quad c \leq t \leq d\}.$$

The advantage in having an elementary set S as the domain of integration is that the integral $\iint_S f$ can be evaluated as two successive—or iterated—Riemann integrals, one in each coordinate. To see this we picture a net on some rectangle containing S. The integral $\iint_S f$ is the limit of sums of the form

$$\sum_{\substack{1 \le i \le m \\ 1 \le j \le n}} f(\boldsymbol{\mu}_{i,j}) A(R_{i,j}),$$

where $R_{i,j}$ is the rectangle in the ith row and the jth column. The mn terms in this sum can be totaled by first adding the terms for the $R_{i,j}$'s of each column (that is, for a fixed j), then adding the n column-subtotals to get the value of the sum. Thus

$$\sum_{\substack{1 \le i \le m \\ 1 \le j \le n}} f(\boldsymbol{\mu}_{i,j}) A(R_{i,j}) = \sum_{j=1}^{n} \left\{ \sum_{i=1}^{m} f(\boldsymbol{\mu}_{i,j}) A(R_{i,j}) \right\}. \tag{1}$$

For a fixed value of j, consider the inner sum

$$\sum_{i=1}^{m} f(\boldsymbol{\mu}_{i,j}) A(R_{i,j}) = (t_j - t_{j-1}) \sum_{i=1}^{m} f(\boldsymbol{\mu}_{i,j})(s_i - s_{i-1}), \tag{2}$$

where

$$R_{i,j} = \{\mathbf{x}: t_{j-1} \le x_1 \le t_j \quad \text{and} \quad s_{i-1} \le x_2 \le s_i\}.$$

The right-hand member of (2) is very nearly a Riemann sum for the function $f(x_1, s)$ on an interval $u(x_1) \le s \le v(x_1)$, where x_1 is fixed in the interval $[t_{j-1}, t_j]$. This sum is not exactly a Riemann sum for $f(x_1, t)$ because the points $\boldsymbol{\mu}_{ij}$ need not all have the same first coordinate x_1. This difference tends to zero, however, if f is continuous. Thus the sum in (1) is approximated by

$$\sum_{j=1}^{n} \left\{ \int_{u(t)}^{v(t)} f(t, s)\,ds \right\} (t_j - t_{j-1}). \tag{3}$$

By similar heuristic reasoning we see that the expression (3) is approximated by a Riemann sum for an integral over the interval $a \le t \le b$. Hence we find that

$$\iint_S f = \int_a^b \left\{ \int_{u(t)}^{v(t)} f(t, s)\,ds \right\} dt. \tag{4}$$

Our reasoning leading up to (4) is admittedly lacking in details, but the interested reader can verify the assertions of this paragraph by invoking the continuity of f (see Exercise 16.7.1). We are now prepared to prove the result that connects the concepts of the line integral and the double integral.

THEOREM 16.12: Green's Theorem. Suppose that S is an elementary set with boundary curve C, and let P, Q, P_2, and Q_1 be continuous functions on the domain D that contains S. If $C = \{(g(t), h(t))\}$, then

$$\iint_S [Q_1 - P_2] = \int_C [P\,dg + Q\,dh]. \tag{5}$$

Proof. Each of the integrals in (5) can be written as the sum of two integrals, so we wish to prove

$$\iint_S Q_1 - \iint_S P_2 = \int_C P\,dg + \int_C Q\,dh.$$

We prove here that

$$\iint_S P_2 = -\int_C P\,dg, \tag{6}$$

and the other integrals can be treated similarly. Using (4) we can evaluate the left-hand integral as an iterated integral:

$$\iint_S P_2 = \int_a^b \int_{u(t)}^{v(t)} P_2(t, s)\,ds\,dt.$$

By the Fundamental Theorem of Calculus, the inner integral is given by

$$\int_{u(t)}^{v(t)} P_2(t, s)\,ds = P(t, v(t)) - P(t, u(t)).$$

As we saw above, $C = C_1 \cup (C_2)$, where C_1 and C_2 are given by

$$C_1 = \{(t, u(t)): a \le t \le b\}$$

and

$$C_2 = \{(t, v(t)): a \le t \le b\}.$$

Thus we have

$$\iint_S P_2 = \int_a^b P(t, v(t))\,dt - \int_a^b P(t, u(t))\,dt. \tag{7}$$

Substituting $g(t) = t$ into the right-hand member of (7), we get

$$\iint_S P_2 = -\int_{-C_2} P\,dg - \int_{C_1} P\,dg$$

$$= -\int_{C_1 \cup (-C_2)} P\,dg$$

$$= -\int_C P\,dg.$$

This completes the proof.

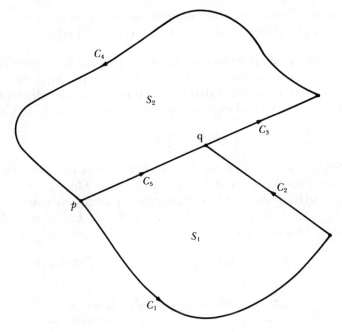

Figure 16.6

It is not hard to see that Theorem 16.12 can be extended to the case where S is the union of a finite number of elementary sets. We demonstrate this by proving the result for the union of the two elementary sets shown in Figure 16.6. Let P, Q, P_2, and Q_1 be continuous on $S = S_1 \cup S_2$. The boundary of S is given by $C_1 \cup C_2 \cup C_3 \cup C_4$, and by introducing the curve C_5 from **p** to **q**, we have

$$b[S_1] = C_1 \cup C_2 \cup (-C_5) \quad \text{and} \quad b[S_2] = C_5 \cup C_3 \cup C_4.$$

Then

$$\int_C [P\,dg + Q\,dh] = \int [P\,dg + Q\,dh]$$

$$= \int_{C_1 \cup C_2 \cup (-C_5)} [P\,dg + Q\,dh] + \int_{C_5 \cup C_3 \cup C_4} [P\,dg + Q\,dh]$$

$$= \iint_{S_1} [Q_1 - P_2] + \iint_{S_2} [Q_1 - P_2]$$

$$= \iint_{S_1 \cup S_2} [Q_1 - P_2]$$

$$= \iint_S [Q_1 - P_2].$$

Green's Theorem makes it particularly simple to evaluate integrals of exact differentials, as is seen in the next result.

THEOREM 16.13. Suppose that P, Q, P_2, and Q_1 are continuous functions on the domain D, and let $C = \{g(t), h(t)\}$ be a sectionally smooth curve in D that is the boundary of the elementary set S. If the differential expression $Pdg + Qdh$ is exact in D, then

$$\int_C [Pdg + Qdh] = 0.$$

Proof. The exactness of $Pdg + Qdh$ implies the existence of a function f on S such that $f_1 = P$ and $f_2 = Q$. Then $f_{12} = P_2$ and $f_{21} = Q_1$, and by Theorem 15.6 these mixed partials are equal. Thus $Q_1 - P_2$ is identically zero throughout S (see Exercise 16.5.5). Therefore, by Green's Theorem,

$$\iint_C [Pdg + Qdh] = \iint_S [Q_1 - P_2] = \iint_S 0 = 0.$$

16.7. Analogues of Green's Theorem

In this final section we discuss theorems whose form is similar to that of Green's Theorem. We do not prove these results, although they will be recognized by students of multivariable calculus. Our present purpose is merely to present an overview of the general context of which Green's Theorem is a part. We begin with an examination of the form of that result. Suppose the function f is defined on an open domain D that contains a set S, and let df denote a "differential expression" involving the derivative or partial derivatives of f. If we wish to integrate df over S, then according to Green's Theorem, with the appropriate assumptions about f and S, we get the same value by integrating f itself over the boundary of S. In formal symbols this is

$$\iint_S df = \int_{b[S]} f. \tag{1}$$

Of course, we are using the verb "to integrate" in a deliberately vague way. The left-hand and right-hand members of (1) are different types of integrals on domains of different dimensions, but the form of the assertion is readily observed: The integral of a function over the boundary of a set equals the integral of some differential expression over the enclosed set. A similar statement has been seen before. If the dimensions are reduced by 1, f becomes a function on an interval $I = [a, b]$ and $df = f'$; then $b[I]$ is the two-point set $\{a, b\}$ and according to the Fundamental Theorem of Calculus

$$\int_I df = f(b) - f(a) = \int_{b[I]} f. \tag{2}$$

The right-hand member of (2) is an undefined symbol that we can define to equal the middle member of (2). The point is that the values of f on the boundary of I determine the integral of df on the enclosed set.

Equation (1) can also be formulated in a three-dimensional setting. In this situation the set S can be either a surface Σ in E^3 or a solid figure V. In the first case we have *Stokes's Theorem,* in which the line integral

$$\int_{C=b[\Sigma]} [Pdg + Qdh + Rdv]$$

is equal to the surface integral

$$\iint_{\Sigma} [(Q_3 - R_2) + (R_1 - P_3) + (P_2 - Q_1)]d\sigma,$$

for "sufficiently nice" surfaces and functions. (The boundary $b[\Sigma]$ is a curve that encloses the surface S.) In the second case, the surface Σ encloses a solid figure V, and $f(\mathbf{x}) = (P(\mathbf{x}), Q(\mathbf{x}), R(\mathbf{x}))$ is a transformation from E^3 into E^3 where P, Q, and R are functions on E^3. When the dot product $f \cdot d\sigma$ is integrated over the boundary surface, the result is the same as the volume integral of $df = [P_1 + Q_2 + R_3]dV$ taken over V:

$$\iint_{\Sigma=b[V]} f \cdot d = \iiint_V df. \tag{3}$$

This is *Gauss's Divergence Theorem,* and in Equation (3) it is plain that once again the integral of a function over the boundary of a set is equal to the integral of a differential expression over the enclosed set.

A full discussion of Stokes's Theorem and Gauss's Divergence Theorem is best presented as part of the subject of vector analysis. In that setting the full power of vector notation and concepts can be used to simplify the descriptions of the various functions, curves, and surfaces, and their relationships can be verified more efficiently.

Exercises 16.7

1. Prove assertion (4) of section 16.6. (*Hint:* Use the uniform continuity of f to get points v_{ij}, . . . , v_{mj} that have the same second coordinate and satisfy

$$\sum_{\substack{1 \le i \le m \\ 1 \le j \le n}} |f(\mathbf{\mu}_{ij}) - f(\mathbf{v}_{ij})| < \frac{\varepsilon}{A(S)}.$$

2. Complete the proof of Green's Theorem by proving that $\iint_S Q_1 = \iint_C Q dh$.

In Exercises 3–7, use Green's Theorem to evaluate the line interval $\int_C [P dg + Q dh]$ for the given functions and curves. Assume the curves are oriented so that the set S lies to the left as the curve is traversed.

3. $P(\mathbf{x}) = 2x_1 - x_2$; $Q(\mathbf{x}) = x_1 + 3x_2$; C is the rectangular path from $(0, 0)$ to $(0, 3)$ to $(3, 2)$ to $(0, 2)$ to $(0, 0)$.

4. $P(\mathbf{x}) = 2x_1 x_2$; $Q(\mathbf{x}) = x_1^2$; C is the boundary of the bounded set enclosed by the line $x_2 = x_1$ and the parabola $x_2 = x_1^2$.

5. $P(\mathbf{x}) = 2x_1 + x_2$; $Q(\mathbf{x}) = 3x_1 - x_2$; C is the boundary of $N_2(\mathbf{0})$.

6. $P(\mathbf{x}) = 2x_1 \sin x_2$; $Q(\mathbf{x}) = x_1^2 \cos x_2$; C is the ellipse $5x_1^2 + 3x_2^2 = 1$.

7. $P(\mathbf{x}) = x_1 - x_1^2 x_2$; $Q(\mathbf{x}) = x_2 + x_2^2 x_1$; C is the boundary of the quarter circle:

$$S = \{\mathbf{x} \in E^2 : \|\mathbf{x}\| \le 1, x_1 \ge 0, x_2 \ge 0\}.$$

8. Show that Green's Theorem can be applied to an integral over a Z-shaped region as shown below.

APPENDIX A

MATHEMATICAL INDUCTION

This discussion is not intended as a general development of the subject of mathematical induction. A student who has reached the level of this book will have encountered the concept in a more basic and perhaps more complete presentation. The purpose of this appendix is to show, by one typical proof, how this very fundamental tool of mathematics is used many times throughout the theory presented in this book.

PRINCIPLE OF MATHEMATICAL INDUCTION. Suppose that T is a subset of $\mathbb{N}$ that satisfies

  (i) 1 is in T, and
  (ii) if n is in T, then $n + 1$ is in T;

then T is all of $\mathbb{N}$.

The letter T is chosen in the preceding statement of the Principle of Mathematical Induction because of the way that this principle is most frequently applied. We let $\{S_n\}_{n=1}^{\infty}$ be a sequence of statements, and let T be the *truth set* of $\{S_n\}_{n=1}^{\infty}$: that is,

$$T = \{n \in \mathbb{N}: S_n \text{ is true}\}.$$

Then we can infer that S_n is true for every n by showing that T satisfies (i) and (ii).

Throughout the development of the theory of calculus we introduce collections of functions that satisfy some particular definition, and then we prove that the collection is closed under some operation of combining two of its elements; that is, whenever two elements of the collection are combined, they produce an element that is also in that collection. The collection may consist of the convergent sequences or, perhaps, the integrable functions; and the manner of combining two of its elements may be an arithmetic operation such as addition or multiplication, or it may be some other combination like the composition of functions. All that we need to assume here is that the combining operation is associative:

$$(f * g) * h = f * (g * h).$$

In such an instance, once we have proved that the collection is closed under an associative combining of any two of its elements, then the Principle of Mathematical Induction allows us to conclude immediately that the collection is closed under the combination of any finite subset of its elements. That is the one result that we prove here. For convenience, we use a plus sign to indicate the binary operation on the collection C, and we use the letters f and g to denote elements of C. But the proof is not limited to the case in which the operation is addition and C consists of functions. We assume only that the operation is associative and satisfies this condition:

$$\text{for any } f \text{ and } g \text{ in } C, \quad f + g \text{ is in } C. \qquad (1)$$

THEOREM A1. Suppose C is a collection with an associative combining operation $+$ that satisfies property (1). If $\{f_0, f_1, \ldots, f_n\}$ is any finite subset of C, then the sum $f_0 + f_1 + \cdots + f_n$ is in C.

Proof. For each n in $\mathbb{N}$ let S_n be the statement "for any subset $\{f_0, f_1, \ldots, f_n\}$ of C, $f_0 + f_1 + \cdots + f_n$ is in C." We wish to prove that the truth set T of $\{S_n\}_{n=1}^{\infty}$ is $\mathbb{N}$. First note that S_1 is precisely property (1), so T satisfies (i). Now assume that n is in T; that is, for any set $\{f_0, \ldots, f_n\}$ in C, $f_0 + \cdots + f_n$ is in C. We must show that $n + 1$ is in T, so consider an arbitrary subset $\{f_0, \ldots, f_n, f_{n+1}\}$ of C. By associativity we have

$$f_0 + \cdots + f_n + f_{n+1} = (f_0 + \cdots + f_n) + f_{n+1}.$$

By our assumption, $f_0 + \cdots + f_n$ is in C, so (1) implies that the right-hand side is in C. Hence $n + 1$ is in T; so by the Principle of Mathematical Induction, T is all of $\mathbb{N}$.

APPENDIX B

COUNTABLE AND UNCOUNTABLE SETS

From our elementary experiences with sets of numbers, it is natural to infer that if two sets are both infinite then they are "equally large." It is extremely advantageous, however, to use a more sophisticated concept for comparing large sets. This theory of cardinal numbers and equivalence of sets was developed in the latter 1800s by Cantor, and it has vastly enriched mathematics. We do not go into it very far in this discussion, but the reader is encouraged to seek other sources and pursue this further.

DEFINITION B1. The set S is *countable* provided there is a sequence whose range is all of S. If a set is not countable, it is said to be *uncountable*.

This definition limits the number of elements in a countable set S in the sense that there is at least one positive integer that is associated with each element of S. Thus there cannot be "more" elements of S than there are elements of $\mathbb{N}$. Here are some examples of countable sets that are easy to verify:

- (i) the set $\mathbb{N}$ itself
- (ii) the set of even positive integers $\{2n : n \in \mathbb{N}\}$
- (iii) any subset of $\mathbb{N}$
- (iv) any finite set

The third example can be extended to a more general assertion, which we prove next.

PROPOSITION B1. Any subset of a countable set is itself a countable set.

Proof. Suppose that S is a countable set, and let T be a subset of S. Then there is a sequence s such that

$$s = \{s_n : n \in \mathbb{N}\}.$$

Let t be the subsequence of s corresponding to those integers $n(k)$ such that $s_{n(k)} \in T \subseteq S$. Thus $T = \{s_{n(k)} : k \in \mathbb{N}\}$, and we have shown that T is the range of the sequence t.

Although Proposition B1 is very useful, it is more striking to observe what happens when we enlarge countable sets by combining several of them. For example, suppose $S = \{s_n\}_{n=1}^{\infty}$ and $T = \{t_n\}_{n=1}^{\infty}$ are countable sets. We assert that their union is also a countable set; for $\{s_1, t_1, s_2, t_2, s_3, t_3, \ldots\}$ is clearly a sequence whose range is $S \cup T$. Similarly, the union of three countable sets S, T, and U is shown to be countable by the sequence $\{s_1, t_1, u_1, s_2, t_2, u_2, \ldots\}$. It is easy to prove (by using the Principle of Mathematical Induction) that the union of any finite collection of countable sets is itself a countable set. The next result takes this conclusion beyond the access of mathematical induction.

THEOREM B1. The union of a countable collection of countable sets is itself a countable set.

Proof. For each n in $\mathbb{N}$ let $S^{(n)}$ be the countable set $\{s_{nk} : k \in \mathbb{N}\}$. We wish to show that the set $S = \bigcup_{n=1}^{\infty} S^{(n)}$ is countable, and we can display S in the array shown in Figure B1. The arrows indicate how the elements of S may be written as the terms of a sequence, namely,

$$\{s_{11}, s_{21}, s_{12}, s_{31}, s_{22}, s_{13}, s_{41}, s_{32}, s_{23}, s_{14}, \ldots\}.$$

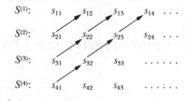

Figure B-1

Note that the sum of the two subscripts of each term is constant along any diagonal. Thus we are listing the terms in groups according to the sum of their subscripts:

in the first diagonal the sum is 2,
in the second diagonal the sum is 3,
.
.
.
in the jth diagonal the sum is $j + 1$,
.
.
.

For every term s_{nk} in S, the sum $n + k$ is equal to an integer j, so it follows that every term of S appears somewhere in the above sequence. Hence S is the range of that sequence, so S is a countable set.

EXAMPLE B1. The set $\mathbb{Q}$ of rational numbers is countable; for if $S^{(n)} = \{k/n: k \in \mathbb{N}\}$, then $\cup_{n=1}^{\infty} S^{(n)}$ consists of all the positive rational numbers, and it is countable by Theorem B1. Similarly, the set of negative rationals is also countable, and so is the singleton set $\{0\}$. Since $\mathbb{Q}$ is the union of these three countable sets, it follows that $\mathbb{Q}$ itself is countable.

By now it appears that virtually all sets are countable, and this would mean that we have gained nothing over our earlier viewpoint that all infinite sets are "equally large." But this is not the case, as we prove in the next theorem.

THEOREM B2. The set $\mathbb{R}$ of real numbers is uncountable.

Proof. In order to show that there is no sequence whose range is all of $\mathbb{R}$, we assume that $\{x_n\}_{n=1}^{\infty}$ is an arbitrary sequence of real numbers and show that there is at least one real number that is not a term of this sequence. Let us assume further that each x_n is in the interval $[0, 1)$. Each x_n can be written in decimal form, say,

$$x_n = .d_{n1} d_{n2} d_{n3} \ldots ,$$

where each d_{nk} is a digit (that is, 0, 1, . . . , or 9). The sequence $\{x_n\}_{n=1}^{\infty}$ can be displayed as in Figure B2. In Figure B2 the sequence

$$x_1 = \boxed{.d_{11}} \quad d_{12} \quad d_{13} \quad \ldots$$

$$x_2 = .d_{21} \quad \boxed{d_{22}} \quad d_{23} \quad \ldots$$

$$x_3 = .d_{31} \quad d_{32} \quad \boxed{d_{33}} \ldots$$

$$x_n = .d_{n1} \quad d_{n2} \quad d_{n3} \quad \ldots \quad \boxed{d_{nn}} \ldots$$

Figure B-2

of nth digits of the nth number has been enclosed in boxes because it is from these digits that we construct a number $y = .\delta_1\delta_2\delta_3 \ldots$ that is different from each of the x_n's. The number y is defined by describing its nth digit δ_n. We want to have $\delta_n \neq d_{nn}$, so we define

$$\delta_n = \begin{cases} 7, & \text{if } d_{nn} \leq 5, \\ 3, & \text{if } d_{nn} > 5. \end{cases}$$

Thus y is a number in the interval $[0, 1)$. Also, y cannot equal any of the x_n's, because for each n, $\delta_n \neq d_{nn}$, so y and x_n differ in at least their nth decimal place. (Note that y cannot be "rounded off" to equal an x_n in the way that .49999 . . . is equal to .50000 . . . , because y has no 9's in its decimal expansion.) Hence the sequence $\{x_n\}_{n=1}^{\infty}$ does not contain among its terms all of the numbers in $[0, 1)$. This proves that the set $[0, 1)$ is uncountable. Since $\mathbb{R}$ contains $[0, 1)$, it follows from Proposition B1 that $\mathbb{R}$ is also uncountable.

COROLLARY B2. The set of irrational numbers is an uncountable set.

Proof. If the irrational numbers X were countable, then we could write $\mathbb{R} = \mathbb{Q} \cup X$, and Theorem B1 would imply that $\mathbb{R}$ is countable.

APPENDIX C

INFINITE PRODUCTS

The concept of an infinite product is the multiplicative analogue of infinite series. It is a topic that is usually omitted from texts at this level. Although infinite products are quite useful in some areas of mathematics such as number theory and complex function theory, we show here that, for the purpose of studying convergence of infinite products, it is sufficient to study infinite series.

Let $\{a_k\}$ be a number sequence and define the related sequence $\{p_n\}$ by the formula

$$p_n = a_1 a_2 \ldots a_n = \prod_{k=1}^{n} a_k.$$

Then $\{p_n\}$ is the *sequence of partial products* of the "infinite product Πa_k." (We use capital pi, Π, for "product" to reinforce the analogy with a series in which we write capital sigma, Σ, for "sum.") Because of the special multiplicative properties of the number zero, it is often necessary to assume that a_k is nonzero for every k. Although we do not make that general stipulation here, it is a necessary consequence of the convergence that we now define.

DEFINITION C1. The infinite product Πa_k is *convergent* provided that its sequence of partial products is convergent to a nonzero limit. In the case of convergence, we write

$$\prod_{k=1}^{\infty} a_k = \lim_n \left\{ \prod_{k=1}^{n} a_k \right\},$$

and this limit number is called the *value* of the infinite product.

In case the sequence of partial products converges to zero, we say that Πa_k is *divergent to zero*. This is obviously the result if any term a_k is zero. Although it may seem an arbitrary exclusion to disallow the number zero as a limit value, it is a necessary assumption in order to prevent some very erratic sequences from yielding a convergent infinite product. For example, if $a_{k*} = 0$ for some $k*$, then $\{a_k\}$ could be an unbounded sequence, or have any values we wish for $k \neq k*$, and still

$$p_n = \prod_{k=1}^{n} a_k = 0 \quad \text{for } n \geq k*.$$

This is not, however, the only way that an infinite product can fail to be convergent. Examples C1 and C2 show other types of non-convergence, and Example C3 gives a convergent infinite product.

EXAMPLE C1. If $a_k = (-1)^k$, then $p_n = (-1)^n$, and Πa_k is nonconvergent.

EXAMPLE C2. If $a_k = (k + 1)/k$, then

$$p_n = \frac{2}{1} \cdot \frac{3}{2} \cdot \frac{4}{3} \cdots \frac{n + 1}{n} = n + 1.$$

Therefore $\lim_n p_n = \infty$, and Πa_k is nonconvergent.

EXAMPLE C3. If $a_{2k-1} = (k + 1)/k$ and $a_{2k} = k/(k + 1)$, then

$$\{a_k\} = \left\{ \frac{2}{1}, \frac{1}{2}, \frac{3}{2}, \frac{2}{3}, \frac{4}{3}, \frac{3}{4}, \cdots \right\}$$

and

$$p_{2n-1} = \frac{n + 1}{n} \quad \text{and} \quad p_{2n} = 1.$$

Therefore $\prod_{k=1}^{\infty} a_k = 1$.

The next result is immediately recognized as the analogue of Proposition 9.1 on infinite series.

PROPOSITION C1. If Πa_k is convergent, then $\lim_k a_k = 1$.

Proof. Suppose $\prod_{k=1}^{\infty} a_k = L \neq 0$. Then

$$\lim_n a_n = \lim_n \frac{p_n}{p_{n-1}} = \frac{\lim_n p_n}{\lim_n p_{n-1}} = \frac{L}{L} = 1.$$

We close this discussion with a simple theorem that connects the convergence of infinite products with that of infinite series.

THEOREM C1. If Πa_k is convergent, then $\Sigma \log|a_k|$ is convergent; if $\Sigma \log|a_k|$ is convergent and $a_k > 0$ for n greater than some N, then Πa_k is convergent.

Proof. If Πa_k is convergent, then by Proposition C1 the general term a_k is positive for sufficiently large k. Therefore we may assume without loss of generality that $a_k > 0$ for all k. Now

$$\lim_n \left\{ \sum_{k=1}^{n} \log a_k \right\} = \lim_n \left\{ \log \prod_{k=1}^{n} a_k \right\}$$

$$= \lim_n \{\log p_n\},$$

and the continuity of $\log x$ plus the Sequential Criterion for Continuity guarantee that the limits exist.

To prove the converse, we may again assume $a_k > 0$ for all k. If

$$s_n = \sum_{k=1}^{n} \log a_k,$$

then we have

$$s_n = \log p_n \quad \text{or} \quad e^{s_n} = p_n$$

Now the convergence of $\{s_n\}$ and the continuity of e^x imply that $\{p_n\}$ is convergent.

INDEX

A

Abel, N., 154, 162, 226
Abel's Test, 162
absolute value, 11, 112
absolutely convergent, 161
accumulation point, 25
addition in E^n, 282
alternating harmonic series, 163, 229 (Ex. 5), 239
alternating series, 163
Alternating Series Theorem, 163
analytic function, 231
Approximation Property, 294
area, 315
  inner, 317
  outer, 317
Asymptotic Comparison Test, 152

B

basis vector, 282
Bernstein, S., 212
Bernstein polynomials, 212
biconditional statement, 2
binomial coefficient, 212
Bliss's Theorem, 119 (Ex. 7)
Bolzano-Weierstrass Property, 263
Bolzano-Weierstrass Theorem, 30
boundary, 319
Bounded Sequence Property, 263
bounded set, 8, 261
bounded variation, 177
bracket function, 17, 45

C

Cantor, G., 343
Cauchy-Bunyakovsky-Schwarz Inequality, 117, 246
Cauchy Condensation Test, 153
Cauchy Criterion, 32
  for Series, 150 (Ex. 9)
  for Uniform Convergence, 211 (Ex. 11)
  as $x \to \infty$, 55 (Ex. 13)
Cauchy form of the remainder, 236

Cauchy product of series, 171
Cauchy sequence, 30, 259
Cauchy's Law of the Mean, 85
Chain Rule, 76, 298
characteristic function, 280
closed curve, 325
closed interval, 33
closed set, 251
closure of a set, 254
closure under multiplication, 6
cluster point, 25
comparison test for integrals, 128
Comparison Test (for series), 151
complement of a set, 252
complete metric space, 262
complete system, 32
composition of functions, 47
conclusion, 2
conditional statement, 2
conditionally convergent series, 161
connected set, 255
constant sequence, 13
continuous
  from the left, 48
  from the right, 48
continuous function, 38
continuous transformation, 270
contradiction, 3
contrapositive argument, 3
convergent function sequence, 194
convergent improper integral, 127
convergent infinite product, 347
convergent number sequence, 13–18
convergent point sequence, 258
convergent series, 147
converse statement, 2
coordinate, 246
coordinate sequences, 258
corollary, 2
countable set, 343
curve, 324
cut point, 12 (Ex. 14)

351

remainder term, 234
removable discontinuity, 51
Riemann integral, 98
Riemann-Stieltjes integrability of
    continuous functions, 191
Riemann-Stieltjes integral, 183
Riemann-Stieltjes Integration by
    Parts, 190
Riemann sum, 98
right-hand limit, 54
Rolle's Theorem, 79
Root Test, 157

S

second derivative, 74
sectionally smooth curve, 325
sequence, 13
sequence of partial products, 347
sequence of partial sums, 146
Sequential Criterion for Continuity,
    43, 275
Sequential Criterion for Function
    Limits, 51
Sequential Criterion for Uniform
    Continuity, 67
shifting property, 143
simple curve, 325
smooth curve, 325
step function, 106 (Ex. 6), 324
    (Ex. 6)
Stieltjes sum, 183
Stokes's Theorem, 339
subcover, 35
subsequence, 23
summation by parts, 162
supremum, 176
symmetry property, 242

T

taxicab metric, 247–248
Taylor series, 231
Taylor's Formula with Remainder, 88
terminal point, 325
theorem, 2
topology, 248
total variation function, 181
transformation, 270
Triangle Inequality, 242–243, 246–
    247
truth set, 341

U

uncountable set, 343
uniform continuity, 62
    negation of, 64
uniform convergence, 197
    negation of, 200
uniform convergence of power se-
    ries, 225
uniformly bounded function se-
    quence, 211 (Ex. 10)
unit circle, 306
universal comparison test, 153–154
upper bound, 8

V

variation of a function, 177

W

Weierstrass Approximation Theo-
    rem, 212
Weierstrass M-Test, 217
Well-Ordering Principle, 8 (Ex. 7)

Z

zero of a function, 60 (Ex. 2)